Comprehensive Virology 8

Comprehensive Virology

Edited by Heinz Fraenkel-Conrat
University of California at Berkeley

and Robert R. Wagner
University of Virginia

Comprehensive

Edited by

Heinz Fraenkel-Conrat

Department of Molecular Biology and Virus Laboratory
University of California, Berkeley, California

and

Robert R. Wagner

Department of Microbiology
University of Virginia, Charlottesville, Virginia

Virology

8

Regulation and Genetics

Bacterial DNA Viruses

SPRINGER SCIENCE+BUSINESS MEDIA, LLC

Library of Congress Cataloging in Publication Data

Fraenkel-Conrat, Heinz, 1910-
  Regulation and genetics.

(Their Comprehensive virology; v. 8)
Includes bibliographies and index.
1. Viral genetics. 2. Bacteriophage. 3. Genetic regulation. I. Wagner, Robert R.,
1923-   II. Title. III. Series.
QR357.F72 vol. 8 [QH434]   576'.64'08   [576'.6482]
                                                              76-22208

ISBN 978-1-4684-2717-2   ISBN 978-1-4684-2715-8 (eBook)
DOI 10.1007/978-1-4684-2715-8

© 1977 Springer Science+Business Media New York
Originally published by Plenum Press, New York in 1977
Softcover reprint of the hardcover 1st edition 1997

Foreword

The time seems ripe for a critical compendium of that segment of the biological universe we call viruses. Virology, as a science, having passed only recently through its descriptive phase of naming and numbering, has probably reached that stage at which relatively few new—truly new—viruses will be discovered. Triggered by the intellectual probes and techniques of molecular biology, genetics, biochemical cytology, and high-resolution microscopy and spectroscopy, the field has experienced a genuine information explosion.

Few serious attempts have been made to chronicle these events. This comprehensive series, which will comprise some 6000 pages in a total of about 22 volumes, represents a commitment by a large group of active investigators to analyze, digest, and expostulate on the great mass of data relating to viruses, much of which is now amorphous and disjointed, and scattered throughout a wide literature. In this way, we hope to place the entire field in perspective, and to develop an invaluable reference and sourcebook for researchers and students at all levels.

This series is designed as a continuum that can be entered anywhere, but which also provides a logical progression of developing facts and integrated concepts.

Volume 1 contains an alphabetical catalogue of almost all viruses of vertebrates, insects, plants, and protists, describing them in general terms. Volumes 2–4 deal primarily, but not exclusively, with the processes of infection and reproduction of the major groups of viruses in their hosts. Volume 2 deals with the simple RNA viruses of bacteria, plants, and animals; the togaviruses (formerly called arboviruses), which share with these only the feature that the virion's RNA is able to act as messenger RNA in the host cell; and the reoviruses of animals and plants, which all share several structurally singular features, the most important being the double-strandedness of their multiple RNA molecules. This grouping, of course, has only slightly more in its favor than others that could have been, or indeed were, considered.

Volume 3 addresses itself to the reproduction of all DNA-containing viruses of vertebrates, a seemingly simple act of classification, even though the field encompasses the smallest and the largest viruses known. The reproduction of the larger and more complex RNA viruses is the subject matter of Volume 4. These viruses share the property of lipid-rich envelopes with the togaviruses included in Volume 2. They share as a group, and with the reoviruses, the presence of enzymes in their virions and the need for their RNA to become transcribed before it can serve messenger functions.

Volumes 5 and 6 represent the first in a series that focuses primarily on the structure and assembly of virus particles. Volume 5 is devoted to general structural principles involving the relationship and specificity of interaction of viral capsid proteins and their nucleic acids, or host nucleic acids. It deals primarily with helical and the simpler isometric viruses, as well as with the relationship of nucleic acid to protein shell in the T-even phages. Volume 6 is concerned with the structure of the picornaviruses, and with the reconstitution of plant and bacterial RNA viruses.

Volumes 7 and 8 deal with the DNA bacteriophages. Volume 7 concludes the series of volumes on the reproduction of viruses (Volumes 2–4 and Volume 7) and deals particularly with the single- and double-stranded virulent bacteriophages.

In the present volume, the first of the series on regulation and genetics of viruses, the biological properties of the lysogenic and defective phages are covered, the phage–satellite system P2–P4 described, and the regulatory principles governing the development of selected typical lytic phages discussed in depth.

Volume 8 will be followed by others dealing with the regulation of gene expression and integration of animal viruses, the genetics of animal viruses, and regulation of plant virus development, covirus systems, satellitism, and viroids. In addition, it is anticipated that there will be two or three other volumes devoted largely to structural aspects and the assembly of bacteriophages and animal viruses, and to special virus groups.

The complete series will endeavor to encompass all aspects of the molecular biology and the behavior of viruses. We hope to keep this series up to date at all times by prompt and rapid publication of all contributions, and by encouraging the authors to update their chapters by additions or corrections whenever a volume is reprinted.

Contents

Chapter 2

Bacteriophage λ: The Lysogenic Pathway

R. A. Weisberg, S. Gottesman, and M. E. Gottesman

Chapter 3

Defective Bacteriophages and Incomplete Prophages

A. Campbell

Chapter 4

The P 2–P 4 *Trans*activation System

Richard Calendar, Janet Geisselsoder, Melvin G. Sunshine, Erich W. Six, and Björn Lindqvist

CHAPTER 1

Regulation of Gene Action in the Development of Lytic Bacteriophages

Dietmar Rabussay and E. Peter Geiduschek

Department of Biology
University of California at San Diego
La Jolla, California 92093

1. GENERAL INTRODUCTION

In this chapter, we shall analyze the ways in which gene activity is regulated in the development of the larger DNA viruses of bacteria. The kind of analysis that we write about has now been carried out for about 20 years, with increasing sophistication and insight as the understanding of genetic regulation has improved and as the methods of genetics, molecular biology, and enzymology have developed. The immediate goal of this search is to understand, in the molecular terms of protein–nucleic acid interaction, all the strategies of genetic regulation that are encompassed in the development of the phages. That goal is far from realization, although great strides toward it have been taken during the last few years. In fact, we originally hoped to organize our presentation of this subject around a classification of the mechanisms of action of regulatory elements and, in this way, to draw together the common principles of phage development. We eventually decided against the scheme because, with the notable exceptions that we shall emphasize, the mechanisms of gene regulation in phage development are not sufficiently worked out. We have instead chosen several prototypes of lytic phage development in order to analyze the diversity of regulatory mechanisms while avoiding excessive repetition of quasi-equivalent detail. However, one should stress that, although regulatory schemes are wondrously diverse, the general features of the development of the lytic phages fit a common general pattern. We return to a consideration of these common features in the final section of this article, but they are worth noting at the outset.

1.   The propagation of the phages that we shall analyze involves time-ordered sequences of gene expression. In general, two or three stages of the development can be distinguished. The earlier stages involve: (a) partial or complete inactivation of host-cell macromolecular metabolism and its replacement by viral genome-directed nucleic acid and protein synthesis; (b) inactivation of host-cell functions that would prevent viral multiplication; (c) preparation for, and initiation of, viral DNA replication and recombination. The late stage of viral gene expression is principally directed toward the production of the components of mature progeny virus particles and toward virion assembly.

2.   The time-ordered sequences of viral gene activity are generated primarily by regulation at the transcriptional level.

3.   Many different regulatory strategies generate similar temporal sequences of gene action. The common element of these strategies is that at least some viral genes are positively regulated at the transcriptional level by proteins that are coded by the viral genome. These positive regulators are of several kinds. They include genome-specific RNA polymerases and proteins that interact with the RNA polymerase of the host. The latter include some proteins that modify the initiation specificity of RNA polymerase and others that (probably) affect RNA chain termination. In some instances, the regulation of gene activity is partly achieved by DNA processing, i.e., modification of DNA structure.

4.   The lytic phages establish diverse regulatory and metabolic interrelations with their hosts. The pattern of total, rapid, and irreversible host genome inactivation and degradation, which was established for the T series of *E. coli* phages in the early years of phage biology, is not universal. Many lytic phages leave host macromolecular synthesis substantially intact until late in infection.

We have chosen our prototypes principally because they embody distinct regulatory patterns of gene expression but also as reflections of our own research activities: phage T 7 has been chosen because it is the best understood of the lytic DNA phages and because it exemplifies regulation of transcription by a phage genome-specific RNA polymerase. We chose the hmU DNA phages (phages which contain hydroxymethyluridine instead of thymidine in their DNA) of *B. subtilis* because they exemplify regulation of transcription by a phage-specific initiation factor. We chose the T-even phages because of their complexity, because of the great range of regulatory events that occur during their development, and because they permit some discussion of post-transcriptional regulation and replication-transcription coupling. Phage T 5 was included for the sake of its stepwise transfer of DNA. Phages N 4 and PBS 2 were included because of their interesting and only recently studied autarchic regulation and, in the case of N 4, because of the selective shutoff of host genes during viral development. We excluded a detailed analysis of the lytic development of phage λ because it will be treated in Chapter 2 of this volume. We also excluded discussion of phage φ 29 despite the detailed analysis to which it is being subjected (e.g., Pène *et al.*, 1973; Schachtele *et al.*, 1973; Carrascosa *et al.*, 1973) simply because it is not yet clear in which of its regulatory principles it differs from our chosen prototypes.

We have introduced our chapter with a review of closely related aspects of transcription, translation, and their regulation in bacteria. The introduction is deliberately less detailed than the rest of the chapter, though indicating salient features of these subjects. At various stages of our writing, we have been tempted to wait for this or that development which would give us a coherent and simple view of everything that we wanted to write about. At the urging of our editors, we have decided not to wait for the millenium but to write about this subject now, to the best of our abilities. We have been greatly helped by J. Abelson, C. Bordier, J. J. Duffy, S. C. Falco, L. Gold, O. Grau, M. Hayashi, M. H. Malamy, D. Nakada, D. Nierlich, H. Noller, L. B. Rothman-Denes, L. Synder, P. R. Srinivasan, J. A. Steitz, W. Studier, A. Wahba, S. B. Weiss, and C. Yanofsky who provided unpublished material, answered our questions, or provided valuable explanations. We are especially grateful to E. N. Brody, M. Chamberlin, O. Grau, R. Haselkorn, C. K. Mathews, D. J. McCorquodale, D. Nierlich, E. G. Niles, L. B. Rothman-Denes, L. Snyder, J. Steitz, and W. Studier who read and criticized different sections of the text.

1.1. Summary of Certain Aspects of Transcription and Translation Control in Uninfected Bacteria

1.1.1. Regulation of Transcription

Of the three steps in the pathway of genetic information transfer—replication, transcription, and translation—transcription must be the simplest, and translation the most complex. RNA is synthesized on a DNA template by a single enzyme composed of several subunits. This enzyme, RNA polymerase, can execute the initiation, propagation, and, in some instances, termination of RNA chains in the absence of accessory proteins. The expression of all but the simplest genomes is selective, either intrinsically or as a result of regulation. In *E. coli* this selectivity generates abundances of mRNA that vary by over four orders of magnitude (Kennell, 1968*b*). At least a part of this selectivity must be unregulated* and must result directly from the interactions of RNA polymerase with promoter-region nucleotide sequences in DNA which permit a discrimination among different transcription units based on the strengths of their promoters. The principles relating DNA structure to promoter strength remain to be discovered. Nucleotide sequences are the experimental foundation on which these principles must be based, and promoter sequences are now being amassed. Regulated selectivity of transcription in bacteria involves the action of numerous proteins. Some of these transcription elements, like the *lac* repressor and the *ara*C protein, have specific effects through their action at unique DNA-binding sites. Others may have more general activities either because they interact with DNA at many sites or because they modify RNA polymerase. However, for any single regulated gene or operon, the number of transcription elements is not large.

1.1.1a. RNA Polymerase

All the bacterial RNA polymerases bear strong resemblances to each other. *E. coli* RNA polymerase, which is the best studied, is a zinc metalloenzyme containing three polypeptides, α, β, and β' (molecular weights 40, 155, and 165×10^3) in a tightly associated complex or "core" whose composition is $\alpha_2\beta\beta'$ (Burgess *et al.*, 1969; Zillig *et al.*, 1970). This core complex possesses polymerase activity and is the unit active in elongating RNA chains *in vivo* (Chamberlin, 1974). A fourth subunit, σ (molecular weight 90×10^3), functions as an initiation factor. A σ subunit of different molecular weight (51×10^3) but similar properties is found in *B. subtilis* RNA polymerase (Avila *et al.*, 1970, 1971; Losick *et al.*, 1970). A second σ (σ_{II}) has been reported by one laboratory as a minor species, tightly bound to only a fraction of RNA

---

* The great majority of *E. coli* genes are transcribed very infrequently *in vivo*. We guess that most of these genes are not specifically regulated; infrequent transcription is likely to be an intrinsic property of these genes and of their promoters.

polymerase (designated as RNA polymerase II; Fukuda *et al.*, 1974; Iwakawa *et al.*, 1974). σ_{II} and RNA polymerase II are reported to have the same spectrum of transcriptional capacity as σ and RNA polymerase I. *E. coli* and *B. subtilis* σ can be readily dissociated from core, but σ_{II} and the initiation factors of a number of bacterial RNA polymerases are less readily dissociable (Herzfeld and Zillig, 1971; Johnson *et al.*, 1971; Bendis and Shapiro, 1973). In *E. coli* and *B. subtilis* only one core RNA polymerase has been found thus far: A single kind of β subunit is responsible for all bacterial messenger and stable RNA synthesis and for the transcription of several bacterial viruses. Another form of RNA polymerase (III) has been reported to participate in primer synthesis for M13 DNA replication (Wickner and Kornberg, 1974). It has been proposed that RNA polymerase III provides one of the two known modes of primer RNA synthesis for DNA replication, having the antibiotic (rifamycin) sensitivity of RNA polymerase I. Another pathway of RNA primer synthesis involves the product of one of the replication genes, *dna*G (Schekman *et al.*, 1975).

1.1.1b. Initiation, Growth, and Termination of RNA Chains

In bacteria, different segments of the genome are expressed at enormously different rates, and most bacterial genes are transcribed very rarely under any growth conditions. Electron micrographs of lysed bacteria, in which coupled transcription and translation can be examined *in situ*, provide the most dramatic illustration of this relationship (Miller *et al.*, 1970). Variations in the frequency of initiation rather than chain growth rate are responsible for this range of genetic activity. The initiation process has been analyzed in some detail *in vitro*, particularly through the use of phage DNA templates which yield only a relatively small number of different transcripts. One can summarize* the current understanding of this subject, which has recently been reviewed in a particularly lucid way (Chamberlin, 1974) in terms of the following reaction sequence: One can conceive of the binding of RNA polymerase to promoter regions of DNA as occurring in at least two steps: DNA recognition, which provides RNA polymerase access to the double helix, followed by tight binding. Little is known about the tight-binding step and one surmises, mainly on the basis of genetic evidence, that it is distinct from recognition-entry. However, no information

---

* This summary refers only to RNA polymerase I, which has been thoroughly studied. Throughout our text we shall refer to this enzyme simply as "RNA polymerase."

about the chemical nature of promoter recognition by RNA polymerase is as yet available.

1. *Binding.* *E. coli* RNA polymerase holoenzyme binds very tightly to limited numbers of sites on double-helical DNA. This binding is strongly dependent on temperature and ionic strength. The interaction between T 7 DNA and *E. coli* RNA polymerase is characterized by the following properties: At 37°C and low ionic strength, the RNA polymerase–DNA complex is extraordinarily stable (dissociation constant of approximately 10^{-14} M). RNA polymerase also binds much more weakly (dissociation constant approximately 10^{-6} to 10^{-8} M) to DNA at very many sites, perhaps at all nucleotide sequences. The σ subunit evidently determines this dichotomy of binding, since RNA polymerase core binds generally to DNA with a single intermediate affinity (dissociation constant approximately 10^{-11} M). At least some of the tight-binding sites on DNA for RNA polymerase are located at initiation sequences for RNA chains. It seems very likely, therefore, that a tight-binding site for RNA polymerase is a part of any strong promoter.

2. *Promoter Activation.* In order for tight binding to occur, RNA polymerase and DNA must undergo an interaction which leads to some unwinding of the DNA double helix (Saucier and Wang, 1972). This process of forming an "open" promoter or "rapid start" complex has the high temperature-dependence characteristic of a cooperative transition (Walter *et al.*, 1967; Zillig *et al.*, 1970; Mangel and Chamberlin, 1974c). Different transcription units and RNA polymerases have different transition temperatures (Remold-O'Donnell and Zillig, 1969; Pène and Barrow-Carraway, 1972; Travers *et al.*, 1973). This must reflect differences in nucleotide sequence and RNA polymerase–DNA interaction. The ease of opening probably helps determine promoter strength.

It is generally assumed that the initiation factor σ is required for formation of open, rapidly initiating complexes between RNA polymerase and promoters. Chamberlin (1974) has pointed out that, plausible as this notion seems, it is not yet supported by experimental evidence. Specifically, it is not yet known whether an open complex would form spontaneously in the absence of σ. The outcome of this experiment would decide whether σ enhances initiation directly by acting at the promoter, or only indirectly by preventing the sequestering of RNA polymerase at nonpromoter sites. (This ambiguity about the role of σ recapitulates past arguments about the role of the initiation factor IF3 in protein synthesis.) Experiments with very small restriction fragments of double-stranded DNA containing intact promotor regions could help to resolve remaining doubts about the mechanism of action of σ.

The notion that promotor activation involves changes of DNA conformation is consistent with the effects of negatively superhelically twisting (supercoiling) DNA on transcription *in vitro*. Supercoiling increases the template activity of circular double-stranded DNA (Hayashi and Hayashi, 1971; Botchan *et al.*, 1973; Wang, 1974; Zimmer and Millette, 1975). It appears to increase the rate of formation and the stability of "open" promoter complexes between *E. coli* RNA polymerase and phage PM2 DNA, allows these complexes to form at lower temperatures, and increases the number of enzyme molecules that can form open complexes on each DNA molecule. Supercoiling evidently does not increase the rate of RNA chain initiation at open complexes (Richardson, 1975). In a general sense, DNA supercoiling produces effects that are analogous to increasing promoter strength. Indeed it has been proposed that the introduction of single-strand breaks into circular double-stranded phage S 13 DNA deactivates transcription as it initiates replication in the infected cell (Puga and Tessman, 1973a).

3. *Initiation.* RNA polymerase that is bound to DNA in an "open" complex can initiate RNA chains extremely rapidly, within a fraction of a second at sufficiently high nucleotide concentrations (Mangel and Chamberlin, 1974a,b,c). The initiation involves the joining of the 5′ terminal nucleotide pppX to its 3′ first neighbor. X is always a purine, *in vitro* and *in vivo* (Maitra and Hurwitz, 1965; Bremer *et al.*, 1965; Maitra *et al.*, 1967; Jorgenson *et al.*, 1969; Konrad *et al.*, 1975). It is the great rate of RNA initiation from these "open" or "rapid start" complexes which evidently makes them so resistant to inhibition by the antibiotic rifamycin, an inhibitor of RNA chain initiation by procaryotic RNA polymerase (Sippel and Hartmann, 1970). The term "initiation" is used loosely here. In fact, rifamycin has recently been shown to block the first translocation step of RNA synthesis: the addition of the third nucleotide to the initial dinucleotide (Johnston and McClure, 1976). *In vitro* and, probably, *in vivo*, RNA synthesis is kinetically limited by the process of RNA polymerase binding to RNA at sites where a rapid start complex can be formed. For this reason, it has been proposed that binding sites for RNA polymerase might be clustered in a "storage region" near those transcription units that initiate RNA chains most rapidly, about once every one or two seconds (Schäfer *et al.*, 1973a). However, there is as yet no compelling experi-

mental evidence in favor of this attractive notion and, in the case of the
T 7 early promoter there is evidence against it (Bordier and Dubochet,
1974; Chamberlin, 1974). At the tight-binding sites, RNA polymerase
protects approximately 40-nucleotide-long segments of double-stranded
DNA from digestion by endonucleases (Okamoto *et al.*, 1972; Heyden
et al., 1972). The protected segments include the template for approximately fifteen 5′ terminal nucleotides of the transcript. The
transcribed sections of different polymerase-protected segments share
no common sequences; presumably each messenger-generating sequence is merely occluded by an RNA polymerase molecule which is
specifically bound at the adjacent tight-binding site. This tight-binding
site may be specified by a staggered sequence (e.g., xAxxGCx-
xCTAAxTx in one strand) resembling the irregular but specific ridges
on a key rather than by a contiguous sequence of nucleotides. Among
the RNA polymerase tight-binding sites that have been analyzed thus
far, an AT-rich sequence of seven nucleotides is relatively "conserved"
(Pribnow, 1975*a,b*).

The varying strengths of different promoters probably arise in one
of the following ways: (a) by variations in the frequency of promoter
recognition or of the rate with which the transition from the
(hypothetical) promoter recognition state to the tight-binding state of
RNA polymerase–DNA complexes occurs; (b) by variations in the rate
of RNA chain initiation by tightly bound RNA polymerase–promoter
complexes. The location of the known promoter strength mutations in
the lactose promoter, both within and adjacent to the tight-binding
region (Dickson *et al.*, 1975; Gralla, 1976) is consistent with all the
above alternatives. However, it may be significant that the two known
mutations in the tight-binding site increase promoter strength, while the
two adjacent mutations decrease promoter strength.

4. *Elongation.* Bacterial RNA chain elongation, *in vivo*, proceeds at varying average rates for different transcription units:
Messenger RNA chains grow 35–50 nucleotides per second (at 37°C),
while ribosomal RNA chains can grow up to twice as fast, depending
on the physiological state of the cell (for recent data on this subject and
a review of prior work see Dennis and Bremer, 1974; Dougan and
Glaser, 1974). *In vitro* experiments (Darlix and Fromageot, 1972;
Dahlberg and Blattner, 1973; Maizels, 1973) suggest that RNA chain
elongation is pulsatile and that there are unique hold-up points.
Nevertheless, the nucleotide sequences and composition at these hold-
up points near the beginning of the *lac* operon are diverse (Maizels,
1973) and offer no insight into the mechanism of saltatory chain elon-

gation. Hold-up points are, however, not confined to promoter-proximal nucleotide sequences. They can evidently also be distinguished in regions that are far removed from promoters (Darlix and Fromageot, 1972; Darlix and Horaist, 1975).

 5. *Termination.* RNA chains terminate at unique signals. RNA polymerase has the ability to terminate RNA chains, releasing the enzyme from its template in an active form that permits reinitiation *in vitro*. This kind of termination, which we call intrinsic, occurs in the absence of any accessory transcription factors at certain DNA sequences under proper reaction conditions. It is possible that σ may also participate in RNA chain termination by *E. coli* RNA polymerase (Zimmer *et al.*, 1974). σ has been found in termination complexes of RNA polymerase with DNA from which RNA chains have already been released (Schäfer and Zillig, 1973b). It has been argued that σ is not required for RNA chain elongation, since the tightness of its association with the RNA polymerase-DNA transcription complex is decreased once RNA chain elongation is underway (Burgess *et al.*, 1969; Darlix *et al.*, 1969; Krakow *et al.*, 1969; Krakow and von der Helm, 1970; C.-W. Wu *et al.*, 1975). It appears therefore likely that there is a σ-cycle in RNA polymerization. It has also been argued that σ is required for the rapid release of RNA polymerase core from DNA. σ may therefore be involved in two separate aspects of the termination of transcription: RNA chain release and RNA polymerase recycling. However, explicit and definitive experiments on this subject remain to be done. The influence of template secondary structure on RNA chain termination has been appreciated for a long time (Chamberlin and Berg, 1964; Maitra *et al.*, 1966). It appears that the superhelical twisting of double-stranded DNA also influences RNA chain termination by RNA polymerase holoenzyme (Zimmer and Millette, 1975).

 RNA chain termination also occurs *in vitro* at sites that are not recognized by RNA polymerase holoenzyme alone. Three *E. coli* proteins, ρ, κ, and ρ_y, act as accessory RNA chain termination factors *in vitro* (Roberts, 1969; Schäfer and Zillig, 1973a; Yang and Zubay, 1974). Rho has been the most extensively studied of these factors. Rho has been shown to terminate RNA chains at sites that correspond to *in vivo* termination signals which are not recognized by *E. coli* RNA polymerase; the signals t_L and t_R of the L and R transcription units of phage λ (Roberts, 1969; De Crombrugghe *et al.*, 1973; Rosenberg *et al.*, 1975). The mechanism of action of ρ is far from completely understood, but a partial picture *is* emerging. Rho can interact with RNA, DNA, and RNA polymerase. It is an RNA polymerase-binding protein and also binds to DNA in a polymerized form at high concentrations (Oda and Takanami, 1972; Minkley, 1973). However, the latter property may not be relevant to its action *in vivo* since high concentrations of ρ can terminate RNA chains *in vitro* at sites that are not used for chain termination *in vivo* (Dunn and Studier, 1973*a,b*). Rho is also a RNA-dependent NTPase, and the NTPase activity evidently is an integral part of RNA chain termination (Galluppi *et al.*, 1976) although the chemical mechanism is not yet understood. The key observations on the function of ρ appear to be: (1) the recent suggestion that ρ action may occur at sites at which RNA chain elongation slows down (Darlix and Horaist, 1975), (2) the recently presented evidence that the polarity-regulating *su*A gene is the structural gene of ρ (Richardson *et al.*, 1975; Ratner, 1976), and (3) the evidence that ρ action is responsible for the polar effect of peptide-chain-terminating mutations (De Crombrugghe *et al.*, 1973; Das *et al.*, 1976), which is in accord with the theory that *amber* polarity involves premature RNA chain termination (Imamoto, 1970, 1973). These observations may be combined into a model with two requirements for ρ activity: (1) that RNA chain elongation slow down at ρ-dependent termination sites, and (2) that naked RNA be available at these sites. Rho interacts with RNA and with RNA polymerase in effecting RNA chain termination at sites at which conditions (1) and (2) are satisfied. Such sites probably are located in certain intercistronic segments or at specific loci in untranslated segments of genes that are distal to (mutationally introduced) chain-termination signals.* Control of ρ-mediated termination may involve proteins that interact with RNA polymerase or with RNA. Ribosomes fall into the latter category, and ρ is, in fact, known to be relatively inactive in cell-free systems in which coupled transcription and translation takes place (O'Farrell and Gold, 1973*b*). The phage λ gene N protein, to which we refer again below, in Section 1.1.2c, falls into the former category. The gene N protein binds to RNA polymerase (Epp and Pearson, 1976) and blocks ρ-mediated RNA chain termination. The precise mechanism of action is not yet known. Adhya and co-workers (1975) have recently suggested a very similar model.

Until very recently, it was not at all certain that ρ functions as a termination factor *in vivo*. While these doubts are now dispelled, it may yet turn out that ρ functions primarily as a *regulatory* element of gene expression in uninfected bacteria and that it only secondarily plays a role in determining the limits of constitutive units of bacterial

---

* See also Notes Added in Proof, p. 196, Note 1.

transcription. We return to this conjecture in Section 1.1.2c. Rho is evidently an essential gene product of *E. coli* (Das *et al.*, 1976; Inoko and Imai, 1976). But is it the transcription-terminating activity of ρ which is essential to cell survival? That is by no means certain. The defective phenotype of conditional-lethal ρ mutants is complex. The complexity might reflect either the pleiotropic effect of defective chain termination, or the multiplicity of rho-mediated functions.

Intrinsic and ρ-mediated terminations by *E. coli* RNA polymerase occur at common nucleotide sequences of the type $CU_{6\text{-}8}A_{0\text{-}1}(OH)_2$ *in vivo* and *in vitro*. (For a compilation and references to the literature, see Rosenberg *et al.*, 1976.) This family of sequences must be recognized by RNA polymerase. The auxiliary nucleotide sequences that determine efficiency and regulation of termination at these more-or-less common sequences must now be sought.

1.1.1c. RNA Processing

The stable ribosomal and transfer RNA's of cells are fashioned from their primary precursor transcripts by sequences of steps that include various kinds of nucleotide modification and endonucleolytic cleavage (Littauer and Inouye, 1973; Dunn, 1975). Some of these cleavages occur rapidly on nascent RNA chains so that the primary transcripts cannot normally be found *in vivo*. It has now been shown that *E. coli* RNase III is the enzyme which executes the cleavage of the 30 S ribosomal precursor RNA into smaller precursors of 23 S and 16 S ribosomal RNA (Kossman *et al.*, 1971; Dunn and Studier, 1973*b*; Nikolaev *et al.*, 1973*a,b*). Other processing RNases are involved in tRNA maturation (Schedl *et al.*, 1975; Ozeki *et al.*, 1975). RNase III recognizes nucleotide sequence (Kramer *et al.*, 1974) and conformation (Robertson *et al.*, 1968*b*). This enzyme also cleaves T 7 mRNA precursors into their constituent messages, as we shall relate in Section 2. However, it is not yet certain whether a substantial fraction of bacterial mRNA is processed by this enzyme (Apirion, 1975). Effects of secondary structure interactions on mRNA translation are postulated for the RNA phages (Weissmann *et al.*, 1973; Eoyang and August, 1974). The secondary structure of a segment of R17 RNA is also important for its interaction with phage coat protein, which acts as a cistron-specific translation inhibitor (Gralla *et al.*, 1974; see Section 1.1.3c). It is conceivable that processing RNases and proteins that protect RNA from these RNases might act as positive and negative posttranscriptional control elements. Regulated posttranscriptional processing of RNA by nucleases may be a major control device of higher eucaryotes, but there is, as yet, no evidence of posttranscriptional regulation of mRNA function by processing nucleases in uninfected bacteria.

The ribosomal and transfer RNAs of bacteria are subjected to numerous modifications; the variety of these modifications is particularly rich in tRNA (Littauer and Inouye, 1973; Hall, 1971). In contrast, *E. coli* mRNA is almost devoid of such modification. It is true that some 3′ terminal posttranscriptional addition of adenylic acid occurs in *E. coli* and that some of these polyA ends are relatively long (25 nucleotides on the average, Nakazato *et al.*, 1975). However, most *E. coli* mRNA lacks the extensive 3′ adenylylation of eucaryotic mRNA, and bacterial mRNA also lacks the modified 5′ termini of eucaryotic mRNA (Rottman *et al.*, 1974). [On the other hand, the fraction of *Caulobacter crescentus* mRNA that bears relatively long polyA tracts is quite large (Ohta *et al.*, 1975).] There is evidence that the 5′ terminal guanylylation and methylation of eucaryotic mRNA is required for initiation complex formation (Both *et al.*, 1975) and that there is a close connection between 5′ terminal modification of eucaryotic messages and their monocistronic translation. In contrast, procaryotic messages have ribosomal binding sites that are recognized as unmodified nucleotide sequences.

1.1.1d. Regulation and Discrimination

At this point we emphasize the precise meanings that we attach to two terms that have already been used: discrimination and regulation. Regulation and discrimination can occur either at the transcriptional or at the translational levels. We regard a gene as being subject to regulation only if, during normal genome function, it exhibits more than one discrete level of function. Transcriptional discrimination is the term that is used to designate those mechanisms which allow different transcription units to be read at different frequencies constitutively, that is, in the absence of regulation. The assumption that transcriptional discrimination must exist was stated at the outset (Section 1.1.1). Transcriptional discrimination must arise from divergencies in the interactions of different promoters with one or more RNA polymerases. The interactions which might have significance for constitutive promoter strength are (1) binding and promoter activation and (2) initiation of synthesis. We believe that the former is more important in this connection than the latter. From an evolutionary point of view the distinction between discrimination and regulation is less clear: We guess that the fitness of *E. coli* is at least as dependent on its ability to transcribe most of its genes constitutively only very rarely (Section 1.1.1) as it is on the ability to shut off the catabolite-sensitive genes when their products are not required (Section 1.1.2a). The evolution of transcriptional discrimination must involve changes of promoter sequence and of polymerase binding, and this is precisely the objective of transcriptional regulation. On the time scale of bacterial and viral evolution, the distinction between regulation and discrimination therefore is blurred. However, on the time scale of cell function and viral development which concerns us, the distinction between discrimination and regulation is useful, and we shall also use it in the context of post-transcriptional aspects of gene expression.

1.1.1e.   Stability of mRNA

Bacterial mRNA is functionally and chemically unstable. Its chemical degradation, ultimately to mononucleotides, surprisingly involves different mechanisms in *B. subtilis* and *E. coli*, being phosphorolytic in the former and hydrolytic in the latter (Duffy *et al.,* 1972; Chaney and Boyer, 1972). Transcription and translation are concurrent and dynamically compatible. Messages are probably protected from degradation by ribosomes. Yet, inactivation of messenger function and chemical degradation of RNA are not necessarily coincident in time (Kennell and Bicknell, 1973); the functionally inactivating event which blocks further initiation of peptide chains is presumed to be some nucleolytic cleavage near the 5′ ends of messages or near internal ribosome-binding sites. The functional lifetimes of different *E. coli* messages and even of different segments of the same polycistronic message are not identical (Blundell *et al.,* 1972; Kennell and Bicknell, 1973). Certain messages, including mRNA for some *E. coli* membrane proteins and for an abundant protein of sporulating *B. thuringiensis,* are relatively stable (Glatron and Rapoport, 1972; Petit-Glatron and Rapoport, 1975; Huang, 1975). The mechanism of mRNA degradation is still not firmly established (for a review and one specific hypothesis, see Apirion, 1973). The quantity of polypeptide that different genes are capable of yielding is determined by the functional stability of their conjugate messages and by the frequency of peptide chain initiation. Gene-specific regulation of protein synthesis might, therefore, involve regulation of mRNA lifetime. This appears not to be the case for those regulated bacterial genes that have so far been studied. On the average,

E. coli messages are translated approximately 15–40 times each (the larger number corresponding to more rapid growth in rich media; Dennis and Bremer, 1974). For tryptophan synthetase mRNA the translational yield is comparable (Baker and Yanofsky, 1972), but translational yields of messages for all *E. coli* proteins probably are not the same. The RNA and DNA phages also show wide variations of message-specific translational yields, which are at least partly due to differences of translational initiation rate. The RNA phages have already been discussed by Eoyang and August (1974) in Volume 2 of this series. In the case of phage λ, all the late genes form part of a single, extremely long transcription unit (Herskowitz and Signer, 1970; see also Chapter 2 of this volume). Different segments of this unit are transcribed with the same frequency and have the same chemical stability (Ray and Pearson, 1974), yet the peptide gene product yields of different late genes vary over a very wide range (Hendrix, 1971; Murialdo and Siminovitch, 1972). Some of the phages also make mRNA that has a distinctly higher functional stability than host mRNA (see Section 2.3.6 and 4.2.4). No gene-specific posttranscriptional regulation of gene expression in uninfected *E. coli* has yet been proven to exist although posttranscriptional discrimination clearly does exist as, for example, in the lactose operon whose messages yield several (3.5 to 7) times as many β-galactosidase as transacetylase polypeptide chains (Zabin and Fowler, 1970). However, such regulation is part of the repertoire of the phages (Sections 1.1.3c and 4.2.4).

1.1.1f. Control of Stable RNA Synthesis

The larger bacteria like *E. coli* and *B. subtilis* derive their great capacity for ribosome synthesis partly from their multiple copies of the rRNA genes. In *E. coli*, these rRNA transcription units are distributed about the genome rather than contiguous. Each ribosomal transcription unit contains a segment for 16 S, 23 S, and 5 S RNA, arranged in such a way that a primary transcript

$$5' \cdots (16\,S) \cdots (23\,S) \cdots (5\,S) \cdots 3'$$

is processed by means of nucleolytic cleavage and methylation to yield mature 16 S, 23 S, and 5 S RNA in ribosomal precursor particles and ultimately in ribosomes. At least some ribosomal transcription units contain a tRNA gene in the spacer region between the 23 S and 16 S genes (Lund *et al.*, 1976).

The six ribosomal transcription units of *E. coli* yield up to 60% of the total synthesized RNA depending on bacterial growth rate. At the highest rates of rRNA synthesis, 60 or more rRNA precursor chains are simultaneously synthesized on a single rRNA transcription unit, implying rRNA initiation rates of up to one per second at each ribosomal promoter. The rate of elongation of these rRNA transcripts is rapid relative to message transcripts, being 70 to 100 nucleotides per second at relatively low (0.7 generations per hour) and high (2.1 generations per hour) growth rates, respectively.

Ribosomal RNA synthesis is regulated to the growth and nutritional state of *E. coli*. In more slowly growing bacteria, the absolute rate of total RNA synthesis is lower and the proportion of RNA synthesis that is devoted to stable species is also lower (reviewed by Kjeldgaard and Gausing, 1974, and by Nierlich, 1974). Deprivation of a required amino acid results in a rapid cessation of stable RNA synthesis and a redistribution of gene activities in which some genes are shut off and some are turned on (Cashel and Gallant, 1974; O'Farrell, 1975*a*). This "stringent" response to amino acid starvation is controlled by genes that are designated as *rel;* *rel*A is the most studied of these genes. The *rel*A function is required for synthesis of guanosine 3′ diphosphate 5′ diphosphate (ppGpp). The *rel*A gene also controls the response of general cellular metabolism to amino acid deprivation, presumably because ppGpp is a general effector in this metabolic integration (reviewed by Cashel and Gallant, 1974). ppGpp is synthesized on ribosomes from GDP and ATP in a reaction that requires uncharged tRNA bound to the aminoacyl tRNA acceptor site of the ribosome and also requires the so-called "stringent factor," which is the product of the *rel*A gene and is a ribosome-binding protein (Block and Haseltine, 1974).

The search for the mechanism of the regulation of rRNA synthesis through *in vitro* experiments has thus far been inconclusive. Experiments with crude cell-free systems which produce relatively high proportions of ribosomal transcripts have not yet shown the appropriate response to ppGpp (Winsten and Huang, 1972; Murooka and Lazzarini, 1973). The assignment of elongation factor T as the positive regulator of rRNA transcription (Travers, 1973; 1974) has been disputed (Haseltine, 1972).

1.1.2. Bacterial Transcriptional Control Elements

The activity of many transcription units in bacteria is subject to positive and negative control. As far as is known, all such control involves proteins. Some of the control proteins, like the *E. coli* catabolite gene activator protein (designated as CGA, CRP, or CAP), affect many transcription units; others, like the λ, *hut,* and *arg* repressors, the *ara* regulatory protein (*ara* C), and the *deo* regulatory protein affect small numbers of transcription units; and still others, like the *gal, lac,* and *trp* repressors, only affect the transcription of single sets of genes located in single transcription units. In this section we shall briefly review the mechanisms of action of four of the best-understood transcription elements.

1.1.2a.   Transcriptional Control Elements Acting on DNA

DNA-binding proteins exert positive and negative control on transcription. The negatively controlling *lac, trp,* and λ repressors are among the best-studied of these. We shall also discuss the CAP protein which exerts positive control and the *ara* regulatory protein (product of gene *ara* C) which exerts dual, that is, positive *and* negative, control.

The *lac* repressor DNA-binding protein is a tetramer of a 37,000 molecular weight subunit (Beyreuther *et al.,* 1973) probably arranged around a twofold axis (Steitz *et al.,* 1974). It binds preferentially to a single operator sequence in *E. coli* DNA (Riggs *et al.,* 1970; Fig. 1) which also has a partial twofold symmetry (Gilbert and Maxam, 1973). (Two nearby weaker binding sites have also been found.) There is evidence that this symmetry is functionally significant (Sadler and Smith, 1971). The *lac* repressor binds sufficiently tightly to the operator to protect a 24-nucleotide-long sequence of double-stranded DNA from nucleases. However, the repressor also has an approximately 10^6-fold lower affinity for DNA in general (von Hippel *et al.,* 1975). Thus, only about one-half of the repressor-operator standard binding free energy is due to their *specific* interaction, which must take place in the grooves of helical operator DNA, since the *lac* repressor does not substantially unwind the operator (Wang *et al.,* 1974). The *lac* repressor blocks the binding of RNA polymerase to the tight-binding site. The overlapping DNA segment of the repressor and RNA polymerase binding sites codes for the 5′ terminal segment of *lac* message; thus, the *lac* operator is transcribed (Maizels, 1973; Majors, 1975b; Dickson *et al.,* 1975; Fig. 1). When the *lac* repressor binds to its operator, *lac* transcription and, consequently, protein synthesis is lowered by between three and four orders of magnitude (Jacob and Monod, 1961). But in *E. coli* growing on lactose, an inducer, allolactose (1-6-*O*-β-D-galactopyranosyl-D-glucose; Jobe and Bourgeois, 1972), binds to the repressor as it sits on its DNA binding site and alters its conformation in such a way as to desta-

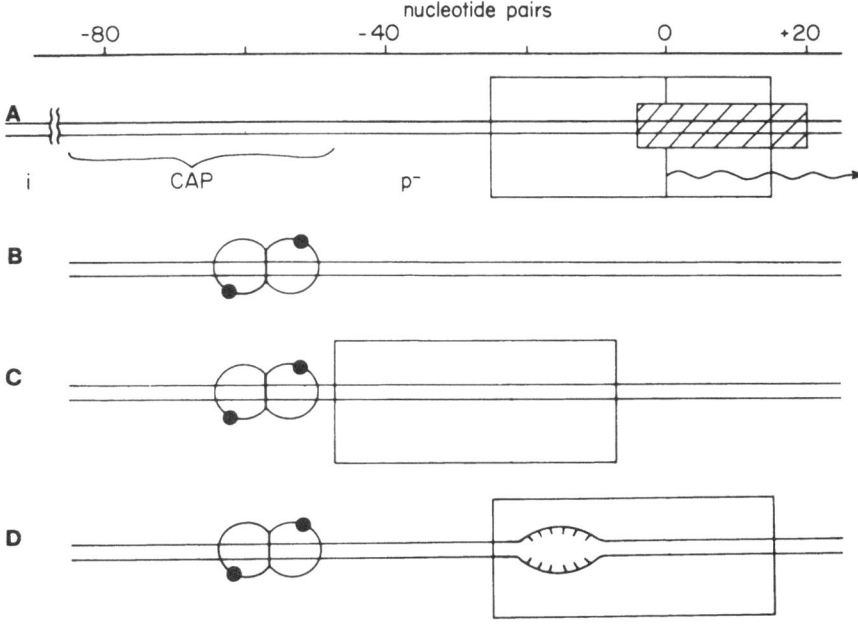

Fig. 1.   The lactose operon regulatory region (after Dickson *et al.*, 1975). Line A shows the segment of DNA protected by the *lac* repressor, ▨▨▨, which overlaps the RNA polymerase tight-binding site ⬚⬚⬚ and the start of the *lac* message (wavy line). Nucleotides 39–41 of that message are the initiating AUG codon for β-galactosidase (gene *z*). Promoter mutations, p⁻, are known to occur to the left of the RNA polymerase tight-binding site. The CAP site is also indicated. The structural gene (i) for *lac* repressor is to the left of the CAP site. The scale is marked in nucleotide pairs from the mRNA initiating nucleotide as origin. Lines B, C, and D present a model of the initiation of *lac* operon transcription. Line B shows the complex of cAMP (small black circle) with the CAP protein (large circle) binding to the CAP site as a dimer and thus facilitating the entry of RNA polymerase [cf. Section 1.1.1b(1)] which is shown on line C. The RNA polymerase then drifts to its tight-binding site (line D) in which a cooperative conformational change of template and protein has occurred. (It is not established that RNA polymerase "drift" is a part of promoter activation. If RNA polymerase were to span a greater stretch of DNA than shown in line A, the transition from promoter recognition to activation might occur without drift. For the sake of being concise, this alternative picture of promoter activation has not been represented.)

bilize the repressor–inducer–DNA complex. RNA polymerase can then bind, initiate, and transcribe the operator sequence and the operon. The maximum rate of *lac* operon transcription is also dependent on the catabolite gene activator protein, CAP, in its activated, cAMP-bound form. The cAMP–CAP protein complex binds to the regulatory region of the *lac* operon on the repressor-distal side of the promoter (Fig. 1). The mechanism by which CAP–DNA binding promotes *lac* RNA synthesis is not well understood. cAMP-activated CAP has a relatively high affinity for all kinds of DNA (Riggs *et al.*, 1971), but a preferential binding to the *lac* promoter region and to a CAP-sensitive site in that region has recently been demonstrated (Majors, 1975). It is not known how CAP binding affects RNA polymerase and whether these two components interact directly or only through DNA. The specific transcription-stimulating activity of CAP might be due to a synergistic interaction with RNA polymerase (Anderson *et al.*, 1974). Whether CAP, like RNA polymerase, unwinds DNA is not yet known. If CAP were an unwinding protein, then it might act on a segment of promoter-contiguous nucleotide pairs that would otherwise depress the formation of an open promoter (rapid-start) RNA polymerase–DNA complex.

The λ repressor also is an oligomeric DNA-binding protein (Chadwick *et al.*, 1970; Brack and Pirotta, 1975). Its interaction with two operator sites on λ DNA is remarkable in two respects: Each operator binds varying numbers of λ repressor oligomers depending on the relative proportions of repressor and operator. The polymerization of repressor on each operator has an origin ($O_R I$ in Fig. 2) and a polarity ($O_R I \rightarrow III$). Depending on the number of repressor molecules bound at each operator, between 30 and 110 contiguous nucleotide pairs of DNA are covered by this binding (Maniatis and Ptashne, 1973; Brack and Pirotta, 1975). The nucleotide sequence of this operator DNA segment is not, itself, precisely repetitive. As mentioned, the λ repressor binds to the operator without unwinding the DNA double helix. But the sites for RNA polymerase and repressor binding in the L and R operator regions overlap: At each locus, the approximately 40-nucleotide-pair segment of DNA protected by RNA polymerase includes a segment of approximately 20 nucleotides also covered by the λ repressor. This implies that the λ repressor and RNA polymerase probably have different effects on the same nucleotide sequence, which the latter unwinds and the former does not. λ repressor therefore blocks RNA polymerase–promoter binding. The λ operators are message-distal to their promoters and are therefore not transcribed from the λ_L and λ_R promoters. The spatial relationship of the *lac* operator and promoter is the opposite (compare Figs. 1 and 2), yet the mechanism of action of these two repressors (and of the *trp* repressor) appears to be the same. Once the repressors were shown to be DNA-binding proteins, it seemed likely that they would function in one of three different ways: (1) by blocking RNA chain elongation very close to the 5′ end of the message; (2) by blocking RNA polymerase binding or entry; or (3) by altering RNA polymerase–promotor interaction, for example, by sta-

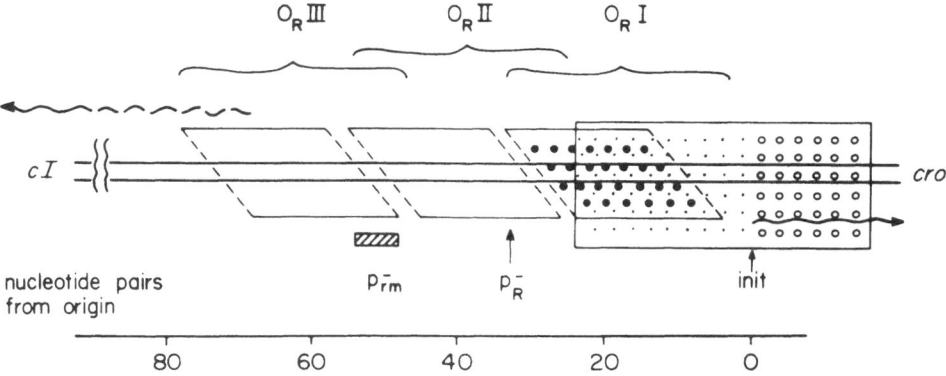

Fig. 2.   Structure of the transcription-regulatory site of the major rightward early (R) operon of bacteriophage λ (after Walz and Pirotta, 1975). The RNA polymerase tight-binding site [:::88] overlaps the major repressor binding site, O_RI \\•••••\\. The rightward transcript starts at zero and continues through the regulatory gene *cro*. Tightly bound RNA polymerase protects approximately 15 transcribed nucleotide pairs of DNA, [o o], as well as approximately 25 untranscribed nucleotide pairs, [:::::::::], of DNA from nuclease digestion. A promoter mutation, P_R^-, is located out-side this protected region (Allet and Solem, 1974) in a DNA segment that is postulated to regulate RNA polymerase entry. The λ repressor is an oligomeric protein that binds to the two operator regions, O_R and O_L, in repressing the major rightward and leftward λ operons, respectively. At both operators, increasing concentrations of λ repressor protect increasing lengths of DNA from nuclease digestion (Maniatis and Ptashne, 1973). The segments covered by a unit parallelogram might be protected by repressor tetramer. Three such tetramers are shown bound at sites I → III. The lambda repressor, which is the product of gene *c*I, stimulates its own synthesis (Heinemann and Spiegelman, 1970; Reichardt and Kaiser, 1971; Echols and Green, 1971). There is evi-dence that this positive regulatory activity also involves repressor-operator binding in the O_R region (Reichardt, 1975b). Indeed, a "promoter" mutation, p_{rm}^-, for the transcription unit that transcribes the λ repressor gene, *c*I, in its maintenance mode has been found (Hedgepeth and Smith, quoted by Walz and Pirotta, 1975). The 5′ end of the transcript that contains the *c*I message and starts near p_{rm} has not yet been mapped and is therefore indicated by a broken wavy line. This regulatory region also includes one or more sites at which the *cro* gene regulatory protein blocks its own expression (Reichardt, 1975a) and blocks synthesis of λ *c*I repressor (Calef and Neubauer, 1968; Reichardt, 1975b). The sites have not yet been precisely mapped.

bilizing DNA against unwinding and thereby preventing the formation of open, rapid-start complexes. All the hitherto analyzed repressors block the tight-binding site of the promoter region. It transpires that the λ repressor also is a positive regulator—of its own transcription from the "maintenance" or p_{rm} promoter (Reichardt, 1975b; Figure 2 legend). We omit further discussion here because the wonderful intricacies of λ repressor regulation are discussed in Chapter 2 of this volume.

The regulatory protein of the *E. coli* arabinose operon (*ara* genes B, A, D) is both a repressor and an activator. It is a repressor in the absence of its inducer, L-arabinose, or in the presence of its antiinducer, D-fucose. The *ara* repressor acts at an operator site (o) to limit expression of the contiguous three *ara* genes, almost certainly at the transcriptional level. When the *ara* regulatory protein binds L-arabinose, it undergoes a conformational change that converts it to an activator. The activator–L-arabinose complex promotes expression of the contiguous *ara* genes by binding to an initiator region (i) of the *ara* BAD operon which also holds sites for the activated CAP protein and for RNA polymerase. The order of sites and genes in the direction of transcription is oiBAD. The detailed structure of the i region is not yet known, and it is also not yet known whether the *ara* repressor blocks RNA polymerase entry or tight binding (reviewed by Engelsberg and Wilcox, 1974).

DNA-binding proteins which stimulate transcription *in vitro* have been isolated from *E. coli*. Three of these, which have been characterized to some extent, are named H_1, D, and HU (Cukier-Kahn *et al.*, 1972; Ghosh and Echols, 1972; Rouvière-Yaniv and Gros, 1975). Protein H_1 can act selectively on *lac* transcription at a site that must be close to the CAP-binding site of the *lac* promoter region but also stimulates transcription *in vitro* generally (Crepin *et al.*, 1975). However, there is no genetic evidence regarding the function of these proteins *in vivo*. Protein HU might have a structural role in DNA folding within the nucleoid (Rouvière-Yaniv and Gros, 1975; Griffith, 1976).

1.1.2b. Transcription Control Elements Acting on RNA Polymerase

It has long been surmised that proteins interacting with RNA polymerase might modify its specificity. When σ factor was first separated from RNA polymerase core and shown to be released in transcription, it was postulated that varieties of σ factors would be found in normal and virus-infected cells, where they would act as transcription-selecting elements. With the notable exception that we discuss in Section 3, this prediction has not been borne out thus far, although it is far from certain that it is wrong. A very interesting recent experiment (Ratner, 1974a) shows that many *E. coli* proteins interact with RNA polymerase. Among the numerous transcription-enhancing proteins that have been isolated from *E. coli* in recent years one, the M protein, binds to RNA polymerase holoenzyme (Davison *et al.*, 1970; Ramakrishnan and Echols, 1973), while the mode of binding of several others is not established but may include RNA polymerase attachment

(Mahadik and Srinivasan, 1971; Ghosh and Echols, 1972; Murooka and Lazarini, 1972). The physiological roles of all these proteins are entirely unknown at present. However, if all the RNA polymerase-binding proteins of *E. coli* modulate transcription, then we still have a great deal to learn about RNA synthesis in bacteria and its regulation.

Only one transcription protein is known to be capable of interacting with DNA and RNA polymerase: the termination factor ρ, which we have already mentioned above.

1.1.2c.   The Attenuator: Regulation of RNA Chain-Termination

The regulatory element of transcription that we describe next has only recently been discovered in two *E. coli* operons; in fact, it is only the sites of action of the as-yet-unknown regulatory substances that are clearly identified (Kasai, 1974; Bertrand *et al.*, 1975). The properties of the *trp* attenuator are as follows: There is a 160-nucleotide *leader* region between the 5′ terminus of *E. coli trp* operon mRNA and the initiating AUG codon of the *trp E* (anthranilate synthetase component I) gene. Deletions of a portion of this region lead to a cis-dominant increase of operon expression while retaining regulation by the *trp* repressor (Fig. 3). These properties of the *trp* leader region indicate the presence of an attenuator site within approximately 30 base pairs of the nucleotides that code for the amino terminus of the *trp E* gene. The *trp* attenuator is a site at which RNA chain elongation *in vitro* is blocked, yielding 130-nucleotide-long transcripts that are not released from the template. The 3′ termini of these chains contain the characteristic sequence $CU_{7-8}(OH)_2$, which is one of a family of closely related se-

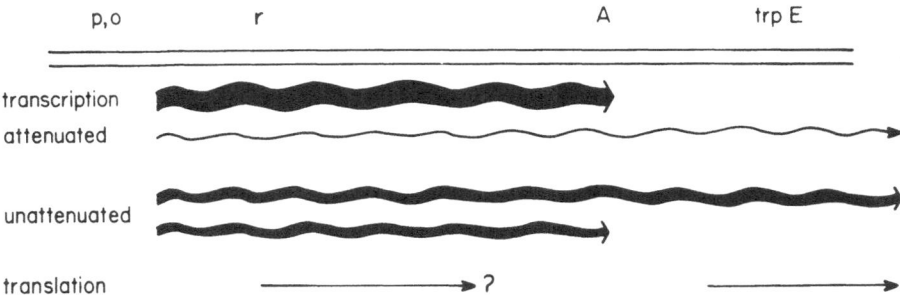

Fig. 3.   The regulatory region of the tryptophan operon (after Bertrand *et al.*, 1975). The order of the promoter, p, and operator, o, has not yet been determined. The letter A designates the attenuator, r designates the ribosomal binding site, and *trp E* the anthranilate synthetase component I gene. The model shows regulated arrest of transcription occurring at A. A small peptide may be coded by the leader of *trp* mRNA.

quences that are shared by intrinsic and ρ-mediated terminations of
RNA synthesis (compare Section 1.1.1b). *In vivo,* the transcription
block at the atternuator also responds to tryptophan levels in the cell.
Certain mutations which reverse the polarity of nonsense or frameshift
mutations also act at or near the attenuator site to increase *trp* mRNA
levels. Several of these mutations map near *su*A, the putative structural
gene of the termination factor, *rho* (see Section 1.1.1b), and probably
are located in the *su*A gene itself (Yanofsky, personal communication).
Thus, RNA chain termination at the attenuator appears to be subject
to regulation (Bertrand *et al;.* 1975). The existence of a similar regulatory site in the histidine operon leader region had been previously inferred (Kasai, 1974). Regulation at the *trp* operon attenuator involves
tryptophanyl tRNA synthetase and correlates with the level of charged
trp tRNA (Morse and Morse, 1976). While the molecular species
directly responsible for attenuation control is not known at this writing,
it may act by antagonizing ρ-mediated RNA chain termination.

   The *trp* operon presents us with an example of regulated termination of transcription in uninfected *E. coli.* However, such a mechanism
had previously been proposed as operating in phage development. The
regulation of RNA chain termination is thought to be the role of the
phage λ gene *N* product which controls the transcription of distal genes
in the two major viral early operons (Roberts, 1969, 1970; Thomas,
1971; Echols, 1971; Dottin and Pearson, 1973). The gene *N* product
also regulates the polar, transcription-terminating effect of chain-
terminating mutations in bacterial genes so long as these genes are
fused to the two major viral early transcription units (Franklin, 1974;
Adhya *et al.,* 1974). This peculiar cis-action of the *N* gene product on
chain termination has received an intriguing explanation (Friedman *et
al.,* 1975): that the capacity for *N*-mediated antitermination anywhere
in a transcription unit is set by an interaction that occurs in the region
of the promoter. The phage λ gene *N* product may act on (or interact
with) the bacterial gene *su*A$^+$ gene product which, as we have already
mentioned, is most likely the RNA chain-termination factor, ρ. The
gene *N* protein also interacts with RNA polymerase (Epp and Pearson,
1975).* But how these dual interactions lead to the peculiar spatial
duality of *N*-mediated antitermination remains to be worked out.

   The *trp* operon of *E. coli* presents us with still another

----

* The first evidence suggesting that the gene *N* protein might interact with RNA
polymerase was the discovery of bacterial (*gro N*) mutations affecting phage λ
development, which could be suppressed by phage λ mutations in gene *N*. The *gro N*
bacterial mutations were located by transduction in the region of the gene coding for
the RNA polymerase β-subunit (Georgopoulos, 1971; Ghysen and Pironio, 1972).

phenomenon with potential regulatory significance. The leader segment of *trp* mRNA also contains a complete ribosome binding site with an initiation codon. Protein synthesis from the RNA segment that lies between this site and the next downstream termination signal could generate a 40-amino-acid peptide, but no such component of the *trp* operon is yet known. Such a peptide might conceivably have a regulatory role in *trp* operon expression.

1.1.3. Regulation of Translation

We intend, in this section, to summarize those steps of normal bacterial translation which could be the targets of regulatory and discriminatory processes either in virus-infected or in uninfected cells. We use the term *translational discrimination*, as distinct from *regulation*, to designate those mechanisms which allow different equally long-lived messages to be translated constitutively, i.e., in the absence of regulation, with different frequency, thereby yielding different quantities of protein. We proceed on the assumption that when regulation or discrimination occurs, it involves initiation of peptide chains and preceding steps of protein synthesis rather than the subsequent peptide-chain elongation. Excellent textbook summaries of protein synthesis are available (e.g., Davis *et al.*, 1973; Lewin, 1974; Stryer, 1975), as is a recent and detailed monograph (Nomura *et al.*, 1975) and several excellent reviews (Haselkorn and Rothman-Denes, 1973; Ochoa and Mazumder, 1974; Lucas-Lenard and Beres, 1974; Tate and Caskey, 1974). We shall not attempt to summarize all of protein synthesis but confine our attention to initiation of peptide synthesis and specifically to those topics that recur in the context of phage development in the rest of this chapter.

1.1.3a. The Initiation Complex

The 30 S initiation complex unites the 30 S ribosome, *fmet* tRNA$_f^{met}$, initiation factors, GTP, and the initiating AUG (or GUG) codon of the message, together with a contiguous noncoding segment of mRNA that serves as a ribosomal binding site. The 30 S initiation complex then reacts with a 50 S ribosome, cleaving GTP, releasing initiation factors, and forming the 70 S initiation complex which is ready to join the first peptide linkage and enter the peptide elongation cycle.

When 30 S ribosomes are bound in initiation complexes, they protect segments of mRNA from RNase digestion. The protected segments include the initiating AUG codon and contiguous leading sequences which are evidently essential for ribosomal binding; no message with an initiation-functional 5'-terminal AUG triplet is known. In the multicistronic RNA phage messages that have been analyzed, the leading ribosome binding sequences are located in untranslated intercistronic regions between contiguous functioning termination and initiation codons. The lengths of these intercistronic regions vary from approximately 25 to several hundred nucleotides. The ribosome binding sequences and leading sequences probably contain structural information for translational discrimination and for regulation. They should, accordingly, be diverse and yet contain some common elements. In fact, their diversity is more readily discernible than their commonality. The ribosome binding sites share no extended sequences or secondary structure but contain a purine-rich oligonucleotide segment approximately 10 nucleotides ahead of the initiator codon. mRNA can bind to 30 S ribosomes in the absence of initiation factor; such binding is dependent on the 30 S ribosomal protein S1 (Szer and Leffler, 1974). Thus, S1 is implicated in mRNA-ribosome binding (Van Duin and Kurland, 1970; Noller et al., 1971; Fiser et al., 1974). However, mRNA-ribosome binding is strengthened by initiation factors, particularly by IF3. Factors IF2, IF3, and protein S1 occupy contiguous positions on the 30 S ribosome (Kurland, 1974; Traut et al., 1974), and all three proteins are relatively easily detached from the 30 S ribosome.

There is evidence that the 3' end of 16 S RNA is important for mRNA binding and for initiation: (1) Kasugamycin inhibition of peptide-chain initiation (Okuyama et al., 1971) is associated with two dimethylated adenine nucleotides near the 3' end of 16 S RNA (Helser et al., 1972). (2) The site of action of streptomycin, which also inhibits peptide-chain initiation, also lies near the 3' end of 16 S RNA and involves ribosomal protein S 12 which is a neighbor of S1 (reviewed by Traut et al., 1974). (3) Colicin E3 exerts its cytotoxic effect by specifically cleaving 16 S RNA near the 3' end. This endonucleolytic cleavage blocks initiation of protein synthesis as well as chain elongation (Tai and Davis, 1974; reviewed by Nomura et al., 1974). (4) The 3' sequences of 16 S RNA from diverse bacteria have recently been determined, and the correlation between these sequences and the 30 S ribosome-species-specific binding and translation of phage RNA (Leffler and Szer, 1973) has been persuasively presented (Shine and Dalgarno, 1974, 1975). (5) An RNA–RNA complex containing the 3' end of 16 S RNA and the initiator region of a message has recently been isolated from a 30 S initiation complex (Steitz and Jakes, 1975).

The model of Shine and Dalgarno identifies RNA–RNA interactions as a component of translational discrimination. Since all *E. coli* 16 S rRNA has the same 3′-terminal nucleotide sequence, the diversity of translational discrimination in this model depends on the diversity of mRNA sequences. To the extent that these discriminatory RNA–RNA interactions are specific for ribosomes from different bacteria, they represent specific adaptations of viral messages to the ribosomes of their respective hosts.

As we have already pointed out, mRNA-ribosome binding involves more than RNA–RNA interactions. Protein S1 and initiation factors play determining roles in messenger-ribosome binding. Indeed, S1 probably interacts with the 3′ end of 16 S RNA (Kenner, 1973). S1 is a protein which binds to single-stranded nucleic acid (Miller and Wahba, 1974; Carmichael *et al.*, 1975) and which also displays a nucleic acid unwinding activity (Noller, 1975). S1 might unwind ribosomal or messenger RNA or, perhaps, both kinds of RNA.

Only two 30 S ribosomal proteins have so far been noted in connection with the formation of the initiation complex. The detailed topography of the ribosome is of course not yet known, but something is known about proximities or neighborhoods on the ribosome. At least nine proteins are situated near the 3′ end of 16 S RNA or the initiation factors IF2 and IF3. Proteins S11, 12, 13, 14, and 19 are near IF2 and IF3. Protein S7 binds to the 3′ terminal one-third of 16 S RNA (but not to the 3′ end), and proteins S9, 10, 13, and 19 are associated with S7 as well as with each other. Proteins S3, 9, 10, and 14 probably lie near the 3′ terminal segment of 16 S RNA (Traut *et al.*, 1974).

If we give *fmet* tRNA and its initiation-factor-dependent and GTP-dependent binding to the 30 S ribosome-mRNA complex short shrift, it is because this step seems unlikely to be involved in cistron-specific discrimination.

1.1.3b. Initiation Factors and *i*-Factors

It is in principle possible that translational discrimination might be built into the translational machinery as well as the collection of messages. We shall next consider this question and the view that the capacity for some discrimination *in vivo* is associated with translational initiation and interference factors, or *i*-factors. (Interference factors, or *i*-factors, are proteins which selectively inhibit protein synthesis *in vitro.*) It is clear that such translational discrimination can be exhibited *in vitro* by different inhibitory *i*-factors from uninfected *E. coli*; according to a widely held view, two classes of the initiation factor IF3 also exhibit such discrimination *in vitro*. Two *i*-factors, i_α and i_β, have been purified and relatively well studied. Both bind strongly to synthetic polynucleotides. i_α is the 30 S ribosomal protein S1 and incorporates into Qβ RNA replicase as subunit I (Groner *et al.*, 1972*b*; Kamen *et al.*, 1972; Wahba *et al.*, 1974); i_α-S1 inhibits polyU and polyC translation and the translation of the coat-protein cistron of RNA phage RNA (Lee-Huang and Ochoa, 1972; Revel *et al.*, 1973*a*; Miller *et al.*, 1974; Miller and Wahba, 1974), yet its presence in ribosomes is required for translation (Van Duin and Kurland, 1970; Noller *et al.*, 1971). Strong translational selectivity *in vitro* only occurs when relatively large proportions of S1 to mRNA are used (Lee-Huang and Ochoa, 1972; Jay and Kaempfer, 1974; Miller and Wahba, 1974; Van Dieijen *et al.*, 1975). An excess of certain other ribosomal proteins (S3, 14, and 21) also inhibits polyU-directed polyphenylalanine synthesis (Van Dieijen *et al.*, 1975). Inhibition by S1 and by the other ribosomal proteins is relieved by excess mRNA (Jay and Kaemfer, 1974; Miller and Wahba, 1974; Van Dieijen *et al.*, 1975). Probably the i_α-factor activity of S1 is associated with its binding to pyrimidine-rich nucleotide sequences on mRNA, and this binding may be in competition with the functional interaction of ribosome-bound 16 S RNA and protein S1 with mRNA. It is entirely possible that S1 might play a role in translational discrimination *in vivo*. However, the mechanism would most probably involve S1 as a part of the ribosome and not as a ribosome-free *i*-factor. We return to S1 in a related role below.

It has been suggested that a second *i*-factor, i_β, is identical with host factor I of Qβ replicase (Carmichael *et al.*, 1975). i_β preferentially inhibits late T4 mRNA translation *in vitro* (Groner *et al.*, 1972; Lee-Huang and Ochoa, 1972), and host factor I inhibits polyA translation *in vitro* (Carmichael *et al.*, 1975).

According to one view, IF3 is a single protein with general initiation activity towards RNA phage RNA, endogenous *E. coli* mRNA, as well as early and late T4 mRNA (Schiff *et al.*, 1974). Another view is that two closely related IF3 proteins, IF3$_\alpha$ and IF3$_\beta$, exist and that they have slightly different molecular weights. One of these could be a cleavage product of the other (Schiff *et al.*, 1974). IF3$_\alpha$ and IF3$_\beta$ are said to assert translational preferences that are conjugate with i_α and i_β: IF3$_\alpha$ preferentially translates MS2, *E. coli*, and T4 early RNA, while IF3$_\beta$ preferentially translates T4 late RNA (Ochoa and Mazumder, 1974).

1.1.3c. Translational Repression

It is probably significant that every bacterial regulatory system that has so far yielded to analysis involves regulation at the transcriptional rather than the translational level. We must turn to the RNA phage messages to find examples of translational repression. Since the regulation of RNA phage development has already been discussed in Volume 2 of *Comprehensive Virology* (Eoyang and August, 1974), we shall merely summarize very briefly so as to refocus that review for our purposes. The coat proteins and the RNA replicases of the *E. coli* RNA phages are site-specific repressors of viral protein synthesis. The coat protein of R 17 phage represses the synthesis of R 17 replicase *in vitro* (Gussin 1966; Sugiyama and Nakada, 1967; Robertson *et al.*, 1968a) by binding to RNA at a specific site which includes the segment of the coat-protein gene that codes for the C-terminal peptide, the intercistronic divide, the ribosomal binding site of the replicase gene, and its initiating codon (Bernardi and Spahr, 1972). A single molecule of coat protein probably recognizes the specific secondary structure of this RNA segment (Gralla *et al.*, 1974).

The $Q\beta$ replicase is an inhibitor of $Q\beta$ coat-protein-synthesis initiation and of *fmet* tRNA binding to $Q\beta$ RNA–ribosome complex. It exerts this effect by binding to the RNA at a site that overlaps with the ribosome binding site of the coat-protein gene. $Q\beta$ replicase does not inhibit completion of initiated coat-peptide chains (Kolakofsky and Weissman, 1971; Weber *et al.*, 1972 Weissman *et al.*, 1973). $Q\beta$ replicase is a complex of three bacterial nucleic-acid-binding proteins with one virus-specified subunit. The three bacterial protein subunits are S1-i_α, T_u, and T_s (Blumenthal *et al.*, 1972; Wahba *et al.*, 1974; Inouye *et al.*, 1974). Thus S1, in conjunction with the other proteins of the $Q\beta$ replicase, can act as a site-selective regulator of translation. The role of this translational repression in the development of the RNA phages has already been discussed in Volume 2 of this series. The recruitment of three RNA- and nucleotide-binding proteins of the host's translational apparatus for viral RNA replication is a wonder-inspiring manifestation of parasitism at the molecular level.

2. BACTERIOPHAGES T 7 AND T 3

2.1. Introduction

We have chosen bacteriophage T 7 as the first prototype for our discussion of viral gene regulation. This is neither a difficult nor a subtle choice, for the analysis of T 7 gene expression teaches us two basic lessons. The first lesson concerns the relationship between promoters, genes, transcription units, messages, and translation units. The second lesson concerns the elaboration of a genome-specific RNA polymerase as a strategy of viral gene regulation.

T 7 is a virulent phage of *E. coli,* first isolated, together with its relative T 3, by Demerec and Fano (1945) and independently by Delbruck (1946*a,b*). It has a polyhedral head with a small tail. Its entirely double-stranded, linear DNA has a molecular weight of 25 × 10^6 daltons and contains approximately 38,000 base pairs (Dubin *et al.,* 1970; Freifelder, 1970). The DNA is nonpermuted but has a terminally repetitious double-stranded segment of approximately 270 nucleotide pairs (Ritchie *et al.,* 1967). T 7 is the most-studied representative of a group of phages, including the *E. coli* phages T 3, ϕ I, ϕ II, W 31, and the *Yersinia pestis* phage H, whose genetic and regulatory homologies have been studied to some extent (Hausmann and Härle, 1971; Hausmann, 1973; Davis and Hyman, 1971; Hyman *et al.,* 1973, 1974; Brunovskis *et al.,* 1973). These coliphages multiply in female *E. coli* but are restricted by strains carrying the F episomes. T7 phage infection of these male *E. coli* is aborted within minutes, and the infected cells die. We shall discuss the roots of this nonpermissiveness in Section 2.6. The virulence of bacteriophage T 7 has two particular manifestations: The phage first shuts off host-gene expression and then, independently, degrades host DNA so as to provide a major source of nucleotides for viral DNA replication.The characteristic time scale of viral development (at 30°C) is that viral DNA replication begins approximately 8 min after infection, complete phage accumulate from about the 23rd minute, and bacterial lysis begins 25–30 min after infection, releasing progeny virions into the medium.

2.2.   The Physical and Genetic Map of T 7; General Features of Regulation

The physical and genetic map of phage T 7 which is presented in Fig. 4 is probably almost complete, at least with respect to the location of genes (Studier, 1972). All T 7 proteins are coded by the same strand, the *r*-strand of T 7 DNA (Summers and Szybalski, 1968; Studier and Maizel, 1969). The T 7 proteins are synthesized in a temporal sequence that bears a particularly simple relationship to the genetic map. The protein products of genes 0.3 to 1.3, designated as group I, are synthesized first, from 4 to 8 min after infection at 30°C. When these proteins cease to be made, host-protein synthesis is also shut off. The group I genes are the early T 7 genes (Studier, 1972). Genes 1.3 to 6, designated as group II, start to be read 6 to 8 min after infection and are shut off about 15 min after infection. Their products include enzymes essential for T 7 DNA replication. The proteins of genes 7 to 19, which constitute group III, start their synthesis almost together with the group II proteins but continue to be synthesized until lysis. The group II and III genes are the late genes. Only gene product (gp) 1.3,* the T 7 DNA ligase, needs to be included in more than one of these groups: It is synthesized as a group I and as a group II protein. We next turn to an analysis of what is known about the underlying mechanisms of this regulation. Two regulatory principles are relatively well understood (see reviews by Studier, 1972; Summers, 1972; Hausmann, 1973): (1) The group I genes, which constitute the left-hand 20% of the genome (Fig. 1), are injected first into the bacterium (Pao and Speyer, 1973) and are read by unmodified bacterial RNA polymerase. No viral protein is required for the expression of these genes. (2) The group II and group III late genes are read by T 7 RNA polymerase, which is the product of T 7 gene 1; the positive control of the expression of these genes probably is exclusively transcriptional and is independent of DNA replication.

2.3. The Early Genes and *E. coli* RNA Polymerase

The T 7 early genes code for five proteins: gp 0.3, 0.7, 1, 1.1, and 1.3. The RNA polymerase (gp 1; molecular weight 107,000) accounts for almost half of the coding capacity of this chromosome segment. Gene product 0.3 functions in the inhibition of host DNA restriction enzymes (Studier, 1975a). Gene product 0.7 is a protein kinase (Rahmsdorf *et al.*, 1974) and functions in shutting off host-gene expression, and gp 1.3 is a DNA ligase. The function of gp 1.1 is not yet understood. Only the function of gene 1 is absolutely required for T 7 multiplication on wild-type nonrestricting F⁻ *E. coli*; consequently, a number of deletion mutants in this region have been isolated. Their accurate mapping (Studier, 1973a,b; Simon and Studier, 1973; Minkley, 1974a) has helped to align RNA and protein molecules with

---

* We use the following convention for designating the connection between genes and proteins: gp n is the primary polypeptide product of gene n. If this primary product is further processed, for example, by proteolytic cleavage, the processed protein is designated as gp n*.

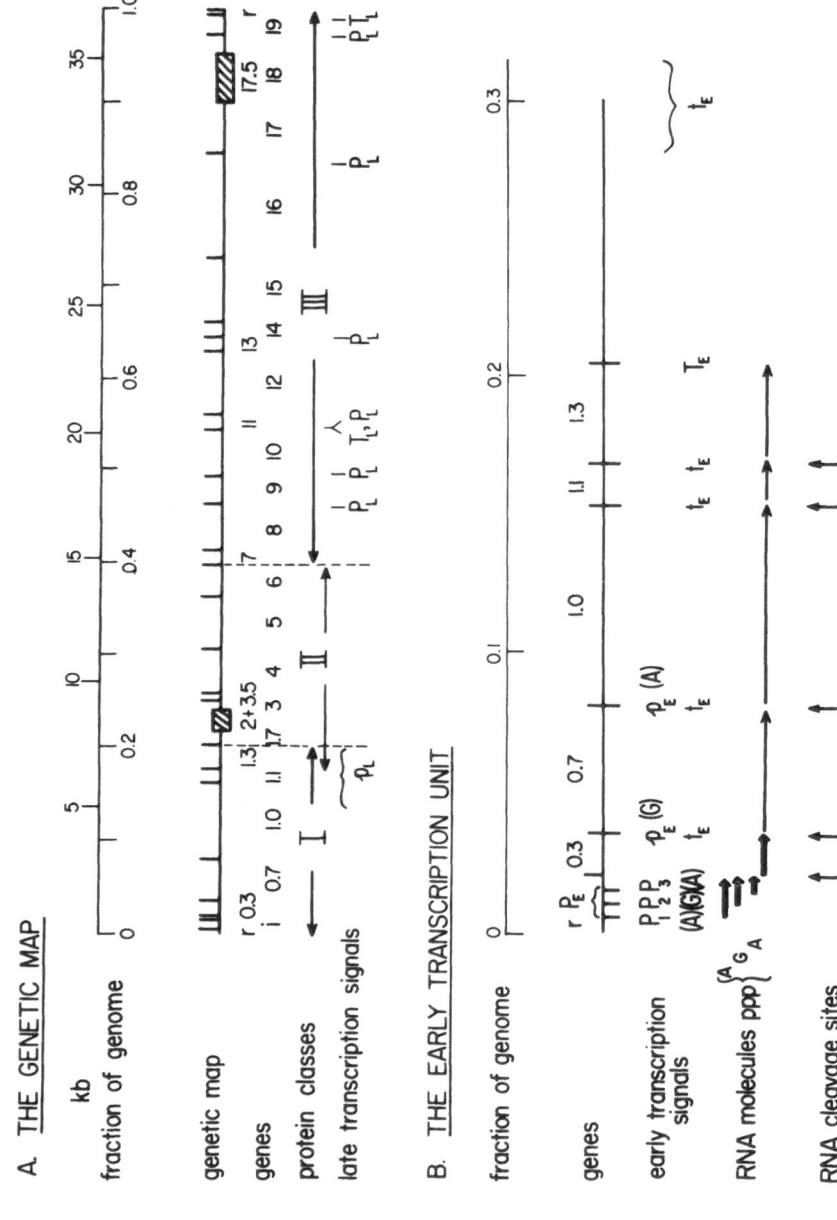

Fig. 4. The genetic map of Bacteriophage T 7. (A) The entire map. (B) The early genes. Map positions are presented in units of thousands of nucleotide pairs (kb) and in fractions of the genome. The boundaries of genes are presented, except at two locations ($\boxed{///}$). One of these incompletely mapped regions (2+) may code for a newly isolated DNA-binding protein (Reuben and Gefter, 1974). The boundary region of genes 0.3 and 0.7 also codes for two small peptides (Steitz, 1975; Studier, 1975b). The letter r designates the redundant ends of the DNA, and i the promoter region of the early transcription unit. Signals for the initiation (P,p) and termination (T,t) of early (E) and late (L) transcription are designated according to strength (P and T, strong). The initiating nucleotides for the early transcripts are given in parentheses [e.g., p(A), p(G)]; all late transcripts are initiated with G. In Fig. 4B cleavage sites of the early transcripts are shown on the bottom line by vertical arrows, and the RNA molecules that these cleavages generate are shown on the line above. Molecules in relatively great abundance are indicated by heavy arrows. Data compiled from Studier (1972); Dunn and Studier (1973a,b), Minkley and Pribnow (1973), Simon and Studier (1973), Golomb and Chamberlin (1974a), Skare et al. (1974), and Niles and Condit (1975). Several additional P$_L$ sites have recently been tentatively mapped into region II. All the conjugate units of transcription terminate at the common T$_L$ site shown in the map near position 0.55 (Studier, 1975b).

the genetic map and the nucleotide sequence in an elegant and accurate way (Hyman, 1971; Studier, 1973a,b; Simon and Studier, 1973; Summers *et al.*, 1973; Dunn and Studier, 1973a; Minkley and Pribnow, 1973; Minkley, 1974a).

The early genes 0.3, 0.7, 1.1, and 1.3 are commonly called "nonessential," but this catchy name deserves some explanation. It is common to lump into this category all genes whose products are not absolutely required on standard laboratory strains of bacteria under standard conditions of growth. Many of these viral genes either quantitatively affect yield, or they broaden the range of natural bacterial hosts. In the evolutionary sense, these genes must be "essential." It remains for us to understand virus–host relationships well enough to be able to appreciate and analyze the selective advantages that derive from all of these "nonessential" gene products. The gene 0.3 protein is already understood in this context, as previously mentioned. It acts to overcome the DNA restriction system of the host (Studier, 1975a).

2.3.1. T 7 Early RNA Is Monocistronic and Chemically (Relatively) Stable

Summers first found that T 7 RNA is chemically sufficiently stable for entire molecules of viral messenger to accumulate after infection (Summers, 1969; 1970; Siegel and Summers, 1970; Summers and Siegel, 1969). The sizes of RNA and polypeptide chains can be relatively accurately measured by gel electrophoresis, and the combination of deletion analysis with RNA sizing has taken the original microscopic mapping (Hyman, 1971) very far (Hyman and Summers, 1972; Studier, 1973a,b; Simon and Studier, 1973; Minkley, 1974a; Summers *et al.*, 1973). The alignment of T 7 early RNA molecules on the genetic map is presented in Fig. 4B. The predominant *in vivo* RNA species are monocistronic* and can be assigned to the genes coding for the five early proteins. Three small RNA chains come from the leftmost end of the T 7 DNA template but apparently do not code for proteins. The 5′ termini of these small RNA chains are near, but not at, the 3′ end of the T 7 DNA*r* strand. The left-hand, terminally redundant end of the *r* strand is not transcribed (Davis and Hyman, 1970; Dunn and Studier, 1973a).

---

* The designation is not entirely accurate. Additional small peptides are also coded by the messages which yield gp 0.3 and 0.7 (Steitz, 1975; Studier, 1975b).

2.3.2. T 7 Early RNA and Posttranscriptional Processing

There is direct and clear evidence that the left-hand 20% of the T 7 DNA molecule serves as a single unit of transcription *in vivo* to yield a primary transcript of approximately 7500 nucleotides, which is subsequently cleaved by *E. coli* RNase III. The termination point for this large unit of transcription is located near 0.20 on the genetic map of Fig. 4. At this site, termination of RNA synthesis by the host RNA polymerase *in vivo* is the general, but not absolute, rule. There is some read-through at this site for all strains of T 7, but the relative rate of read-through differs considerably for strains of T 7 from which the mutant collections of Studier and Hausmann have been derived (Issinger and Hausmann, 1973).

The discovery of T 7 early RNA processing (Dunn and Studier, 1973*a*) has helped to simplify at the same time that it has enlarged our understanding of transcription and of the T 7 early transcription unit. The evidence for this processing is comprehensive: (1) Only very short A and G 5′ triphosphate-terminated T 7 early RNA chains accumulate *in vivo* (Dunn and Studier, 1973*a*; Minkley and Pribnow, 1973). (2) The nucleotide sequences of the 5′ termini of three separable species of T 7 early mRNA are identical; moreover, this same sequence is generated *in vitro* by *E. coli* RNAse III acting on *in vitro* synthesized T 7 early RNA (Kramer *et al.*, 1974; Rosenberg *et al.*, 1974). (3) The 3′ termini of four of the T 7 early messages and of two initiator RNA species are identical (Kramer *et al.*, 1974) but differ from the 3′ termini of RNA chains generated by *E. coli* RNA polymerase *in vitro* at the intrinsic termination site at location 0.20 (T_E; Fig. 4; Millette *et al.*, 1970). (4) A large precursor primary T 7 early transcript accumulates *in vivo* in RNase III-deficient *E. coli* (Dunn and Studier, 1973*b*). (5) Although RNase III is known to cleave double-stranded RNA nonspecifically, all internal cleavages of the T 7 early transcript occur at a specific, repeated site

$$5' \cdots pCpCpUpUpUpApU \!\uparrow\! pGpApUp \cdots$$

Evidently the internal processing is a sequence-specific event.

The T 7 early messages are posttranscriptionally processed in one further way. One or more adenylic acid residues can be added to their 3′ termini (Kramer *et al.*, 1974).

The complete processing of the early T 7 transcript into monocistronic messages* is apparently unique within our current experience

---

* See footnote on p. 32.

with procaryotic transcription. Equally long or longer bacterial transcripts (of the *lac, trp,* and *his* operons) are either not cleaved at all or are certainly not as completely cleaved. However, as we note in Section 2.3.6, the functional significance of the monocistronic translation of this multicistronic transcription unit is still obscure.

2.3.3. The Major Promoter of the Early Transcription Unit

E. coli RNA polymerase transcribes T 7 DNA *r*-strands relatively asymmetrically *in vitro* (Summers and Szybalski, 1968) especially at low ratios of added enzyme to DNA (Dausse *et al.,* 1972*b*; Minkley, 1974*b*). (The partial symmetry of T 7 RNA synthesized *in vitro* with *E. coli* RNA polymerase is especially prominent at higher RNA polymerase:DNA molar ratios, at lower ionic strength, or when repeated initiation at the principal promoter is hindered by rifampicin (Dausse *et al.,* 1972*a*; Dunn *et al.,* 1972*a*; Minkley, 1974*b*; Schäfer *et al.,* 1973*a*; Iida and Matsukage, 1974). We would guess that the *in vitro* transcription of the "wrong" T 7 DNA *l*-strand probably involves specific sites, but the significance of these sites for cellular processes is obscure.) Under suitable circumstances, *in vitro* transcription is substantially confined to class I transcripts initiated near the left-hand end of the DNA template and terminating at 0.20 on the genetic map to yield an RNA chain containing approximately 7500 nucleotides (Millette *et al.,* 1970; Maitra *et al.,* 1970). The termination signal 0.20 (T_E) is evidently recognized by RNA polymerase alone both at high and low ionic strengths *in vitro* (Millette *et al.,* 1970; Chamberlin, 1974*a*). Release of RNA polymerase from the DNA template after RNA chain termination is promoted by increased ionic strength.

RNA chain termination *in vitro* is induced at internal sites of the early transcription unit by the *E. coli* factor ρ (Davis and Hyman, 1970; Goff and Minkley, 1970). However, the RNA distributions generated in this way do not correspond in any simple manner with *in vivo* T 7 early RNA (Darlix, 1973; Dunn and Studier, 1973*a*), and it was this discrepancy which led Dunn and Studier to the discovery of T 7 messenger processing.

The synthesis of the large, major T 7 early transcript is initiated *in vitro* at three discrete but closely linked sites (P_1, P_2, P_3; Fig. 4B); at two of these (P_1 and P_3), RNA chains start with A and at the third (P_2) with G (Studier, 1973*b*; Dunn and Studier, 1973*a*; Minkley and Pribnow, 1973; Minkley, 1974*b*). The complex of RNA polymerase with DNA at these three sites has been visualized in the electron microscope (Bordier and Dubochet, 1974; Vollenweider *et al.*, 1975), and the nucleotide sequences of the DNA segment that are protected from nuclease digestion when RNA polymerase binds tightly to P_2 and P_3 have been determined (Pribnow, 1975*a,b*). Cleavage by RNase III of the large *in vitro* T 7 transcripts initiated at these three sites yields fragments containing γ-^{32}P-labeled 5′ termini which are approximately the same size as three small RNA molecules found *in vivo* (Fig. 4B; Dunn and Studier, 1973*a*). The three sites constitute the strong promoters (P_E) of T 7 transcription by bacterial RNA polymerase *in vitro*; they are most probably the principal sites of RNA chain initiation *in vivo*. At each of these three sites, RNA polymerase undergoes a cooperative interaction with a segment of DNA (Mangel and Chamberlin, 1974*c*; Sarocchi and Darlix, 1974). The resultant complex initiates RNA chains very rapidly, within a fraction of a second of ribonucleoside triphosphate addition (Mangel and Chamberlin, 1974*a,b*; Chamberlin, 1974*a*).

It has also been proposed that this region contains a storage site for additional RNA polymerase molecules, but the evidence on this subject is not entirely clear. The argument for storage sites rests on experiments with heparin, a polymer anion inhibitor of RNA polymerase which sequesters loosely bound and free RNA polymerase. A single T 7 DNA chain can "store" six to eight molecules of RNA polymerase in a heparin-resistant state for initiation of early RNA chains (with ATP only; Schäfer *et al.*, 1973*a*). The most convincing confirmation of this interesting proposal would come from direct electron-microscopic observation. The three discrete sites for single molecules of RNA polymerase at one end of T 7 DNA have been observed (Bordier and Dubochet, 1974; Vollenweider *et al.*, 1975), as have two weaker sites. However, no direct evidence for storage sites bearing adjacently arranged bound RNA polymerase could be found (Bordier and Dubochet, 1974).

These are the known properties of the T 7 early promoter. It is a collection funnel for RNA polymerase molecules—a modest but useful weapon for this phage as it starts its contest with the resident genome of the host cell.

2.3.4. Other Signals in the Early Transcription Unit

We next examine the abundance distribution of early transcripts as a way of looking for other features of transcriptional regulation. A primary large transcript that is processed by a ribonuclease should yield its stable fragments in equimolar proportions. Deviations from equimolarity might indicate subsidiary initiation or termination sites of RNA synthesis. The early T 7 RNA chains are not present in equimolar proportions (Summers *et al.*, 1973; Minkley, 1974*a*; Studier, 1973*b*); the small fragments as well as the 600-nucleotide-long messages (coding for gp 0.3 and/or gp 1.1) are in excess. We have already mentioned that the small fragments come from the 5′ terminal segment of the transcription unit. Termination sites (t_E) at the right-hand and left-hand ends of gene 0.3 (Fig 4B) might yield excess quantities of these RNA chains. The relative abundance of the smallest RNA chains is difficult to measure with any accuracy. If these chains still contain the 5′ triphosphate termini, then their specific radioactivities during short periods of ^{32}P labeling would differ considerably from those of RNA chains labeled only in internucleotide ^{32}P, and this would make the estimation of relative abundance from ^{32}P distribution even less reliable (Lutkenhaus *et al.*, 1973).

There are two other kinds of singularities in the early transcription unit whose probable location is itself remarkable. Hold-up points for RNA chain elongation *in vitro* by *E. coli* RNA polymerase have been mapped by electron microscopy approximately into the intercistronic divides of early genes (Darlix and Horaist, 1975). Under suitable circumstances of reduced RNA chain-elongation rate, *rho*-induced termination of RNA chains can occur in these same regions.

Additional initiation (p_E) and termination (t_E) sites for *E. coli* RNA polymerase in the early T 7 transcription units have been mapped *in vitro* by Minkley and Pribnow (1973). Surprisingly, these also fall near the cistron boundaries and near sites of processing of the primary transcript by RNase III. Two subsidiary initiation sites $p_E(G)$ and $p_E(A)$ use GTP or ATP, respectively, as the RNA chain-initiating nucleotides (Fig. 4B). These initiation sites are relatively weak compared with the strong promoter at positions 0.005–0.015. Presumably, they are also weak *in vivo* since the complete transcript of the entire early genes is the major T 7 early RNA species accumulating in RNase III-defective *E. coli* (Dunn and Studier, 1973*b*). Minor RNA chain-termination sites (t_E) recognized by *E. coli* RNA polymerase *in vitro* are located at positions 0.034, 0.080, 0.153, and 0.167 (Minkley, 1974*b*). Together, these initiation and termination sites generate a large set of transcripts. If these diverse RNA chain termini are really generated by *E. coli* RNA polymerase, then one should anticipate that their 3′ termini and RNase III-generated 3′ termini would have different nucleotide sequences. The comparison has not yet been made.

In Fig. 4B we present a model of the T 7 early transcription unit which contains the features discussed in the preceding sections. Lines 2 and 3 of that figure show the positions of the genes, of the terminally redundant end (r), and of the promoter region (P_E) with its three strong initiation sites [P_1(A), P_2(G), and P_3(A)]. Subsidiary initiation sites (p_E) are located elsewhere in the early region. The principal termination signal is located 20% of genome length from the left end, at position 0.20 (T_E). Another important termination signal (t_E) is located near 0.035. This signal is responsible *in vivo* for a greater abundance of initiator and gene 0.3 RNA chains (heavy horizontal arrows on line 4). Other, subsidiary termination signals have also been detected. Post-transcriptional cleavage at sites designated on line 5 (↑) generates the RNA molecules that are found *in vivo*.

2.3.5. A Giant T 7 Transcription Unit

The termination signal of the early T 7 transcription unit is located near the 3′ end of gene 1.3, which codes for the nonessential T 7 DNA ligase. We have already mentioned that strains of T 7 vary in the efficacy of this stop signal (Issinger and Hausmann, 1973). The stop signal can also be deleted, and then synthesis of the group II and III proteins is partly relieved from dependence on the function of T 7 RNA polymerase (gp 1). Transcription of the T 7 DNA *r*-strand then encounters a second stop signal recognized by *E. coli* RNA polymerase at position 0.30 *in vitro*, but some transcription continues through this signal *in vitro* and *in vivo* (Studier, 1972; Peters and Hayward, 1974) along an extraordinary 38,000-nucleotide-long transcription unit (Bräutigam and Sauerbier, 1973). The T 7 proteins now make their first appearance in a time sequence that corresponds with the order of genes along the chromosome. RNA synthesis must be rate-determining for this progression, which consumes about 15 min at 30°C, corresponding to an average RNA chain-elongation rate of 50 nucleotides per second. This is about two to three times the average rate of elongation of *E. coli* *trp* mRNA and T 4 early mRNA chains (Rose *et al.*, 1970; Bremer and Yuan, 1968).

2.3.6. Functional Stability of T 7 mRNA

Although T 7 mRNA is chemically stable, it is functionally inactivated *in vivo*. The functional stability of global T 7 early RNA and of the messages for T 7 RNA polymerase and lysozyme is considerably greater than that of *E. coli* mRNA (Schleicher and Bautz, 1972; Hagen and Young, 1973; Yamada *et al.*, 1974*a*), but much less than the total chemical stability of T 7 mRNA. The loss of functional activity of T 7 and φ II early RNA is perhaps due to nucleotide removal. In the case of T 7, this may occur at the 5′ end, thus inactivating the mRNA for binding initiator tRNA to ribosomes (Linial and Malamy, 1970*a,b*; Yamada *et al.*, 1974*a*).

It is conceivable that processing of the early T 7 transcript by RNase III might significantly affect translation by altering the possibilities for RNA secondary-structure interactions. The selective synthesis of the *E. coli* RNA phage A protein on nascent or fragmented viral RNA messages would be a prototype for relationships of this kind (Weissmann *et al.*, 1973). However, only the synthesis of P 0.3 is significantly affected (increased) by RNase III cleavage of the T 7 transcript (Dunn and Studier, 1975*a,b*). Since the gene 0.3 message is only in part generated by cleavage of the large early transcript (Section 2.3.4 and Fig 4B), RNase III action on the latter is probably of relatively little significance even for the synthesis of gp 0.3. While RNase III processing might affect the functional stability of early messages and might thus be involved in early gene shutoff, the functional role that is served by the monocistronic translation of the T 7 early genes is still unclear.

2.4. The T 7 Late Genes and T 7 RNA Polymerase

2.4.1.   T 7 RNA Polymerase

The transcription of all the genes from DNA ligase to the right-hand end of the T 7 chromosome is coordinately initiated and controlled. We now know that this is due to the activity of T 7 specific RNA polymerase coded by T 7 gene 1 (Chamberlin *et al.*, 1970). The discovery of this enzyme followed the finding that gene 1 mutants of T 7 are pleiotropically defective for T 7 DNA replication and for the synthesis of many viral proteins (Hausmann and Gomez, 1967; Studier and Maizel, 1969; Studier, 1969) and that a failure of transcription is responsible (Siegel and Summers, 1970).

T 7 RNA polymerase is composed of a single polypeptide chain of molecular weight 107,000. It has a remarkably restricted range of effective DNA template utilization including double-stranded and single-stranded T 7 DNA, polydG:dC, and, less effectively, T 3 DNA. The enzyme does not transcribe other single-stranded DNAs and is inhibited by monovalent salts. T 7 RNA polymerase, like *E. coli* RNA polymerase, is a zinc metalloenzyme (Coleman, 1974) that also requires Mg^{2+}. The T 7, unlike the *E. coli* enzyme, does not permit substitution of Mn^{2+} for Mg^{2+}. It is resistant to inhibition by rifampicin and streptolydigin, two antibiotics that inhibit bacterial RNA polymerase, but is inhibited by relatively high concentratons of certain rifamycin derivatives. T 7 RNA polymerase selectively transcribes the *r*-strand of T 7 DNA, initiates RNA chains only with pppG, and elongates these chains very rapidly *in vitro*; an average rate of 230 nucleotides per second has been measured at 37°C, which is approximately five times faster than the rate of transcription of the left-hand end of T 7 DNA by *E. coli* RNA polymerase (Chamberlin *et al.*, 1970; Chamberlin and Ring, 1972, 1973*a,b*).

2.4.2. *In Vitro* and *in Vivo* Transcription

The T 7 RNA that is made *in vitro* by T 7 RNA polymerase is in many ways as remarkable as the enzyme that catalyzes its synthesis. The major products of transcription are seven polynucleotide chains ranging in length from approximately 700 to 15,000 nucleotides and synthesized in equimolar proportions at various enzyme:DNA mole ratios. The seven promoters at which these chains are initiated must be functionally identical; it will be important to confirm that they are also structurally identical. Six of the transcription units that generate these abundant late RNA molecules have been mapped (Fig. 4). Four units have been mapped by hybridization analysis using chemical deletions of the ends of T 7 DNA and T 7–T 3 hybrid phage DNA. These units (1, 2, 3b, 6) share a common terminus near, but not at, the right-hand end of the genetic map (Golomb and Chamberlin 1974*a,b*). Two more units (4, 5) have been mapped by *in vitro* synthesis of specific proteins with isolated single transcripts. These units share a common terminus near the middle of the T 7 DNA molecule (Niles and Condit, 1975). The six mapped transcripts originate in the right-hand 55% of the T 7 map and thus correspond to those genes which code for the class III proteins. Minor RNA species are transcribed from the region of the class II genes, and one or more additional transcripts overlap with the right-hand end of the early transcription unit (Skare *et al.*, 1974; Niles and Condit, 1975). A provisional map of transcription units in this segment of the genome has just recently been constructed (Studier, 1975*b*). Perhaps its most interesting feature is that it shows several overlapping transcription units with a common termination site (T_L, at 0.55 in Figure 4; the weak promoters in region II are not shown in Fig. 4). Gene 10, which codes for the major phage head protein, is actually transcribed from six different promoters! One of the minor transcripts is more than 15,000 nucleotides long and could represent the entire right-hand 80% of T 7 DNA (Golomb and Chamberlin, 1974a,b). The existence of more and less abundant late transcripts implies the recognition of more than one kind of promoter by T 7 RNA polymerase.

The largest of the late T 7 transcripts (1, 2, and one of the 3 species) are also subjected to processing by RNase III *in vivo* and *in vitro* (Dunn and Studier, 1975a), which accounts for the failure to find these transcripts intact in T 7-infected cells (Niles *et al.,* 1974). The other transcript 3 and transcripts 4 and 5 are evidently not cleaved by RNase III (Dunn and Studier, 1975a), and T 7 endolysin mRNA (gene 3.5, class II) is found as multiple molecular species, suggesting that this message is either processed incompletely or not at all (Hagen and Young, 1974). Individual T 7 transcripts produce unequal (molar) yields of their conjugate peptides *in vitro*. In other words, the translational efficiency, *in vitro,* of different segments of these late T 7 messages is nonuniform. Since the peptide abundance distributions generated *in vitro* by unfractionated *in vitro* RNA corresponds reasonably well with the *in vivo* abundances of T 7 proteins, these variations of *in vitro* translational yield may be a biologically significant property of the transcripts (Niles and Condit, 1975). However, for most of the late proteins, as for early proteins, RNase III cleavage does not affect translational yield (Dunn and Studier, 1975b).

2.4.3.   How Is the Synthesis of Class II Proteins Shut Off?

This is the aspect of T 7 regulation which is least understood. Is the shutoff regulated, in the sense that it involves an effector-dependent time sequence of late gene expression? Or, might it result from the intrinsic properties of the late genes and the late messages? The facts for deciding between these alternatives are not at hand. The suggestion has been made that the shutoff might not be regulated by an effector, but that it might be due to competition between the less abundant class II and the more abundant class III messages (Niles and Condit, 1975). Alternatively, if the class II messages were functionally less stable than the class III messages, then the general decline of viral RNA synthesis that normally sets in about 12 min after infection would result in a selective shutoff of the synthesis of class II proteins.

2.4.4.   T 7 Early–Late Genes

Are other T 7 and T 3 genes controlled together with the viral early (group I) genes? This problem had been surrounded by persisting confusion but appears now to have been solved (Issinger and Hausmann, 1973): The T 7 early genes proper are genes 0.3, 0.7, 1, 1.1, and 1.3 and no others; a correspondingly located set of T 3 genes constitute that phage's early transcription unit. Genes located to the right of these on the genetic map are accessible to transcription by host RNA polymerase only if the early transcription unit's termination signal can be breached. This occurs in certain wild strains of T 7 *in vivo*; it also happens *in vitro* when viral DNA is used for coupled transcription-translation in a cell-free system. Unfortunately, these bacterial cell-free systems exhibit a relatively low but significant level of read-through of transcription even at termination signals that are well recognized *in vivo*. Such read-through is irrelevant to *in vivo* regulation but generates the *in vitro* synthesis of the T 4 lysozyme and nucleotide kinase, the SPO1 dCMP deaminase, and T 7 endolysin with the cell-free system from uninfected bacteria (Salser *et al.*, 1967; Schweiger and Gold, 1970; Schweiger *et al.*, 1971; Scherzinger *et al.*, 1972; Shub, 1975). Thus, the demonstration that a T 7 enzyme can be made in a cell-free system from uninfected bacteria is not an adequate criterion for its classification as an early protein. Three additional criteria for regulatory classification can be usefully imposed: (1) comparison of the relative yields of several proteins in a bacterial cell-free system programmed by DNA with yields of the same system programmed by RNA isolated at various times from virus-infected cells (O'Farrell and Gold, 1973b; Shub, 1975); (2) comparison of promoter proximity *in vivo* and *in vitro* (Milanesi *et al.*, 1970; Brody *et al.*, 1971; Herrlich and Schweiger, 1971; Bräutigam and Sauerbier, 1973); (3) comparison of viral and bacterial RNA polymerases in transcription-translation-coupled cell-free systems (Dunn *et al.*, 1972b; Niles and Condit, 1975).

It appears that at least one of the early proteins, DNA ligase (gp 1.3), continues to be synthesized together with the group II proteins, probably because T 7 RNA polymerase transcribes this gene. Evidently, the early transcription unit contains at least one internal promoter recognized by T 7 RNA polymerase (Fig 4, p_L; Skare *et al.*, 1974). The status of gene 1.1 is unclear.

2.5.   T 3 and the Other Related Phages

Up to this point we have concentrated on phage T 7 as a representative of its class. We now turn to several aspects of the development of T 7 and its congeners which are less clearly understood. The available fund of information on these topics is much less concentrated on T 7 so that it becomes useful to be more catholic in our consideration.

The homology of the phages related to T 7 has been assessed at the DNA level (Davis and Hyman, 1971; Hyman *et al.*, 1973; Brunovskis *et al.*, 1973; Beier and Hausmann, 1973) and at the level of functional interactions (Hausmann and Härle, 1971; Issinger *et al.*, 1973). Phages T 7, ϕ 1, ϕ II, W 31, and H are much more closely related to each other than to T 3. T 7, H, and ϕ II are closely homologous along more than 80% of the nucleotide sequence (by the criterion of thermal stability of heteroduplex DNA segments relative to homoduplex DNA). H and ϕ II are also closely homologous to each other along the remaining 20% of the sequence, but T 7 is essentially heterologous to H and ϕ II along almost 10% of the nucleotide sequence. According to the same criteria, T 3 and T 7 are closely homologous along only approximately 30% of their nucleotide sequence, are partly homologous along 50% of the sequence, and are heterologous along the remaining 20%. The regions of partial and nonhomology are interspersed among the early genes and the genes of group II.

In this context, it is interesting to summarize relationships between T 3 and T 7, which are considered in more detail below. T 3 codes for an RNA polymerase whose properties are similar to those of the T 7 RNA polymerase (there is close homology between T 3 and T 7 DNA along a segment which probably includes a part of T 3 and T 7 genes 1). T 3, like T 7, shuts off host macromolecular synthesis but may use different mechanisms, and regulates the expression of the genes that code for shutoff products differently from T 7. A T 3 gene coding for an S-adenosyl methionine cleaving enzyme is located to the left (Fig. 4) of gene 1. Its counterpart in T 7 is a gene (0.3) whose product interferes with bacterial DNA restriction but without cleaving S-adenosyl methionine. Here, perhaps, is another interesting example of genes in *homologous* loci of chromosomes that code for *heterologous* proteins which perform *equivalent* functions (Botstein, 1974).

2.5.1. Comparison of T 3 and T 7 RNA Polymerases

These similar but clearly nonidentical enzymes are composed of single polypeptide chains of molecular weight approximately 100,000 (Chamberlin *et al.*, 1970; Chamberlin and Ring, 1973a; Chakraborty *et al.*, 1973). Each enzyme preferentially transcribes the homologous double-stranded DNA with essentially perfect asymmetry, initiating and terminating RNA chains repeatedly *in vitro* without intervention of other factors (Chamberlin *et al.*, 1970; Summers and Siegel, 1970; Dunn *et al.*, 1971; Maitra, 1971; Maitra and Huang, 1972; Chakraborty *et al.*, 1974; Niles *et al.*, 1974). Both enzymes have template preferences for their homologous double-stranded DNA. They share the properties of chain growth and salt and antibiotic sensitivity already described for T 7 RNA polymerase in Section 2.4.1 (Chamberlin and Ring, 1972; McAllister *et al.*, 1973; Küpper *et al.*, 1973; Salvo *et al.*, 1973; Maitra *et al.*, 1974; Chakraborty *et al.*, 1974). Both enzymes lose selectivity on denatured DNA templates (Salvo *et al.*, 1973).

Both enzymes initiate RNA chains on double-stranded DNA with GTP; differences are apparent at the second nucleotide. The T 3 RNA polymerase initiates RNA chains with $pppG(pG)_m(pA)_n$ ($n,m \geq 1$) on T 3 DNA and with pppGpA on T 7 DNA, while the T 7 RNA polymerase initiates chains with pppGpG on both templates (Chakraborty *et al.*, 1973; McAllister *et al.*, 1973; Chamberlin and Ring, 1973*a,b*; Maitra *et al.*, 1974). The effect of ppGpp, which accumulates in T 7 infection (Friesen and Fiil, 1973), on RNA synthesis with either of these enzymes has not been studied.

Why is *E. coli* RNA polymerase so much larger and more complex than these enzymes? We think that the answer must have to do with the contrast between the dynamic and regulatory versatility of bacterial transcription and the relatively monotonous burst of virus-specific gene expression that produces the late period of phage development. We guess that bacterial RNA polymerase is complex because of its diverse interactions with other *proteins*, interactions about which we still know far too little.

2.6. Other Features of Regulation

2.6.1. Nonpermissiveness and the Absence of Positive Translational Control

Nonpermissiveness of *E. coli* strains for bacteriophage T 7 is not uncommon (Hausmann, 1973). The molecular mechanisms of this host–virus incompatibility must be diverse in T 3 and T 7 (Hausmann *et al.*, 1968; Hausmann, 1968, 1973; Chamberlin, 1974*b*), as they are with other viruses (Mathews, 1970; Pulitzer and Yanagida, 1971; Georgopoulos, 1971; Georgopoulos *et al.*, 1972; Takano and Kakefuda,

1972; Coppo *et al.*, 1973; Montgomery and Snyder, 1973; Sternberg, 1973*a,b*; Coppo *et al.*, 1975*a,b*; Takahashi *et al.*, 1975). Here we shall discuss only one restriction system, which involves genes on the F episome of *E. coli* (Mäkelä *et al.*, 1964; Linial and Malamy, 1970*a,b*; Morrison and Malamy, 1971).

ϕ II and T 7 inject their DNA normally into F$^+$ *E. coli* and make early proteins, including the virus-specified RNA polymerases. Very soon after the normal time for starting late gene expression, macromolecular synthesis ceases (Condit and Steitz, 1975), leaving some residual functional T 7 mRNA in these cells (Morrison and Malamy, 1971; Summers and Jakes, 1971; Blumberg and Malamy, 1974), which does not become translated *in vivo* (Morrison and Malamy, 1971; Morrison *et al.*, 1974). What basic defect produces these consequences? Soon after infection by T 7, F$^+$ *E. coli* develop a leaky plasma membrane, so that cells suffer a rapid loss of nucleotides and are unable to accumulate nutrients (Condit, 1975; Britton and Haselkorn, 1975). According to one hypothesis, the breakdown of protein and other macromolecular synthesis is a direct consequence of this lesion, and the particular distribution of T 7 RNA and proteins that is found in the abortively infected cell reflects the particular time at which macromolecular syntheses stop (Condit and Steitz, 1975; Britton and Haselkorn, 1975). However, there still is disagreement over whether a separate defect of protein synthesis also develops in the infected nonpermissive cell. The experiments on this issue involve cell-free protein-synthesizing systems. It is agreed that cell-free extracts from uninfected F$^+$ and F$^-$ bacteria can translate both early and late T 7 messages (Yamada *et al.*, 1974*a,b*; Condit and Steitz, 1975; Niles and Condit, 1975; Young and Menard, 1975; Yamada and Nakada, 1976; Blumberg *et al.*, 1976). This finding argues against positive translational control of T 7 gene expression. However, there is some disagreement about the existence of a general debility of cell-free systems isolated from nonpermissive, infected F$^+$ bacteria (relative to uninfected F$^+$ or F$^-$ and particularly to infected F$^-$ bacteria). Such a defect has been found by some (Blumberg *et al.*, 1976) but denied by others (Condit and Steitz, 1975). If such a defect does exist, it is not selective for T 7 mRNA or for T 7 late relative to early mRNA. Thus, to repeat, there is no evidence for positive translational regulation of viral gene expression in normal or defective T 7 development. At least a part of F$^+$ restriction appears instead to be an example of a general phenomenon: Phage development is concomitant with interactions between virus- and host-determined components of the plasma membrane. Some of these interactions lead to a loss of membrane integrity: The affected cell is nonpermissive for viral development. Compensatory changes of the membrane proteins of the host or of the phage may circumvent membrane dysfunction. Events of this general nature probably underlie the incompatibility of T 4 rII mutants with *E. coli* carrying the prophage *rex* gene (which we do not discuss) and the incompatibility of T 5 with *E. coli* carrying the wild-type *col*Ib plasmid (which is discussed in Section 5.3.2a).

2.6.2. Host and Early Viral RNA Shutoff

T 3 and T 7 phages shut off the RNA and protein synthesis of their hosts and also shut off the expression of their own early genes. There is evidence that these effects are generated by the products of "non-essential" viral genes. The simplest expectation is that each genome might harbor genes, one or more of which might code for negative transcriptional control elements while others might code for negative translational elements. Evidence in support of both of these notions has been developed recently.

The shutoff of T 7 early RNA and host RNA synthesis involves an early viral function. As a consequence, T 7 early gene expression is self-limiting (Brunovskis and Summers, 1971). The T 7 gene 0.7 product is associated with this process (Brunovskis and Summers, 1972; Rothman-Denes *et al.*, 1973), and recent systematic studies with the early T 7 deletion and amber mutants begin to define other roles more clearly (Studier, 1975*b*, personal communication): (1) T 7 gene action is required for host shutoff and killing. (2) In the absence of functional gp 0.7, a late (class II or III) protein can shut off host and viral early protein synthesis, perhaps by drastically reducing mRNA stability. (3) Gene product 0.3, by itself, has little or no effect on host shutoff.

A protein kinase activity has been associated with gp 0.7 (Rahmsdorf *et al.*, 1974). The targets of this kinase activity include ribosomes (Rahmsdorf *et al.*, 1973) and the β' subunit of RNA polymerase (Zillig *et al.*, 1975). RNA polymerase phosphorylation probably is the gp 0.7-mediated transcriptional control activity. The role of the other phsophorylations in phage development is not yet known. The RNA polymerase phosphorylation must be labile under at least some conditions of purification, because prior searches for this modification had failed to uncover it (Rahmsdorf *et al.*, 1973). Detailed analyses of the transcriptional properties of the phosphorylated *E. coli* RNA polymerase have not yet been published.

The same mechanism of transcriptional shutoff may not operate in

T 3 infection; the T 3-induced shutoff of host RNA synthesis is not an early function, since T 3 gene 1 mutants are defective in this shutoff (Srinivasan, personal communication). Moreover, this shutoff operates in the face of the ability of T 3 RNA polymerase to transcribe the entire T 3 genome (Dunn *et al.*, 1972*a*), at least *in vitro*. In contrast, T 7 gene 1 mutants shut off host RNA metabolism with normal kinetics (Brunovskis and Summers, 1971). If such a basic difference between T 3 and T 7 exists, it raises the interesting question of whether the T 3 homolog of T 7 gp 0.7 is also a protein phosphokinase. Transcription-modifying activity in T 3-infected bacteria has been described (Dharmgrongartama *et al.*, 1973) which could be homologous with the T 7-specific phosphorylation. A fraction of bacterial RNA polymerase is modified after T 3 infection. The modification results in changes of chromatographic properties of RNA polymerase holoenzyme and core and lowers the enzyme activity (nonspecifically with respect to DNA templates) by three-fourths. The modified enzyme does not inhibit unmodified RNA polymerase activity (this is a simple test of whether modified RNA polymerase might exert regulatory activity *in vivo* by acting as a general repressor; cf. Yura and Igarashi, 1968; Khesin *et al.*, 1972; Chao and Speyer, 1973). It has been suggested that the modification involves condensation of a large peptide to subunit β'.* The relationship of this RNA polymerase modification to *in vivo* regulation has not yet been clarified, but it is an early function (Srinivasan and Dharmgrongartama, 1975). If this modification is an activity of T 3 gp 0.7, the latter may be a peptidyltransferase rather than a phosphokinase. Bacterial RNA polymerase is subject to the action of another inhibitor, which was first identified after T 3 infection (Mahadik *et al.*, 1972). The T 3-specified protein inhibits transcription by bacterial RNA polymerase holoenzyme, but not by T 3 or by T 7 RNA polymerase. It interacts with *E. coli* RNA polymerase holoenzyme and DNA in the presence of ATP and GTP (Mahadik *et al.*, 1972, 1974). Inhibitor activity is under the control of a T 3 late gene that has not yet been identified. There have been numerous searches for a comparable inhibitory activity in T 7-infected cells, and the first success has recently been reported (Hesselbach *et al.*, 1974). An inhibitor

---

* In Section 4 (Table 5), we criticize the evidence for a similar modification of RNA polymerase after T 4 infection. The linkage of the T 3-specific peptide to the β'-subunit of RNA polymerase has been detected on SDS polyacrylamide gels and is therefore likely to be covalent. However, the homogeneity of the RNA polymerase with respect to the linked peptide has not been established, and other questions about this modification remain to be answered: for example, whether modification occurs during preparation, i.e., *in vitro* rather than *in vivo* (cf. Linn *et al.*, 1973).

that is substantially selective for bacterial RNA polymerase holoenzyme over core and T 7 RNA polymerase has been purified. It is a small protein (Ponta *et al.*, 1974a; Hesselbach and Nakada, 1975). Surprisingly this protein is also a translational inhibitor, resembling i_α-S1 (Groner *et al.*, 1972a; Miller and Wahba, 1974). While the relation of these interesting factors to transcriptional shutoff *in vivo* remains to be clarified, they could constitute the products of class II or III genes that also shut off host transcription, as we have already mentioned.

A selective inhibitor of the initiation of translation of bacterial and certain early viral messages has also been identified in lysates of T 7-infected cells (Herrlich *et al.*, 1974). The significance of this finding remains relatively unclear, partly because the specificity of the inhibitory activity is not yet fully explored and partly because an indirect analysis placed the T 7 gene associated with this activity near gene 0.3. Gene product 0.3 shuts off DNA restriction (Studier, 1975a) and, although it interacts with RNA polymerase (Ratner, 1974a), it appears not to be involved in the shutoff of host protein and RNA synthesis.

3. THE LARGE hmU-CONTAINING PHAGES OF *B. SUBTILIS*

3.1. Introduction

In this section, we analyze a program of viral gene expression in which changes of viral transcription specificity are effected through an alteration of the bacterial host's RNA polymerase. Of course, this strategy is not unique to the hmU phages of *B. subtilis*; T-even coliphages, among others, do the same and are much better studied. Nevertheless, we consider the hmU phages first because at least some of their positive controls of viral transcription are being successfully studied *in vitro* and because we can discuss them in a relatively straightforward way before turning to the intricacies of phage T 4. The properties of these and other phages of *B. subtilis* have recently been reviewed by Hemphill and Whiteley (1975).

The hmU phages of *B. subtilis* are large, tailed phages with icosahedral heads and relatively elaborate tail plates (Eiserling, 1963). They derive their name from their double-stranded DNA (M.W. ca. 10^8) in which hydroxymethyluracil (hmU) replaces thymine. This group of phages includes SP 6, 7, 8, 9, 13, and 82 and SPO 1, ϕe, 2C, 5C, and ϕ 25 (Mandel, 1968). Although lytic, they differ from the T-even and T-odd *E. coli* phages in their host–virus metabolic relationships. These differences concern the shutting off of host-cell macromolecular syn-

thesis and host-genome degradation. Bacterial spores can act as carriers of these viral genomes (Sonenshein and Roscoe, 1969; Sonenshein,
1970), because sporulating bacteria become nonpermissive for the lytic
development of these phages (Yehle and Doi, 1967; Sonenshein and
Roscoe, 1969). For phage φe, nonpermissiveness is a late development
of sporulation and primarily results from inhibition of viral replication
(Kawamura and Ito, 1975) rather than viral transcription (Losick and
Sonenshein, 1969; Losick et al., 1970a,b; Brevet and Sonenshein, 1972;
Losick, 1972; Brevet, 1974). Evidently changes of RNA polymerase
that occur very early in sporulation do not block the program of viral
gene expression. On the other hand, modifications of RNA polymerase
that occur during sporulation most probably are significant for the
regulation of cellular gene expression during this differentiation (see,
for example, Fukuda et al., 1975; Klier et al., 1973; Pero et al., 1975c).

Phages SPO 1 and SP 82 are the most studied of these viruses.
Much of the work on the regulation of gene expression has been done
with SPO 1 (Shub, 1966; Levinthal et al., 1967; Gage and Geiduschek,
1967, 1971a,b; Fujita et al., 1971). However, a comparable analysis of
SP 82 has been initiated, and certain aspects of the physiology of the
hmU phages have been investigated (Roscoe and Tucker, 1966; Pène
and Marmur, 1967; Pène, 1968; Roscoe, 1969; Liljemark and
Anderson, 1970; Kahan, 1971; Marcus and Newlon, 1971; Stewart et
al., 1971, 1972; Lavi and Marcus, 1972; Lavi et al., 1974; Spiegelman
and Whiteley, 1974b). Phage SPO 1 shuts off the replication of its host,
and so do other viruses of this class (Shub, 1966; Pène, 1968; Marcus
and Newlon, 1971; Roscoe, 1969; Liljemark and Anderson, 1970).
SPO 1 also shuts down most host-protein syntheses and substitutes its
own program of gene expression in which several distinct stages can be
recognized either by analysis of proteins or of mRNA (Shub, 1966;
Levinthal et al., 1967; Gage and Geiduschek, 1967, 1971a,b). The
genetic analysis of these phages is, by comparison with T 7 and T 4,
rudimentary. Genetic maps containing less than 40 genes have been
constructed for both SPO 1 and SP 82 (Kahan, 1966; Okubo et al.,
1972; Green and Laman, 1972). It appears that these genomes, like
other viral genomes, are so organized that genes for related functions
(such as replication or viral tail assembly) are clustered. Two of the
SPO 1 genes show little or no genetic linkage with each other or with
the rest of the viral genome (Okubo et al., 1972). A similar lack of
linkage in the T 5 genome is associated with the extensive terminal
redundancy of that DNA (Section 5.2). The hmU phages have not yet
been examined for such terminal redundancy. The nucleotide
homologies of SPO 1 and SP 82 DNA are very extensive as judged by

hybridization analysis (Truffaut *et al.*, 1970), yet the preliminary genetic maps of SP 82 and SPO 1 are not congruent (Green and Laman, 1972; Okubo *et al.*, 1972). Heteroduplex and restriction fragment mapping would solve an important problem of the relationship between SPO 1 and the other phages. The initial purpose of the genetic analysis of SPO 1 was to find genes controlling the program of viral gene expression, and three such genes have, in fact, been identified (Fujita *et al.*, 1971).

When SPO 1 and the other related phages infect their permissive hosts, they shut off macromolecular metabolism, albeit not as completely as the T-even phage. Bacterial DNA synthesis is shut off 6–8 min after infection (at 37°C), but the DNA is not extensively degraded and does not, for example, lose transforming activity (Yehle and Ganesan, 1972). Host mRNA and protein synthesis is severely diminished during the first minutes after infection, but rRNA synthesis continues considerably longer, eventually declining after the onset of viral replication (Shub, 1966; Cocito, 1974). There are also indications that this shut-down of host protein and RNA synthesis is reversible (see next section).

3.2. The Program of Viral Gene Expression during SPO 1 and SP 82 Phage Development

The developmental cycle of these phages is marked by a relatively clearly defined pattern of gene expression which can be discerned by examining the synthesis of viral proteins or RNA (Shub, 1966; Levinthal *et al.*, 1967; Gage and Geiduschek, 1967, 1971*a,b*). SPO 1 mRNA, unlike T 7 mRNA, is chemically unstable.* Accordingly, the concentration of a group of messages decreases as soon as their synthesis stops, so that RNA-DNA hybridization-competition experiments allow the duration of synthesis of different groups of messages to be followed quite closely. Six groups of SPO 1 messages can be distinguished on the basis of their times of first appearance and cessation of synthesis. Their schedule is shown in Table 1. If any collection of SPO 1 transcripts is made throughout the entire time of viral development, they must constitute a small part of viral transcription both at the beginning and at the end of viral development (Gage and Geiduschek, 1971*a,b*). The pattern of viral transcription during infection with phage SP 82 is very

---

* Degradation of SPO 1 mRNA to nucleotides is not absolutely complete; a small part of at least some messages remains in cells as a compositionally heterogeneous collection of 5–7 S fragments (Gage, 1969).

TABLE 1

The Normal Program of SPO 1 Transcription and Its Regulation by Three Viral Genes[a]

Viral RNA	Wild-type phage program (37°C)		Defects in viral mutants[b]: *Sus*F4 and F14 (genes 33, 34) (maturation-defective)		*Sus*F21 (gene 28) (replication- and transcription-defective)	
	On	Off	Synthesis	Shut off	Synthesis	Shut off
e	1 min	3.5–5 min	+	+[c]	+	+[c]
em	1 min	12 min	+	+[c]	+	p
m	4–5 min	12 min	+	–[d]	–	n.a.
m_1l	4–5 min	—	+	n.a.	–	n.a.
m_2l	8–12 min	—	–	n.a.	–	n.a.
l	13–14 min	—	–	n.a.	–	n.a.

[a] This table summarizes the results of extensive hybridization-competion studies with unlabeled and pulse-labeled RNA isolated at various times of infection by wild-type and mutant phage. Times refer to infection at 37°C in an amino acid-supplemented glucose minimal medium in which viral DNA replication starts 8–10 min after infection and the eclipse period is 24–26 min (Wilson and Gage, 1971). Data are taken from Gage and Geiduschek (1971*a,b*) and Fujita *et al.* (1971).
[b] Abbreviations: +, normal; –, defective; p, premature; n.a., not applicable.
[c] However, the timing of *e* and *em* repression has not yet been established for these mutants.
[d] However, the rate of phage transcription is ultimately reduced relative to the wild type.

similar (Spiegelman and Whiteley, 1974*b*). For both of these viruses, it is useful to roughly distinguish early, middle, and late phases of viral gene expression. Many genes are read during two of these periods, and the various classes of SPO 1 messages are named and symbolized accordingly (as *e, em,* m_1l, etc. in Table 1).

Mutants in three SPO 1 genes are defective in the program of viral gene expression. Two of these genes (genes 33 and 34) apparently code for proteins required for late gene expression. Mutants in these genes are "maturation defective" and are pleiotropically lacking in the production of subassemblies of SPO 1 phage as well as the complete phage themselves. They closely resemble T 4 maturation-defective mutants in phenotype and, presumably, in the molecular basis of defective phage development. Mutants in a third gene (gene 28) are pleiotropically defective in middle and late gene expression and lack various enzymes of DNA replication. In the nonpermissive host, gene 28 *sus* mutants can synthesize only early viral RNA. Almost all of this RNA can also be made when infection is in the absence of viral protein synthesis. Evidently, viral proteins determine two successive modifications of the specificity of transcription so that different genes characteristic of the middle and late stages of the viral development can be read (Fujita *et al.*, 1971).

The relationship between viral DNA replication and gene expression is less clear. What is still lacking, surprisingly enough, is the detailed survey of viral protein synthesis throughout infection, with all the available mutants, that has provided so much information about the regulation of T 7 and T 4 gene expression. One replication-defective SPO 1 mutant (in gene 22; Okubo *et al.*, 1972) was originally examined and found to be selectively lacking *l* messages that are normally produced late in infection (Gage and Geiduschek, 1971*b*). A substantial decrease in viral transcription late in infection with this replication-defective phage mutant was also noted (Gage and Geiduschek, 1971*b*; Fujita *et al.*, 1971). However, a replication-defective mutant of SP 82 was found to make appreciable quantities of the major phage immunogen (Stewart *et al.*, 1972). The most extensive analysis of the relationship between replication and viral gene expression thus far is a survey of SPO 1 mutants in all the known genes whose products are essential for viral DNA replication (the DO genes; Añon, 1974; Añon and Grau, 1974). These DO genes are located in two groups on the genetic map (Okubo *et al.*, 1972). In Añon's experiments, nonpermissive cells were infected with these mutants, and RNA was labeled at a time that would correspond to the very end of the eclipse period of a productive infection, a time at which SPO 1 RNA normally is the major transcription product of the infected cell. At this time, nonpermissive bacteria infected with DO mutants in one group (genes 29–32) barely make any SPO 1 RNA and are selectively defective in m_2l transcription while mutants in the other group (genes 21 and 22) transcribe substantially less SPO 1 RNA than the wild type but make at least some m_2 RNA. [There is disagreement about *l* RNA synthesis which Añon (1974) detected at a reduced level and which Gage and Geiduschek (1971*b*) found to be missing.] However, the main conclusion to be drawn from these experiments is that two groups of SPO 1 replication proteins are involved in different ways with viral transcription. The biochemical implications of this interesting analysis remain to be explored. Further work will also be required to clarify the relationship between SPO 1 and SP 82.

The transcription program mutants and the DNA replication mutants of SPO 1 allow us to see an interesting aspect of the relationship between this phage and its host (*B. subtilis* 168): The shutoff of host gene expression is, in large part, reversible. In general, whenever the temporal progression of the transcription program to a new stage is blocked, SPO 1 RNA synthesis decreases, there is a compensating increase in host RNA synthesis, and the synthesis of host proteins resumes (Fujita *et al.,* 1971; Grausz, 1972; Anderson, 1973). The quantitative aspects of this effect, i.e., when the shutoff of viral transcription occurs and how complete it is vary from one mutation to another, but the regulatory and replication mutants all show pronounced effects. Even the MD gene 33 and 34 mutants which are defective in late (m_2l and *l*) viral transcription have markedly depressed total viral transcription during the late periods of their abortive development, while synthesizing large quantities of viral DNA. An almost complete shutoff of viral transcription occurs 8–10 min after infection of nonpermissive bacteria with a mutant defective in middle transcription (F 21, gene 28, Table 1). The timing depends on the multiplicity of infection, and the shutoff is delayed at higher multiplicity of infection (Grausz, 1972). The shutoff evidently requires protein synthesis since it can be prevented by addition of chloramphenicol during the first two minutes after infection (Fuhrman, 1971). This shutoff is reversible by superinfection, and the reversal liberates both the primarily and subsequently entering genomes from repression (Grausz, 1972). Similarly, viral transcription with a maturation defective–replication defective double mutant (genes 33 or 34 and 22, Table 1) is shut off 12–15 min after infection (Fujita, 1971). We cite these facts to emphasize the plasticity that characterizes the relationship between expression of bacterial and viral genes. It is a relationship that would reward further analysis.

3.3. RNA Polymerase Modification and Viral Transcription

The transcription of the SPO 1 and SP 82 early genes *in vivo* occurs without the intervention of any newly synthesized viral proteins. Bacterial RNA polymerase holoenzyme transcribes mature double-stranded viral DNA selectively and asymmetrically *in vitro* to yield this collection of transcripts (Geiduschek *et al.,* 1968; Spiegelman and Whiteley, 1974*b*). The early genes of both these viruses are read with both transcription polarities, i.e., off both strands of the viral DNA (Gage and Geiduschek, 1971*a*; Spiegelman and Whiteley, 1974*b*). The initiation of early SPO 1 transcription *in vitro* occurs at sites at which rapidly initiating complexes of RNA polymerase with DNA can form (Duffy and Geiduschek, 1976). Although lacking the information that a transcriptional mapping of the SPO 1 genome will provide, one can therefore anticipate with some assurance that the SPO 1 early genes, like the corresponding regions of T 7, T 4, and many other viral genomes, contain strong promoters that are recognized by RNA polymerase alone. The SPO 1 early transcription units, like the T 7 early transcription unit, also contain termination signals that are recognized by RNA polymerase alone. As in the case of T 7, termination is not perfect (Shub, 1975), and it is such read-through which allows the synthesis of the viral middle protein, dCMP deaminase, in a bacterial system capable of carrying out coupled transcription and translation *in vitro* (Schweiger and Gold, 1970). We have already indicated the characteristic features of such a leakage synthesis (Section 2.4.4). For SPO 1 dCMP deaminase, as for the T 7 class II proteins, these features include a low rate of synthesis relative to the SPO 1 early proteins and transcriptional initiation from a distant promoter (Shub, 1975). In the main, however, an *E. coli* cell-free coupled system can be made to synthesize the early SPO 1 proteins in the same molar proportions when it is programmed by SPO 1 DNA as when it is programmed by early SPO 1 mRNA produced *in vivo* (Shub, 1975).

It has been known for some time that viral middle RNA could be made *in vitro* with crude fractions from phage SPO 1-infected *B. subtilis* programmed with exogenously added viral DNA (Geiduschek *et al.*, 1969; Grau *et al.*, 1970). RNA polymerases have now been purified essentially to homogeneity from SP 82- and SPO 1-infected cells at different times after infection (Spiegelman and Whiteley, 1974a; Duffy and Geiduschek, 1973a,b, 1975; Pero *et al.*, 1975a). These enzymes have properties that are clearly relevant to *in vivo* transcription since each can be instructed to synthesize classes of viral RNA asymmetrically *in vitro* that are positively regulated by viral genes *in vivo* (Duffy and Geiduschek, 1973b, 1975; Spiegelman and Whiteley, 1974b; Fox and Pero, 1974; Pero *et al.*, 1975a,b; Swanton *et al.*, 1975; Shub *et al.*, 1976).

The significant common ground of all work with these enzymes is that they contain new subunits. In the case of SPO 1, these peptides are known to be synthesized after phage infection, and their insertion into RNA polymerase is controlled by those viral genes that regulate the program of viral transcription (Duffy and Geiduschek, 1973a; 1975; Fox and Pero, 1974; Spiegelman and Whiteley, 1974a; Fox *et al.*, 1975; Tjian *et al.*, 1976). Different RNA polymerase fractions with different compositions can be isolated from SPO 1-infected cells, their relative proportions apparently being a function of the biological state of the system, i.e., the time after infection, as well as the method of purification.

The proteins that have been found associated with RNA polymerase during phage infection are listed in Table 2. The significant enzyme, from the point of view of *in vitro* transcription of SPO 1 mid-

TABLE 2

Phage SPO 1-Specified Proteins Associated with *B. subtilis* RNA Polymerase

Enzyme designation[a]	Association of subunits that are synthesized after infection	Comments
	A. *Tightly bound proteins*	
B,[b,d] B-P[c]	IV[b] (M = 25,000[b] or 26,000[d]); ν^{28} (M = 28,000[c])	Protein IV is gp 28.[e] IV and ν^{28} are the same.
	VI (M = 13,500[d]); ν^{13} (M = 13,000[b])	VI and ν^{13} are the same. Protein VI-ν^{13} is gp 33[f]
C[d]	V (M = 23,000[b] or 24,000[d]); VI	Enzymes C and B are resolved by phosphocellulose chromatography.[d,g] Protein V is gp 34.[f]
	B. *Less tightly associated proteins*	
"Total RNA polymerase"[b]	I (M = 85,000) II (M = 40,000) III (M = 28,000)	Detected by precipitation with antibody to *B. subtilis* RNA polymerase core.[b] Protein VI is not detected by this method, but proteins IV and V are (see above).
A, B[c]	Three proteins (M = 85,000, 38,000, 18,000)	Tend to copurify with RNA polymerase on DNA cellulose. The 85,000 molecular weight protein copurifies with fraction A; the 38,000 molecular weight protein is removed from fraction B by zone centrifugation at high ionic strength; the 18,000 molecular weight protein is not always found with the B enzyme. Fractions A and B are separated by DNA cellular chromatography.

[a] The capital letters A, B, C, designate subfractions of RNA polymerase that can be separated from each other. Different designations are used by different laboratories. We have refrained from replacing these designations with descriptive names (such as "middle" or "late") in advance of a really detailed analysis of the functions of the virus-specified subunits, although plausible designations could be made according to Table 3.
[b] Fox and Pero (1974).
[c] Duffy and Geiduschek (1975).
[d] Pero *et al.* (1975a).
[e] Fox *et al.* (1976).
[f] Tjian *et al.* (1976). (See also Notes Added in Proof, p. 196, Note 2.)
[g] Enzymes resembling B and C have also been isolated from SP82-infected *B. subtilis* (Whiteley, personal communication).

dle RNA, contains two "new" phage-dependent subunits designated as ν^{28} and ν^{13} (Duffy *et al.*, 1975) or IV and VI (Fox and Pero, 1974; Pero *et al.*, 1975a,b), respectively. ν^{28}-IV has a molecular weight in the range of 28,000 (Duffy and Geiduschek, 1975), 25,000 (Fox and Pero, 1974), or 26,000 (Pero *et al.*, 1975a). ν^{13}-VI has a molecular weight of 13,000

(Duffy and Geiduschek, 1975; Pero *et al.*, 1975*a*). Subunits ν^{28} and ν^{13} remain bound to RNA polymerase during phosphocellulose chromatography, which dissociates the bacterial initiation factor σ. Two other subunits, I (molecular weight 80–85,000) and II (molecular weight 38,000 or 40,000) can be resolved (I) or removed (II) from the enzyme fraction of primary interest (Fox and Pero, 1974; Duffy and Geiduschek, 1975; Pero *et al.*, 1975*a*), and a third subunit, III (molecular weight slightly greater than IV), was found only by Pero and collaborators and evidently only in some preparations of RNA polymerase (Fox and Pero, 1974; Pero *et al.*, 1975*a*). These peptides attach to RNA polymerase containing β, β', α, and two smaller subunits ω^1 (molecular weight 11,000) and ω^2 (molecular weight 9,500; Duffy and Geiduschek, 1975; Pero *et al.*, 1975*c*). The SP 82 phage-modified RNA polymerase was originally reported to have a different, characteristic subunit (Spiegelman and Whiteley, 1974*a*), but more recent reanalysis and development of purification resolves peptides with very similar molecular weights in a polymerase fraction from SP 82-infected cells (Spiegelman and Whiteley, 1975). There may, therefore, be a rather complete convergence of the properties of SPO 1- and SP 82-specific modifications of RNA polymerase. The RNA polymerases that contain these virus-specified subunits show distinct template specificities in the absence of bacterial initiation factor σ. Both the SPO 1- and the SP 82-modified enzymes show template preferences for homologous hmU-containing DNA (Spiegelman and Whiteley, 1974*a*; Duffy and Geiduschek, 1973*b*, 1975). The homologous DNA preference of the SPO 1 enzyme is especially striking when tested in terms of the formation of rapidly initiating enzyme–DNA complexes. On several other DNA templates, this enzyme behaves like RNA polymerase core, including its ability to be activated by σ and its inability to form rapidly initiating polymerase–DNA complexes. However, on SPO 1 DNA such complexes are formed in the absence of σ (Duffy and Geiduschek, 1975, 1976). The enzymes also differ from bacterial RNA polymerase holoenzyme in the selective transcription of their homologous, preferred DNA templates. While bacterial RNA polymerase selectively transcribes the early viral genes, as already stated, the RNA polymerases that contain these phage-specified subunits selectively and asymmetrically transcribe other segments of the genome, predominantly viral middle genes (Duffy and Geiduschek, 1973*a,b*, 1975; Pero *et al.*, 1975*a,b*).

A selective preference for middle viral transcription can be conferred on RNA polymerase core from uninfected *B. subtilis* by a protein which resembles subunit ν^{28} (and probably is ν^{28}) alone. The mechanism of selective transcription has been analyzed with a polymerase fraction which contains subunits ν^{28} and ν^{13} in addition to the subunits of the bacterial RNA polymerase core (designated as B-P; see Table 2; Duffy and Geiduschek, 1976). The transcriptional selectivity of enzyme B-P involves template-selective binding and initiation. In view of the fact that enzyme B-P contains the subunits of the bacterial polymerase core in apparently unmodified state and in view of its ability to endow RNA polymerase core with selective transcribing activity for SPO 1 middle RNA in the absence of ν^{13}, we anticipate that the template-selective binding-initiation properties of enzyme B-P are due to subunit ν^{28} alone. We therefore anticipate direct proof that ν^{28} acts as a template-specific and transcription unit-specific determinant of DNA binding and initiation. These are properties that are also attributed to σ (Section 1.1.1b). ν^{28} and ν^{13} evidently do not affect the rate of SPO 1 RNA chain elongation by *B. subtilis* RNA polymerase (Armelin, 1973). ν^{28} has recently been shown to be the product of gene 28 (Fox *et al.*, 1976), which controls middle transcription (Table 1).

Middle viral transcription comprises several classes of RNA that can be distinguished by hybridization-competition: *em, m,* and m_1l (Table 1). No really detailed analysis of transcription units has yet been undertaken, and it is entirely possible that other virus-specified proteins intervene in the regulation of viral middle transcription, particularly during the late period of virus development. The roles of ν^{13} in particular, and of the phage-specified proteins that bind less tightly to RNA polymerase core, remain to be analyzed.

It has been suggested that a bacterial polypeptide, δ, is associated with asymmetric transcription by the viral middle RNA-transcribing polymerase (Pero *et al.*, 1975a,b). Delta is a small protein (molecular weight ca. 21,000) which is associated with RNA polymerase in uninfected *B. subtilis*. Like σ, it is released from the core enzyme during phosphocellulose chromatography. However, it does not direct asymmetric RNA chain initiation by unmodified bacterial RNA polymerase core and does not affect the stimulation of unmodified bacterial RNA polymerase core by σ (Pero *et al.*, 1975a). The role that δ might play in bacterial transcription is as yet obscure. It may be analogous to some of the *E. coli* proteins such as M (Ramakrishnan and Echols, 1973) which influence transcription *in vitro* but have as yet uncharacterized functions *in vivo*, or to some of the *E. coli* proteins which bind to RNA polymerase but are as yet uncharacterized with respect to function at any level (Section 1.1.2b). It is also interesting to consider the possibility that bacterial proteins which normally have other functions can be recruited to phage gene regulation. In Section 6, we discuss features of the regulation of phage N4 and PBS2 which very concretely raise the same possibilities. However, asymmetric transcription of SPO 1 viral middle RNA by ν^{28}- and ν^{13}-containing RNA polymerase which lacks δ has already been demonstrated (Duffy and Geiduschek, 1973b, 1975). We surmise that δ is not absolutely required for asymmetric viral middle RNA synthesis. However, it may be required for other viral transcription.

One other subunit, designated as V (molecular weight 23,000), has been found associated in varying proportions with RNA polymerase from phage SPO 1-infected *B. subtilis*. RNA polymerase from nonpermissive bacteria infected with maturation-defective mutants in genes 33 and 34, which are unable to synthesize m_2l or l RNA (Table 1), lacks peptide V (Fox and Pero, 1974). Recently, RNA polymerase core containing peptides V and VI-ν^{13} has been resolved from RNA polymerase containing polypeptides ν^{28}-IV and ν^{13}-VI (Pero *et al.*, 1975a). The RNA polymerase containing peptides V and ν^{13} shows some preference for the transcription of late (m_2l and/or l) SPO 1 RNA (Pero *et al.*, 1975b).* Once again, the specific transcription of this late RNA is said to depend on the bacterial protein, δ, but the nature of this effect remains to be worked out.

Subunits V- and VI-ν^{13} have now been shown to be the gp 34 and gp 33, respectively (Table 1, Tjian *et al.*, 1976). thus, all three of the known transcription-regulating SPO-1 genes code for RNA polymerase-binding proteins. Since the binding of these subunits produces readily detectable changes of transcription specificity *in vitro*, their mechanisms of action in generating the positively regulated transcription is open to detailed biochemical and structural analysis.

The separation of two modified forms of RNA polymerase with two different transcription specificities is a striking result but leaves one, for the time being, with only a static picture of transcriptional regulation. One of the aspects of this regulation that remain to be analyzed is the dynamics of regulatory protein–polymerase core interaction. It is not yet known whether the phage-specified proteins, like σ of *E. coli*, cycle between RNA polymerase molecules during transcription (Section 1.1.1), whether any of the as-yet uncharacterized viral proteins which interact with RNA polymerase also promote interchanges of regulatory subunits, and whether different regulatory subunits compete with each other for binding to RNA polymerase core.

---

* See also Notes Added in Proof, p. 196, Note 2.

Transcription of the SPO 1 and SP 82 early (*e*) genes (Table 1) occurs only during the first four to five minutes after infection at 37°C. The shutting off could be due to the action of a transcription inhibitor or to a switch in the transcription specificity of RNA polymerase. It has been shown that the modified RNA polymerase-containing subunits ν^{28} and ν^{13} do not transcribe the SPO 1 *e* genes. Such a change of specificity is consistent with a mechanism of *e* gene shutoff by RNA polymerase conversion. However, some apparently unmodified RNA polymerase remains in infected bacteria several minutes after early transcription has been shut off, so that transcription-inhibitory control of early transcription is likely to exist. The possibility that modified RNA polymerase itself acts as this negative control element, preventing early transcription by the unmodified enzyme (cf. Khesin *et al.*, 1972; Chao and Speyer, 1972; Bordier, 1974), has not yet been investigated.

In this connection, it is also relevant to describe the properties of a template-specific transcription-inhibitory protein that has been isolated from SPO 1-infected bacteria. This small, basic protein is synthesized at a high rate after SPO 1 infection, and more than 10^5 molecules of TF1 have been estimated to accumulate in an infected cell late in SPO 1 development. Originally named TF1, it was sought as a possible answer to the problem of gene turnoff in viral development (Gage and Geiduschek, 1967; Wilson and Geiduschek, 1969). It has been purified (Johnson and Geiduschek, 1972) and shown to be a binding protein that has a higher affinity for SPO 1 DNA than for non-hmU-containing phage ϕ1 DNA (Johnson, 1972; Johnson and Geiduschek, 1977). Rather than being restricted to a small number of sites, a TF1-saturated complex of SPO 1 DNA contains almost 40% protein. In spite of this general binding, TF1 blocks the initiation of RNA chains specifically and does not stop or even slow their elongation. SPO 1 DNA that is saturated with TF1 is unable to form rapidly initiating complexes of RNA polymerase with DNA. TF1 also reduces the total binding of RNA polymerase to DNA (Wilson and Geiduschek, 1969; Armelin, 1973; Petrusek *et al.*, 1975*b*). TF1 specifically blocks SPO 1 but not T 4 DNA-dependent protein synthesis in a coupled *E. coli* cell-free system (Wilhelm *et al.*, 1972; Shub and Johnson, 1975), which mainly makes the corresponding viral early proteins (O'Farrell and Gold, 1973*b*; Shub, 1975). TF1 does not discriminate among the SPO 1 early proteins—all are repressed proportionately—and it is not an inhibitor of RNA-dependent protein synthesis (Shub and Johnson, 1975).

The function of this protein in the infected cell is still not understood, at least in part because its genetics are not worked out. It appears, for a variety of reasons,* unlikely to be solely responsible for the shutoff of *em* or *m* class viral transcripts. It might, however, be responsible for the decline of all viral transcription that occurs in cells infected with replication-defective (DO) SPO 1 mutants (Fujita *et al.*, 1971; Añon and Grau, 1974), and it could be involved in DNA packaging although it is not part of the mature virion (Johnson, unpublished). Indeed, it has been pointed out that a DNA-binding protein which blocks the initiation of SPO 1 transcription but not the growth of RNA chains would serve to strip viral DNA for subsequent packaging (Wilson and Geiduschek, 1969). In its amino acid composition and size, TF1 resembles small, basic proteins that have been found in three different procaryotes and are postulated to have a role in chromosome folding (Rouvière-Yaniv and Gros, 1975; Searcy, 1975; Haselkorn and Rouvière-Yaniv, 1976; Griffith, 1976).

Where do we stand with respect to a biochemical analysis of the regulation of gene action in the development of the hmU phages? Clearly, an encouraging beginning has been made, in part because at least some of the control elements are stable, interact recognizably with RNA polymerase, and can be isolated. There is concrete evidence that two of the most important positive controls of transcription can be identified with RNA polymerase-binding proteins and that the mode of action of these positively regulating proteins is accessible to conventional biochemical analysis. The remaining steps of the regulatory sequence of viral transcription must be guessed at. In Table 3 we present an updated but still tentative scheme of control elements that act during page SPO 1 development.

4. THE T-EVEN BACTERIOPHAGES

The T-even phages, T 2, T 4, and T 6, first isolated by Delbrück and Demerec (Delbrück, 1946a,b; Demerec and Fano, 1945), have been the objects of the classical work on lytic bacterial viruses (Adams, 1959; Stent, 1963; Cairns *et al.*, 1966; Cohen, 1968). These are large, tailed DNA viruses with elaborate base plates; they are structurally

---

* These reasons are: (1) Bacteria infected with the gene 33 and 34 mutants, which are defective in shutting off *m* transcription, synthesize more TF 1 than wild-type SPO 1-infected cells. (2) Purified phage-modified RNA polymerase that synthesizes viral middle RNA can also be inhibited by TF 1 *in vitro*. (Some of the characteristics of inhibition of unmodified bacterial RNA polymerase and of the phage-modified enzyme by TF 1 are not identical, but these differences probably refer to details of chain initiation kinetics only; Petrusek *et al.*, 1976b). This enzyme also makes some RNA that is not completely shut off late in infection. (3) TF 1 indiscriminately blocks transcription of *e* and *em* RNA *in vitro* (Shub and Johnson, 1975).

TABLE 3

A Tentative Scheme of Positive Regulation in SPO 1 Transcription[a]

Transcription Units	Positive regulation by	Rationale for assignment	Uncertainties and comments
e, em	(σ)	1. *In vitro* transcription by bacterial RNA polymerase (Geiduschek *et al.*, 1968). 2. *In vivo* transcription in the presence of chloramphenicol (Gage and Geiduschek, 1971a).	To regard σ as a positive regulatory element of transcription is perhaps stretching the point (see Sections 1.1.1b and 1.1.2b).
m_l, m	ν^{28}-IV acting as a binding-initiation factor	1. *In vitro* transcription of middle RNA with ν^{28} bound to bacterial RNA polymerase core (Duffy *et al.*, 1975 and to be published). 2. ν^{28} and/or ν^{13} is(are) a template-specific binding-initiation factor for middle transcription (Geiduschek and Duffy, to be published). 3. Subunit IV-ν^{28} is the gp 28 (Fox *et al.*, 1975). 4. Gene 28 controls m and m_l transcription (Fujita *et al.*, 1971; Table 1); m and m_l transcripts appear at the same time after infection and probably are coordinately regulated during the middle period of gene expression (Gage and Geiduschek, 1971a).	1. The *in vitro* experiments so far distinguish middle from late and early RNA synthesis; detailed analysis to distinguish m_l from m transcription *in vitro* by hybridization-competition, transcription unit mapping, or protein synthesis *in vitro* has not yet been carried out. 2. The exclusion of ν^{13}-VI here rests partly on the following indirect evidence: Subunit VI is a middle protein (Fox and Pero, 1974) and appears rather late for the onset of middle transcription; middle RNA synthesis *in vitro* can occur without ν^{13} (Duffy *et al.*, 1975; Petrusek *et al.*, 1976a). 4. It is not excluded that m_l RNA might also be synthesized by RNA polymerase with subunits V and ν^{13}-VI.

m_2l

V together with ν^{13}-VI

1. m_2l and l transcription is separately regulated from middle (m and m_1l) transcription by genes 33 and 34 (Fujita et al., 1971; Table 1).

2. An RNA polymerase fraction containing subunit V and ν^{13}-VI transcribes SPO 1 RNA in vitro that corresponds with in vivo RNA which becomes more abundant after replication starts (Pero et al., 1975b).

3. V is gp 34. VI-ν^{13} is gp 33 (Tjian et al., 1976).

4. Quantitative and selective effects of certain DO mutants on m_2l transcription (Añon, 1974).

2. Discrimination of m_2l from l in vitro transcripts remains to be worked out; reconstitution experiments with isolated subunits V and ν^{13} remain to be done; mechanism of action on transcription (e.g., initiation vs. antitermination) remains to be worked out.

4. Evidence that the replication proteins or their metabolic products (such as a particular form or attachment of viral DNA) positively regulate m_2l transcription is lacking. Quantitative effects of DNA replication on transcription could be, in part, due to gene dosage effects.

l

Some replication proteins in an unspecified role

(V + ?; viral genome dosage?)

1. Although m_2l and l transcription is controlled by genes 33 and 34, onset of transcription is not coordinate (Table 1), implying distinctive regulation of m_2l and l transcripts.

Replication requirement for l transcription in vivo poorly understood (Section 3.2); analysis of in vitro transcription that distinguishes m_2l from l transcription remains to be done; the parentheses in column 2 indicate that this is the most insecure assignment and the least likely to be correct.

[a] This is an updated version of a table from Fujita et al. (1971). The original table contained models of shutoff of e, em, and m transcription which have been omitted, there being little new data against which to refine the models.

homologous, undergo genetic complementation and recombination with each other, and cross-react serologically. Their extensive homologies of nucleotide sequence have been demonstrated by heteroduplex mapping in the electron microscope (Kim and Davidson, 1974). All heteroduplexes show more than 85% homology with the greatest homology in the region of the late genes, except for genes 38 and a part of 37 which determine the diverse host range of these phages (Beckendorf *et al.*, 1973) and which are heterologous; there are not many partially homologous sequences.

T-even DNA is circularly permuted and terminally repetitious (Thomas and Rubenstein, 1964). All the T-even phages package about 170,000 base pairs (170kb) of viral DNA with a precision of approximately 1%, yet the T 4 (166 $\pm$ 2 kb) and T 6 (164 $\pm$ 2 kb) genomes are larger than that of T 2 (160 $\pm$ 2 kb). Correspondingly, the terminal repetitions of T 2 (9 $\pm$ 2 kb) and of T 4 and of T 6 (3 $\pm$ 1 kb each) differ (Kim and Davidson, 1974). Deletions of nonessential genetic material are compensated, in their "head-full" DNA packaging mechanism (Streisinger *et al.*, 1967; Ritchie and White, 1972), by more extensive terminal repetitions. Such a packaging procedure probably does not have the precision of nucleotide sequence-specific DNA cutting (Kim and Davidson, 1974), yet this is without consequence for the integrity and completeness of packaged T-even DNA simply because every normal phage head acquires a generous measure of DNA. Aberrations of packaging generate forms of T 4 which contain less than one genome or up to 44 complete genomes in heads of correspondingly smaller or enormous size (Eiserling *et al.*, 1970; Mosig *et al.*, 1972; Doermann *et al.*, 1973; Bolin and Cummings, 1974). The former, the so-called petite phage, are not singly infectious.

Much of the information about the regulation of T-even phage development has been acquired through analysis of T 4. However, there is every reason to believe that the rules which govern T 4 development apply identically to all the T-even phages.

In this section, we shall show that T 4 regulation is remarkable for its diversity and for the number of regulatory events that are already identified. The interplay of these events increases regulatory complexity as more viral genes are expressed during phage development. We shall show that much of the regulation of gene expression occurs at the transcriptional level but that posttranscriptional regulation clearly exists also. We shall stress the association of much of the phage developmental program with viral regulatory genes. What host regulatory mechanisms also participate in the control of T 4 development? cAMP appears to play no determining role for essential gene expression (Craig *et al.*, 1972); neither, apparently, does ppGpp, since T 4 early RNA synthesis is insensitive to the host cell's stringent response to amino acid starvation imposed before infection (Donini and Edlin, 1972). Most importantly, host RNA polymerase is used throughout infection, albeit with several modifications. We imagine that this preserves for phage gene expression the plasticity and range of interactions of the relatively large bacterial polymerase, a range that the simpler phage-induced T 7 polymerase probably does not have. The host's translational apparatus also is subject to some modifications and additions during T-even phage development. However, those changes that have so far been identified are not required for the expression of the "essential" T 4 genes. Nevertheless, the translation of viral mRNA must be subject to the host's full gamut of translational discrimination. The evolution of gene expression in T-even phage development accordingly must have involved an adjustment to, and exploitation of, the host cell's ability to use different messages at individual and different rates for protein synthesis.

T 4 completely shuts off host-cell gene expression. The shutoff mechanisms which involve posttranscriptional as well as transcriptional inhibition (Kennell, 1968a; 1970; Hattman, 1970; Goldman and Lodish, 1971) were analyzed by Mathews (1976) in his chapter on "Reproduction of Large Virulent Bacteriophage" in Volume 7 of *Comprehensive Virology*. The major regulatory events of T 4 development can be summarized as follows: Within 30 sec after infection, the "immediate early" genes start to be transcribed and translated. About 1–5 min later (at 30°C) "delayed early" RNA and proteins appear. (The regulatory relationships of "delayed early" gene expression are somewhat complicated and will be analyzed in detail.) The "immediate early" and "delayed early" genes, which are collectively called the "prereplicative" genes, comprise roughly one-half of the coding capacity of T 4. The other half of the T 4 genes, the "true late" genes, are activated around 10 min after infection. Transcription of the true late genes requires the function of several phage-coded proteins which interact with the RNA polymerase core and depends on simultaneous DNA replication. The tight coupling of DNA replication and true late transcription is a peculiarity of the T-even phages.

4.1. The T 4 Genome

4.1.1. The Genetic Map

The T 4 genome is about one-twentieth as large as the *E. coli* genome. More than 130 T 4 genes are now identified, and the genetic map which is shown in Fig. 5 is approximately 75% saturated [based on

CONTROL & DNA REPL
CON-TROL
LYSIS
DNA REPL
LYS DEF
expansion I
MAT DEF
DNA ARR
DNA NEG
DNA DEL
tRNA
rI
tk
vs
st I
st III
v
ip II
ip III
e
ip I
nrd C
mod
dtp
dex A
pla AR8
pla CTr5x
rIIA
rIIB
DI
den B
pla 262
stp
ndd
ac
das
frd
td
nrd A
nrd B
den A
wac
imm
reg A
20
40
60
80
100
120
140
160
0
ℓ
r
A
B
C
D
E
CON-TROL
DNA REPL
DNA REPL.
EXPANSION I
distance from gene e
4.0
5.0 kb
relative position
0
0.5
1.0
1.5 kb
psu$_b^+$
arg
ile
thr
ser
pro
gly
leu
gln
1
2
5ε
5δ
5α
3
5β
6
4
5γ
transcription

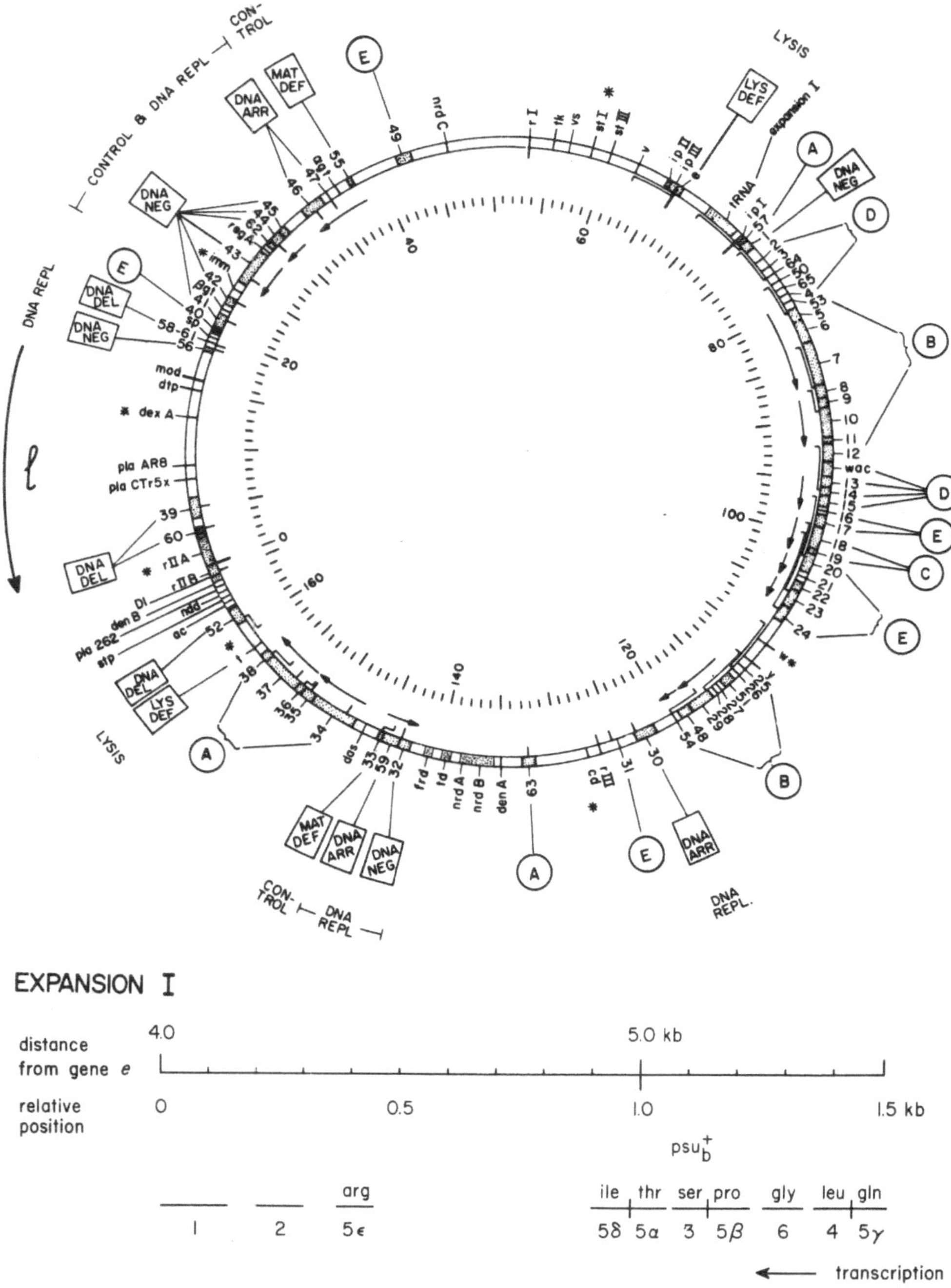

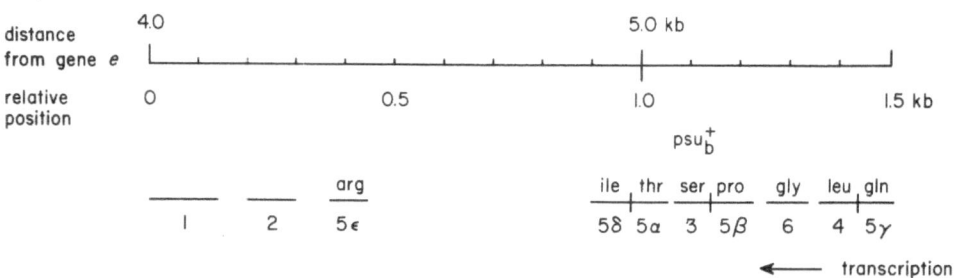

Wood's (1972) prior estimate and on the subsequently identified genes that are shown in Fig. 5]. The circularity of the genetic map results from the circular permutedness of the ends of mature, linear T 4 DNA (Streisinger *et al.*, 1964): Genes which are at the antipodes of some chromosomes are neighbors on others. The majority of the identified genes are essential for T 4 reproduction, as demonstrated by the existence of conditional lethal mutants (Epstein *et al.*, 1963). However, the large T-even genomes include many genes whose products are not essential for, but enhance, T 4 growth in usual laboratory host strains. Among these are genes whose products merely augment cellular metabolic capabilities and can be substituted by corresponding host components (Radding, 1969; Yeh and Tessman, 1972; Chace and Hall, 1973; Capco and Mathews, 1973; Homyk and Weil, 1974). Relatively

---

Fig. 5.   Genetic map of bacteriophage T 4 (based on Wood, 1974). The total map length is 1.66×10^5 nucleotide pairs (Kim and Davidson, 1973). One division on the inner circle represents 10^3 nucleotide pairs (1 kb). The divide between genes rIIA and rIIB, a genetically and physically well-defined site, is arbitrarily chosen as zero. Gene positions are indicated by lines or stippled bars on the circular map. The positions shown by lines on the inside of the circle for some genes were determined by a physical mapping procedure independent of recombination frequencies (Mosig, 1968; Edgar and Mosig, 1970; G. Mosig, quoted by Wood, 1974). Positions of the remaining genes relative to physically mapped loci have been estimated from recombination frequencies (Stahl *et al.*, 1964) or from electron-microscopic heteroduplex mapping of deletions (Homyk and Weil, 1974; Depew *et al.*, 1975). Cotranscribed genes are indicated by arrows showing transcription direction (Stahl *et al.*, 1970; O'Farrell *et al.*, 1973; Vanderslice and Yegian, 1973; Hercules and Sauerbier, 1973). The minimum lengths of stippled genes are based on gene-product molecular weights estimated by discontinuous polyacrylamide gel electrophoresis in the presence of SDS (Laemmli, 1970; Eiserling and Dickson, 1972; O'Farrell *et al.*, 1973; Vanderslice and Yegian, 1974; Homyk and Weil, 1974). Minimum lengths of the two genes, 40 and 60, are estimated from intragenic recombination frequencies (H. Bernstein, quoted by Wood, 1974). Gene functions and references for those loci that have been added to Wood's map are listed in Table 4. All numbered genes and genes *e* and *t* are "essential" (Epstein *et al.*, 1963; Edgar and Wood, 1966; Edgar and Mosig, 1970). Other genes are "nonessential," i.e., their functions are partly or entirely dispensable in common laboratory host bacteria. Encircled capital letters characterize groups of functionally related genes involved in phage morphogenesis; they are categorized in Table 4B. The cistron designated 58-61 represents a single gene (Yegian *et al.*, 1971). Additional genes have been mapped into regions identified by an asterisk, as follows: between genes 24 and 25, *dar;* between genes 31 and 63, *pse*T, *alc;* between genes 38 and 52, *far*I, *mot;* near rII, *m;* between genes 39 and 56, *pse*F, *sud, mb;* between genes 42 and 43, *x;* between genes *tk* and *v*, *far*II. Expansion I: Expansion of the tRNA region (70 to 73 kb from rII), after Fukada *et al.* (1976), shows the disposition of the genes for the various T 4 tRNA (3, 4, 5α-ε, 6) and for two stable low-molecular-weight RNA molecules (1, 2) of unknown function. psu_b^+ is a mutation in tRNA gene 3 which converts T 4 tRNA^{ser} to a suppressor of *ochre* mutations.

large segments of the unmapped portions of the genomes, which probably contain genes that are nonessential in commonly employed laboratory hosts (cf. Kozloff, 1968; Homyk and Weil, 1974; Black, 1974; Black and Abremski, 1974), are homologous in T 2 and T 6 (Kim and Davidson, 1974). If the so-called "nonessential" genes are conserved, they must be presumed to make an essential contribution to the fitness of these viruses. The functions of T 4 genes are listed in Table 4.

4.1.2. Transcription Units

The T 4 genes are arranged in many transcription units containing one or several genes each. The small arrows in Fig. 5 indicate the polarities and lengths of some transcription units: Arrows pointing in a clockwise direction represent transcription from the DNA r-strand (which binds less polyribo(U,G); Guha and Szybalski, 1968); arrows pointing in the opposite direction indicate transcription from the l-strand. The T 4 transcription units are far less completely mapped than bacteriophage λ or T 7 transcription units; many T 4 units must still be totally unmapped. A number of difficulties, none of which is intrinsically insuperable, contribute to this slow development: the size of the genome, the paucity of information about nonessential genes, the absence of lysogens, the apparently complex topography of transcription and of its regulation, and the lack of *in vitro* systems for the expression of certain classes of viral genes. However, it is fair to say that even the currently available methods are far from having been exhaustively employed in the mapping of T 4 transcription units.

4.2. Prereplicative Regulation

The dramatic change in the pattern of gene activity after the onset of DNA replication was one of the first phenomena of T 4 regulation to be recognized (Koch and Hershey, 1959; Wiberg *et al.*, 1962; Khesin *et al.*, 1962; Epstein *et al.*, 1963; Hall *et al.*, 1963). The importance of DNA replication in the regulation of gene activity seems clear, and we shall discuss prereplicative and postreplicative regulation separately. During the prereplicative period (the time span from infection to the start of DNA replication, 5 to 6 min later, at 30°C*), host DNA,

---

* The analysis of T 4 regulation has been done at various temperatures. Unless times of the viral development are otherwise specified in the text, they refer to growth at 30°C. When necessary, results obtained at other temperatures are normalized, taking the ratio of developmental times at 37, 30, and 25°C as 4.6:6.5:10.

TABLE 4

T 4 Genes and Their Functions

Gene[a]	Product	Function[b]	Reference (designated only if not referred to in Wood (1974)
A. T 4 genes and functions other than virion constituents and virion assembly			
rIIA	Membrane protein	Permeability, lysis timing	Sekiguchi (1966); Takacs and Rosenbusch (1975)
60	Unknown	Replication (DD)	
39	Unknown (membrane protein?)	Replication (DD)	Takacs and Rosenbusch (1975)
plaCTr5x	Unknown	Determines ability to grow in a specific bacterial strain	Homyk and Weil (1974)
plaAR-8	Unknown	Determines ability to grow in a specific bacterial strain	Homyk and Weil (1974)
dexA	DNA exonuclease (A)	Bacterial DNA degradation	
dtp	DNA-dependent ATPase	Unknown	Behme and Ebisuzaki (1975)
mod	Unknown	RNA polymerase α subunit modification	Horvitz (1974a)
56	dCTPase-dUTPase	Nucleotide synthesis (DS)	
58 61	Unknown	Replication (DD), recombination	
s(sp)	Unknown	Lysis timing	
41	Unnamed[c]	Replication (DS)	
βgt	β-Glucosyl transferase	DNA modification	
42	dCMP hydroxymethylase	Nucleotide synthesis (DO)	
imm	Unknown	Determines exclusion of superinfecting phage	Okamoto and Yutsudo (1974)
43	DNA polymerase	Replication (DO), recombination	
(Sp62)regA	Unknown	Regulates shutoff of early protein synthesis	Karam and Bowles (1974); Wiberg et al. (1973)
62	Unnamed[c]	Replication (DO)	
44	Unnamed[c]	Replication (DO)	

(continued)

TABLE 4 (Continued)

A. T4 genes and functions other than virion constituents and virion assembly (continued)

Gene[a]	Product	Function[b]	Reference (designated only if not referred to in Wood (1974)
45	Unnamed[c]	Replication (DO), true-late transcription	
46	Exonuclease	Replication (DA), recombination	
47	Exonuclease	Replication (DA), recombination	
αgt	α-Glucosyl transferase	DNA modification	
55	RNA polymerase binding protein	Transcriptional control (MD)	Ratner (1974b)
49	Unnamed	DNA cleavage for phage head assembly	Frankel et al. (1971)
nrdC	Thioredoxin	Nucleotide synthesis	Tessman and Greenberg (1972)
rI	Unknown	Lysis timing	
tk	Thymidine kinase	Nucleotide synthesis	Chace and Hall (1973)
vs[d]	τ peptide	Modification of E. coli val tRNA synthetase	Müller and Marchin (1975)
stI	Unknown	Lysis	Krylov and Yankofsky (1975)
stIII	Unknown	Lysis; locus of suppressor of gene e and t defects	Krylov and Yankofsky (1975)
v	Endonuclease V	UV repair	
ipII[e]	Internal protein iI	DNA packaging	
ipIII[e]	Internal protein III	DNA packaging	
e	Lysozyme	Cell lysis	
RNA 1	A stable RNA of low molecular weight	Unknown	Abelson et al. (1975)
RNA 2	A stable RNA of low molecular weight	Unknown	Abelson et al. (1975)
RNA 3, 4, 5α-ε, 6	tRNAs chargeable with ser, pro, ile, gly, arg, leu, gln, thr	Translation	Abelson et al. (1975)
psu[b+]	Suppressor tRNA (a mutation in tRNA 3)		
ipI[e]	Internal protein I	DNA packaging	
1	Deoxynucleotide kinase	Nucleotide synthesis (DO)	Abelson et al. (1975)

w	Unknown	UV repair and recombination	Hamlett and Berger (1975)
y	Unknown	UV repair	Maynard-Smith and Symonds (1973)
30	DNA ligase	Replication (DA), recombination	
rIII	Unknown	Lysis timing	
cd	Deoxycytidilate deaminase	Nucleotide synthesis	
denA	Endonuclease II	Unknown	
nrdB	Subunit of ribonucleotide reductase	Nucleotide synthesis	
nrdA	Subunit of ribonucleotide reductase	Nucleotide synthesis	
td	Thymidilate synthetase	Nucleotide synthesis	Mathews et al. (1973)
frd	Dihydrofolate reductase	Replication (DO)	
32	DNA binding ("unwinding") protein	Replication (DA), recombination	
59	Unknown	Transcriptional regulation	Horvitz (1973)
33	RNA polymerase binding protein	Suppressor of gene 46 or 47 mutation	
das	Unknown	Lysis	
t	Unknown	Replication (DD)	Takacs and Rosenbusch (1975)
52	Unknown (membrane protein?)	Determines acridine sensitivity	Depew et al. (1975)
ac	Unknown (membrane protein)	Suppressor of the T 4 3′ phosphatase mutant phenotype	
stp	Unknown		Tutas et al. (1974b)
ndd	Unknown	Determines disintegration of bacterial nuclear body after infection	Depew et al. (1975)
pla262	Unknown	Determines ability to grow in a specific bacterial strain	
denB	Endonuclease IV	Unknown	Vetter and Sadowski (1974)
D1	Unknown	Functionally uncharacterized segment adjacent to T 4 rII	
rIIB	Membrane protein	Permeability, lysis timing	Takacs and Rosenbusch (1975); Sekiguchi (1966)
Not precisely mapped or unmapped			
farII[a]	Unknown	Regulates level of synthesis of several early proteins	Johnson and Hall (1974); Chace and Hall (1975)
m[a]	Unknown	Suppressor of gene 30 mutations	Chan and Ebisuzaki (1973)

(continued)

TABLE 4 (Continued)

A. T 4 genes and functions other than virion constituents and virion assembly (*continued*)

Gene[a]	Product	Function[b]	Reference (designated only if not referred to in Wood (1974)
farI[a]	Unknown	Regulates level of synthesis of several early proteins; may control the early regulatory event	Johnson and Hall (1974); Chace and Hall (1975)
mot[a]	Unknown	May control the early regulatory event	Matson et al. (1974)
pseT[a]	a 3′ phosphatase	Unknown	Depew and Cozzarelli (1974)
pseF[a]	a 5′ phosphatase	Unknown	Depew and Cozzarelli; quoted by Homyk and Weil (1974)
sud[a]	Unknown	Locus of suppressors for gene 32 mutations	Little (1973); Homyk and Weil (1974)
mb[a]	Unknown	Required for maturation of several T 4 tRNA's	Wilson and Abelson (1972)
dar[a]	Unknown	Suppressor of gene 59 mutations	Wu and Yeh (1975)
x[a]	Unknown	UV sensitivity; recombination	Harm (1963); Shimizu and Sekiguchi (1974)
alt	Unknown	A change ("alteration") of RNA polymerase α subunit	Horvitz (1974a)
alc[a]	Uncertain (but see Table 5)	Determines DNA nucleotide modification requirement of late transcription	Snyder (1976)
—	RNA-RNA ligase	Ligation of 3′ –OH and 5′ phosphate termini of single and double-stranded RNAs; function in vivo unknown	Silber et al. (1972)
—	endonucleases I, III, and VI	Unknown	Altman and Meselson (1970); Sadowski and Bakyta (1972); Kemper and Hurwitz (1973)
—	Phospholipase	Unknown	Nelson and Buller (1974)

Class	Function	Genes
	B. Genes functioning in virion assembly	
A	Tail fiber and tail fiber joining to tail plate	34, 38, 57, 63
B	Tail plate and tail plate assembly	5, 12, 25, 29, 48, 51, 53, 54
C	Tail core and sheath	3, 18, 19
D	Collar, head-tail joining, and head completion	2, 4, 13, 15, 50, 64, 65, wac
E	Head and head filling	16, 17, 20, 24, 31, 40, 49, ip1,[e] ip11,[e] ip111[e]

[a] Gene order is clockwise from rIIA.

[b] Abbreviations: DO, no replication; DA, replication arrested; DS, replication severely defective; DD, replication delayed; MD, maturation (= "late" function) defective.

[c] These proteins have been purified and used as components of an in vitro system for T 4 DNA synthesis (Alberts et al., 1975). Their properties are briefly described in Section 4.6.3b.

[d] These genes are approximately mapped (see Fig. 5 caption).

[e] The products of these genes are not essential for virion assembly. They are listed in sections A and B of this table.

RNA, and protein synthesis are shut off and the so-called "prereplicative" or early viral gene products appear. The gene products which appear for the first time after viral DNA replication has started have been named "true late." A third group of gene products are already synthesized during the prereplicative period but become more abundant after 10 min of infection; these have been termed "quasi-late." According to this nomenclature of gene products, the corresponding genes, messages, and proteins have been named in the same way (early genes, early mRNA, etc.).

T 4 mRNA, like *E. coli* mRNA, appears generally to be translated *in vivo* almost without delay. Within the prereplicative period, many different species of T 4-coded RNA and proteins are synthesized and appear in a characteristic temporal sequence. In this section we try to analyze this sequence of early gene expression in terms of transcription units and in terms of transcriptional and posttranscriptional control. Before proceeding, however, we want to raise a more general question: What is the functional significance of dividing prereplicative gene expression into several parts with distinctive regulatory features? The simplest intuitive explanation might be that this division serves to create a sequence of viral metabolic activities that allows a more effective staging of virus reproduction. While that explanation probably fits the switch from early to late gene expression, it is not convincing for the early regulatory event, partly because the latter occurs so quickly and partly because most early gene expression is modulated rather than turned on or off by it. Another possibility appeals to us more: that the early regulatory event overcomes a host defectiveness for virus multiplication (which is a collective defense mechanism) and that this event simply occurs at the earliest time that is compatible with postinfection synthesis of a viral regulatory protein. We shall see in Section 6 that some phages may inject their own positive regulatory components into the infected cell. In contrast, T 4 injects an enzymatic actvity which alters the host RNA polymerase, most likely as a first measure to effect shutoff of host RNA synthesis.

4.2.1. Temporal Sequence of the Appearance of Early T 4 mRNA and Proteins

With only one exception, the methods which have been used for the analysis of T 4 RNA and proteins are those which have proved so powerful in general and have already been referred to in the preceding sections: (1) Hybridization of RNA to DNA was first used with T-even

RNA. Hybridization-competition and hydridization to isolated DNA strands were developed subsequently (Hall and Spiegelman, 1961; Bautz and Hall, 1962; Khesin *et al.,* 1962; Hall *et al.,* 1963, 1964; Guha and Szybalski, 1968). Isolation of *gene-specific* T-even messages has, in the past, been confined to the few instances in which suitable deletions were available. However, the advent of restriction enzymes, in principle, removes that requirement (for a review, see Boyer, 1974). When necessary, C-for-hmC-substituted T 4 DNA can be prepared (Bruner *et al.,* 1972; Kutter *et al.,* 1975; Snyder *et al.,* 1976) for restriction analysis (Kaplan and Nierlich, 1975). (2) RNA and proteins can be separated according to size by polyacrylamide gel electrophoresis (Hosoda and Levinthal, 1968; Adesnik and Levinthal, 1970; Laemmli, 1970; O'Farrell *et al.,* 1973; Vanderslice and Yegian, 1974). RNA gel electrophoresis has been successfully used in exploring those viral transcripts that are more stable, like T 7 RNA which we have already discussed. Some T 4 transcripts also are relatively stable, but most T 4 RNA is not, so that the high background level of heterogeneous nascent and degraded T 4 RNA presents major difficulties. These difficulties can be overcome to some extent by controlling initiation of new RNA chains, by choosing appropriate host strains (Sauerbier *et al.,* 1969; Keil and Hofschneider, 1973), and by inhibiting protein synthesis (Adesnik and Levinthal, 1970; Guthrie, 1971; Young, 1970). (3) Cell-free protein synthesis programmed by mRNA yields information on biologically functioning messages (Salser *et al.,* 1967; for a summary, see Gold *et al.,* 1973). (4) A DNA transformation assay for mRNA has been devised. In this method, T 4 *l-* or *r-*strand DNA is protected by hybridized RNA from nuclease digestion and is then used to contribute a specified stretch of DNA as a genetic marker to spheroplasts infected with subviral particles derived from mutant phage. The location of the mutant specifies the DNA segment that must be protected by RNA (Jayaraman and Goldberg, 1969, 1970; Jayaraman, 1972).

A general picture of the time sequence of early RNA and protein synthesis can be conveyed simply: During the first minutes after infection, synthesis of phage-specified RNA, which includes mRNA and some tRNA, is concurrent with residual, but rapidly declining, host RNA synthesis. Thereafter, only phage RNA is transcribed. Early proteins make their first appearance in a somewhat elaborate sequence during the first five to six minutes; no new proteins appear for the next several minutes, and after that come the postreplicative proteins. The early proteins are *not* members of a single coordinately controlled set.

The time sequence of the first appearance of early proteins and their subsequent modulations of synthesis rate are principally the result of three phenomena: (a) The early proteins are the products of multicis-tronic transcription units; a part of the time sequence reflects differ-ences in promoter proximity. The time scale of this sequence is rela-tively drawn out because T 4 early RNA chains grow on the average only half as fast as *E. coli* mRNA chains at the same temperature (Bremer and Yuan, 1968; Guthrie, 1971; Gausing, 1972). (b) A change of transcription specificity, probably including transcription initiation specificity (Section 4.2.6b) occurs within the first few minutes. The transcription of a very few known early genes is entirely dependent on that newly acquired transcription specificity. For many other early genes there are quantitative changes of transcription rate and of the rate of synthesis of the corresponding proteins. (c) The synthesis of some viral proteins ceases after the first few minutes.

The names that have been applied to this collection of RNAs and proteins by now pose something of a problem, mostly because of the promiscuous use of designations for nonidentical or noncorresponding objects. Under the circumstances, we prefer to start with an examina-tion of the basic data and only later to venture into the quasi-swamp of T 4 terminology. The sequential appearance of many T 4 early proteins is shown in Fig. 6. Many of the early proteins are enzymes of the phage-specific pathway of deoxyribonucleotide synthesis and polymeri-zation. Others bind to or enter the cell membrane or modify RNA polymerase and tRNA synthetase; some of these proteins are absolutely required for viral multiplication in all tested laboratory or natural strains of *E. coli*; others are not (Cohen, 1968; Mathews, 1971; Black and Abremski, 1974; Homyk and and Weil, 1974). Only one-third to one-half of the early genes are represented in Fig. 6. The corresponding diagram for T 7 (Section 2) includes all the virus-coded proteins. The difference lies principally with the complexity of T 4; the development of two-dimensional gel electrophoresis provides access to a complete analysis (O'Farrell, 1975), but this remains to be seen.

The synthesis of a number of early messages and their conjugate proteins has been compared *in vivo*. The early messages which have been analyzed in this way code for a number of enzymes of nucleic acid metabolism, the phage internal proteins, gp 57, and the rII proteins (cf. Fig. 6). The first appearance, accumulation, and decay of these func-tional messages has been followed by RNA-programmed synthesis of proteins *in vitro*, and messenger polynucleotide has been followed by hybridization and DNA transformation (Kasai and Bautz, 1969; Jayaraman and Goldberg, 1970; Black and Gold, 1971; Young and Van Houwe, 1970; Wilhelm and Haselkorn, 1971*a,b*; Zetter and Cohen, 1974; Sakiyama and Buchanan, 1972). With only one exception,

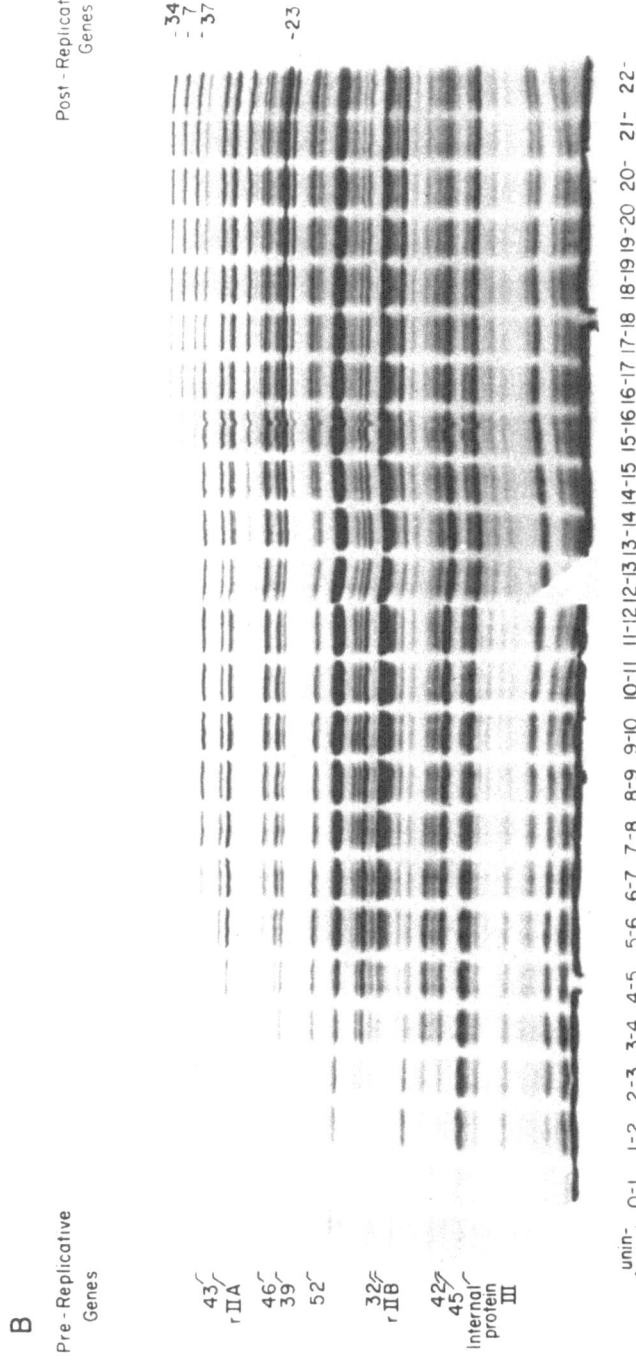

Fig. 6. Sequential appearance of T 4 proteins at 30°C. This picture is taken from Fig. 1 of O'Farrell and Gold (1973a). E. coli cells were UV-irradiated and infected with T 4 at time 0. Aliquots of the infected culture were labeled with ^{14}C-amino acids for 1 min every successive minute. Extracts of each sample were subjected to electrophoresis on SDS polyacrylamide slab gels. The autoradiograms are shown.

messages appear slightly before (or, roughly concurrently with) their respective protein products (Trimble et al., 1972). Lysozyme mRNA is anomalous: although the protein is synthesized together with other late proteins, some lysozme gene RNA, which is not functional as a lysozyme message in vivo or in vitro, is detected during the first minutes after infection (Kasai and Bautz, 1969; Bautz et al., 1966; Jayaraman and Goldberg, 1970).

The early genes are read almost exclusively with one polarity. More than 99% of the earliest T 4 RNA is read off the DNA l-strand (Guha and Szybalski, 1969). At the start of DNA replication, between 98 and 99.8% of RNA is read with that polarity (Notani, 1973; Guha et al., 1971), and 4 min later, there is still less than 5% of transcription from the other strand. Most, but not all, of the late genes (Fig. 5) are transcribed with the opposite polarity.

4.2.2. Shutting Off Early Protein Synthesis

During normal phage development, some early proteins and RNA continue to be synthesized throughout infection. Among these are the phage internal proteins and the phage-specified tRNA. However, the synthesis of many early proteins stops after approximately 12–14 min of infection at 30°C (Wiberg et al., 1962; Hosoda and Levinthal, 1968; Mathews and Kessin, 1967). The synthesis of all early proteins does not stop synchronously (O'Farrell and Gold, 1973a; Trimble et al., 1972). Despite considerable effort, this phenomenon of early gene shutoff is not well understood even at the physiological level. A major problem has been the delayed recognition that the shutoff of early protein synthesis results from a diversity of regulatory phenomena, not from the action of a single mechanism (Cohen, 1970). These regulatory mechanisms exert different influences on different early messages under different circumstances yielding regulatory characteristics that are remarkable for their individuality. Our purpose here is to infer what the regulatory principles might be, but we shall not describe the regulation of all early proteins individually.

The shutoff of early protein synthesis probably results from regulation of transcription (Hall et al., 1964; Khesin et al., 1963; Bolle et al., 1968a; Salser et al., 1970) and also from changes in the posttranscriptional fate of early messages. During the period of its functioning, T 4 early mRNA is quite stable: the chemical half-life of early RNA (measured with rifamycin addition soon after infection) is, on the average, 8–10 min at 30°C (Craig et al., 1972). Yet, when the shutoff of early protein synthesis does occur, mRNA degradation is rapid. For four different early proteins, a functional half-life of approximately 2 min at 37°C has been estimated during this shutoff period (Sauerbier and Hercules, 1973). The rapid decay of early protein synthesis could either be primarily due to competition between messages for translation, with degradation as a secondary event, or it could be the direct result of a change of mRNA (chemical) stability. Competition for translation could be particularly efficient because T 4-infected cells are normally at messenger excess (Sheldon, 1971; Cohen *et al.*, 1972; Gold *et al.*, 1973) so that protein synthesis is not limited by messenger content. Effective competition for translation is evidently not mediated by late messenger-specific (positive) translation factors since late T 4 proteins can be synthesized in cell-free extracts from uninfected *E. coli* programmed with late T 4 RNA (Salser *et al.*, 1967; Wilhelm and Haselkorn, 1971*a,b*).

We guess that the chemical stability of early mRNA first decreases at the time that early protein synthesis is shut off. Such an assumption permits one to reconcile older experiments on continuous labeling of RNA (Haselkorn *et al.*, 1968; Bolle *et al.*, 1968*a*) with measurements of the stability of functioning early mRNA (Adesnik and Levinthal, 1970; Craig *et al.*, 1972; O'Farrell and Gold, 1973*a*; Sakiyama and Buchanan, 1972). However, the evidence on this subject is not strong. Only for the DNA "unwinding" protein, gp 32, is the evidence for posttranscriptional regulation of messenger function really convincing (Gold *et al.*, 1976; Russel *et al.*, 1976). But the autogenous regulation of gp 32 synthesis (Section 4.2.4) is so intimately connected with the unique nucleic acid-binding properties of gp 32 that one cannot automatically generalize to the other early proteins. It would be ideal to measure the functional and chemical stability of several individual early messages concurrently, before and after shutoff, by means of *in vitro* protein synthesis and DNA transformation. Unfortunately, not one complete experiment is available, although the decay of the functional and chemical entities of several early messages after shutoff *in vivo* have been measured (Guthrie and Buchanan, 1966; Greene and Korn, 1967; Bose and Warren, 1967; Bolle *et al.*, 1968*a*; Young and Van Houwe, 1970; Sakiyama and Buchanan, 1971; 1972; Gausing, 1972; Craig *et al.*, 1972; Sauerbier and Hercules, 1973). Most of the half-lives which were measured are greater than 2 min at 37°C, which was cited above for the cessation of synthesis of several proteins *in vivo*, and less than 8–10 min at 30°C, which was cited for the stability of early RNA during the period of early protein synthesis.

The shutting off of the synthesis of at least some early proteins is claimed to occur in advance of the functional inactivation and chemical degradation of their conjugate messages. This is the situation for deoxynucleotide kinase: In the absence of viral replication, or when K^+ is limiting for protein synthesis, accumulation of the enzyme stops while functional message still remains in the cell and general protein synthesis continues (Cohen, 1972; Boloud, 1973; Sakiyama and Buchanan, 1972; Sakiyama, quoted by Wiberg *et al.*, 1973).

In summary, we conclude tentatively that the shutting off of early protein synthesis involves posttranscriptional inactivation of existing messages as well as cessation of their synthesis. In addition, the observations on deoxynucleotide kinase synthesis are consistent with the notion that shutoff might result from messenger competition, although other interpretations involving the specific regulatory properties of this protein are tenable (see Section 4.2.4).

4.2.3. Replication and Early Gene Shutoff

In the absence of DNA replication, the shutting off of early functions is much delayed but ultimately does occur (Wiberg *et al.*, 1962, 1973; Young and Van Houwe, 1970; Sakiyama and Buchanan, 1972, Bolund, 1973) in a time sequence that resembles the normal one (Sauerbier and Hercules, 1973). But for at least some proteins, the relationship between functional mRNA and protein synthesis depends on whether replication does, or does not, occur (Young and Van Houwe, 1970; Sakiyama and Buchanan, 1972). Deoxynucleotide kinase, which we have just discussed, exemplifies this situation. In the wild type, the decay of mRNA activity correlates approximately with the cessation of enzyme synthesis (Cohen, 1972; Sakiyama and Buchanan, 1972). Decay of functional kinase mRNA occurs so long as DNA is replicated but irrespective of whether late T 4 proteins are synthesized (Sakiyama and Buchanan, 1972). Yet the total quantity of nucleotide kinase accumulating after infection evidently depends, conversely, on whether late proteins are synthesized rather than on replication (Bolund, 1973). In the absence of replication, as we have just mentioned, nucleotide kinase synthesis ceases in a cell that continues to harbor functional kinase mRNA.

We use the preceding evidence, imperfect though it is, to argue for the possible existence of posttranscriptional discrimination among T 4 messages. But is that discrimination subject to control? Apparently so, since the product of a recently identified T 4 gene affects the post-transcriptional fate of many early messages and, in that sense, exerts posttranscriptional control. Mutants in this gene (designated as *sp*62 or *reg*A; Fig. 5; Table 4) assert their greatest effect when there is no replication, but show only a slight effect in normal phage development (Wiberg *et al.*, 1973; Karam and Bowles, 1974). Many, but not all, early messages have dramatically greater stability in a *reg*A-replication-defective genetic background than in a *reg*A or wild-type background (Sauerbier and Hercules, 1973; Wiberg *et al.*, 1973; Karam and Bowles, 1974). This replication dependence of the sensitivity of early protein regulation to *reg*A mutations further supports the notion that the repression of T 4 early protein synthesis is multivalent. We have not cited much other evidence that has been offered on the subject of posttranscriptional discrimination or control, mainly in the interest of simplicity and brevity.

4.2.4. Autogenous Regulation of Some Early Proteins

At least two early proteins, T 4 DNA polymerase gp 43 and the T 4 DNA "unwinding" protein gp 32, utilize gene-specific regulatory mechanisms. Each of these proteins negatively autoregulates its *own* synthesis (Russel, 1973; Krisch *et al.*, 1974; Gold *et al.*, 1976; Russel *et al.*, 1976). The basic experimental observations are that *amber* mutants in gene 43 overproduce the corresponding defective polypeptide dramatically. The activity of DNA polymerase as an effector of its own synthesis probably involves the three-dimensional structure of the protein, since many temperature-sensitive gene 43 mutants are also regulation-defective (Russel, 1973) in the sense that they also overproduce the temperature-sensitive T 4 DNA polymerase. In addition, DNA polymerase synthesis shows a dependence on DNA replication, leading to overproduction in the absence of replication. We take this to imply that gene 43 shares some regulatory roles with the other T 4 early proteins. Despite its distinctive regulation, DNA polymerase synthesis normally is shut off roughly concurrently with the early proteins. (On closer examination, DNA polymerase synthesis is actually seen to be shut off later than most early proteins, but roughly together with others.) Here, then, is a clear example of the potential inadequacy of approximate temporal criteria for regulatory classification.

Were it not autoregulated, gene 32 protein could be continuously synthesized until lysis. This is inferred from experiments in which *am* and *ts* mutants in gene 32 have been shown to continuously produce the defective gp 32. The autoregulation of gene 32 protein synthesis has one additional significant feature. Because of its ability to bind gene 32 protein, single-stranded DNA derepresses the synthesis of gene 32 protein. The result is that the production of the DNA unwinding protein, which is required for T 4 DNA replication, recombination, and repair (Epstein *et al.,* 1963; Tomizawa *et al.,* 1966; Warner and Hobbs, 1967; Bernstein, 1968; Berger *et al.,* 1969; Broker and Lehman, 1971; Yeh and Wu, 1973), is coupled to all cellular processes that expose single-stranded DNA. Single-stranded DNA is normally transient and gp 32 is not consumed, inactivated, or permanently tied up by its activities. Consequently, once the pool of replicative DNA reaches its steady size, synthesis of further gene 32 protein normally declines sharply (Krisch *et al.,* 1974). If it were otherwise, if gene 32 protein were bound up or inactivated as a result of replication, then its synthesis and DNA replication might be coupled to each other as histone and DNA synthesis are in higher eucaryotes (Mitchison, 1971).

The autoregulation of gp 32 has recently been shown to involve posttranscriptional regulation. Regulation at the translational level fits with the exceptional stability of gene 32 mRNA which accumulates in the infected cell to a greater degree than any other T 4 mRNA. Remarkably enough, the regulation depends on nonspecific gp 32 RNA binding which probably selectively affects ribosome binding and initiation of gp 32 mRNA (Russel *et al.,* 1976). Figure 7 shows a model of the autoregulation of T 4 gene 32 protein. An important aspect of the model is that it lends itself so readily to elaborations and extensions that might apply to a variety of cellular regulatory processes.

4.2.5. Regulatory Rules for Other Early Genes

We mention the regulatory diversity of the shutoff of other early genes only briefly. The analysis up to this time has been in the framework of correlations with replication, with the function of the transcriptional control genes (especially gene 55), and with continued protein synthesis. The early genes whose shutting off has been analyzed in this context include genes 1, 43, 46, 56, *agt*, IP II, and some unidentified genes whose protein products have been separated and identified on polyacrylamide gels (Sakiyama and Buchanan, 1972; Bolund, 1973; Russel, 1973; Young and Van Houwe, 1970; Young, 1970; Black and Gold, 1971; Adesnik and Levinthal, 1970; Wilhelm and Haselkorn, 1971*a*). We do not list the results in detail because we are not convinced that the proper regulatory principles have yet emerged from the empirical data. For example, analysis of gene 43 or 32 regulation in these terms does not provide the insights into their autoregulation that have been discussed in the preceding section. Specifically, both replication and gene 55 function are likely to act pleiotropically rather than directly in regulating the synthesis of these proteins.

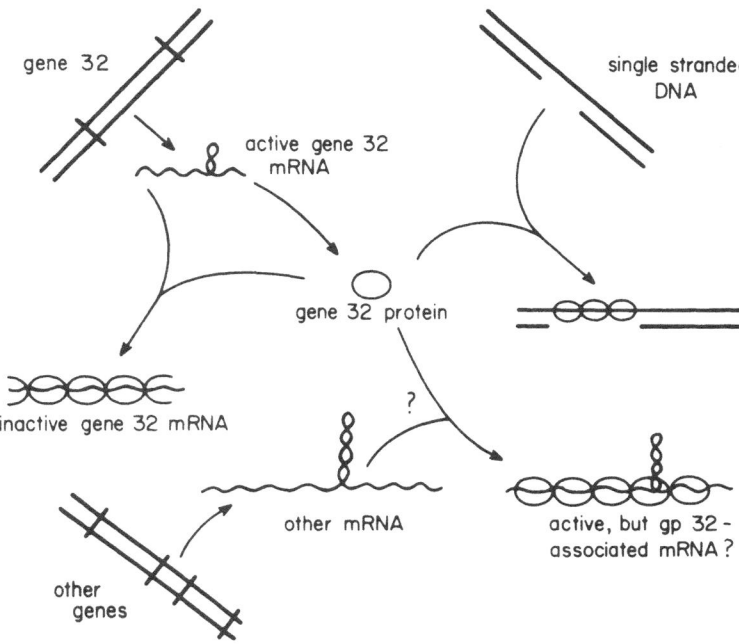

Fig. 7. A model of autoregulation by the nucleic-acid-binding protein, gp 32 (after Russel *et al.*, 1976). The model incorporates the following properties of gp 32: (1) It is a single-stranded DNA-binding protein (Alberts and Frey, 1970; Gold *et al.*, 1976) but also binds to mRNA (Russel *et al.*, 1976); (2) it regulates its own synthesis (Krisch *et al.*, 1974; Gold *et al.*, 1976) at the posttranscriptional level (Russel *et al.*, 1976); (3) translationally inactive gene 32 mRNA is not degraded (Russel *et al.*, 1976). As a result of these properties, single-stranded DNA derepresses the synthesis of gp 32. Conditions which lead to an accumulation of single-stranded DNA prolong the duration of gp 32 synthesis. Temperature-sensitive, labile, mutant gp 32 can also be temperature-sensitive as an autorepressor. The model also incorporates the speculation that gp 32 might be nonselective in its binding to mRNA but selective in its effect on translation. If this is so, it might be related to differences in secondary structure of gene 32 mRNA and other T 4 mRNA. The possibility that gp 32 might also affect transcription has not been disproved but has, for the sake of being concise, been omitted from the model.

4.2.6. Transcriptional Regulation

4.2.6a. Physiological Considerations

T 4 RNA chains grow more slowly than *E. coli* mRNA; at 37°C the average rate is only 25 nucleotides per second. Corrected to 30°C, this corresponds to approximately 1100 nucleotides per minute (Bremer and Yuan, 1968; Guthrie, 1971; Gausing, 1972). *E. coli* ribosomal RNA and T 7 mRNA grow three to four times as fast (Sections 1.2.1b and 2.3.5). The lower rate of T 4 RNA chain elongation occurs without substantial changes of ribonucleotide pools after infection (Mathews, 1972). It may be a property of the glucosylated DNA template (Cox and Conway, 1973), and we shall see below that glucosylation of DNA can, in special circumstances, be significant for transcriptional regulation (Snyder, 1972; Snyder and Montgomery, 1974). However, the rate of elongation of early T 4 polypeptide chains is also slower by a comparable margin (Kleppe and Nygaard, 1969; Gausing, 1972). Since T 4 early messages are relatively stable, and must therefore continue to function after completion of transcription, slow translation might be the cause, but cannot be the consequence, of slow transcription.

The messenger content of the T 4-infected bacterium is at least as high as that of the uninfected cell (Landy and Spiegelman, 1968; Kleppe and Nygaard, 1969; Gausing, 1972); more importantly, as we have already noted, the capacity to synthesize phage proteins is probably limited by ribosomes and translational proteins rather than by messages. It is useful to bear this observation in mind when one tries to piece together an understanding of the controls on phage gene action *in vivo* from experiments with various cell-free systems for protein synthesis. The latter are customarily programmed so as to be limited by mRNA.

4.2.6b. Initiation and Termination of Early RNA Chains

The idea that more than one regulatory class of prereplicative messages exists (Grasso and Buchanan, 1969; Salser *et al.,* 1970) was first proposed on the basis of hybridization-competition experiments, the interpretation of which is now in dispute, but the idea is nevertheless valid. It has been observed that chloramphenicol addition at the time of infection blocks the appearance of a large fraction of early RNA; the sensitivity of early transcription to chloramphenicol disappears within 2–3 min at 30°C. Most of the RNA whose synthesis is blocked by chloramphenicol is that which first appears 2 min after infection and which is relatively distal to the T 4 early promoters used by *E. coli* RNA polymerase (Grasso and Buchanan, 1969; Salser *et al.,* 1970; Brody *et al.,* 1970*a,b*; Milanesi *et al.,* 1969, 1970; Sederoff *et al.,* 1971). Inhibition of protein synthesis by puromycin or K$^+$ starvation has similar effects on early RNA synthesis (Grasso and Buchanan, 1969; Peterson *et al.,* 1972).

It was originally assumed that this finding implied the participation of a virus-specified protein in the transcription of some viral genes.

However, as the information about early RNA synthesis has accumulated, it has become clear that at least this reasoning, if not the conclusion, must be abandoned. First, early phage T4 transcription is not sensitive to the presence of amino acid analogs or to the presence of antibiotics that induce miscoding and therefore should produce nonfunctional proteins (Black and Gold, 1971; Gold, 1975; Brody, 1975). This contradicts the assumption that chloramphenicol acts indirectly on viral transcription by preventing the synthesis of a positive regulatory protein. Secondly, the effects of chloramphenicol on T 4 early transcription are complex and probably involve both promoter recognition and "read-through" from promoter-proximal to promoter-distal genes.

Nevertheless, there is ample evidence that more than one class of prereplicative messages exists: (a) The synthesis of mRNA for a small number of early proteins (gp 32, 40, 41, 1) cannot be initiated within the first 1–2 min after infection (O'Farrell and Gold, 1973a; Gold et al., 1973; Cohen et al., 1974). The synthesis of these proteins is dramatically depressed relative to the other early proteins when rifamycin is added concurrently with, or very soon after, infection.* (b) The relative yield of a number of proteins (e.g., gp 43, gp 45, gp rIIB) depends significantly on that delayed RNA chain-initiation event (Bautz et al., 1969; Schmidt et al., 1970; Gold et al., 1973; Hercules and Sauerbier, 1974). (c) An as-yet-uncharacterized T 4 mutation, named tsG1 (gene mot) selectively affects the synthesis of those T 4 early proteins which depend qualitatively or quantitatively on the delayed RNA chain-initiation event (Matson et al., 1974). Surprisingly, certain mutants in rIIB display a phenotype which is analogous to tsG1 (Nelson and Gold, 1975). The synthesis of these same proteins is quantitatively defective in a cell-free system from uninfected bacteria that carries out coupled RNA and protein synthesis programmed by T 4 DNA (Gold et al., 1973). (An apparent exception, deoxyribonucleotide kinase, gp 1, will be discussed below in the context of transcriptional read-through.) Yet, in vivo generated messages for these same proteins function perfectly well in the same cell-free system. Regardless of the mechanistic basis for this relationship, it does help to define a second regulatory class of

---

* Unless specially treated (e.g., Leive, 1965), most strains of E. coli are poorly permeable to rifamycin and its derivatives. In the experiments of different laboratories, this permeability barrier has been overcome to different degrees, leading to variations of elapsed time between addition of the antibiotic and inhibition of RNA chain initiation. In comparing the experiments of different laboratories, the absolute timing of events is therefore poorly reproduced; however, sequences of events should be comparable.

T 4 genes. (d) The transcription of genes 40 and 41 from mature T 4 DNA by *E. coli* RNA polymerase is defective *in vitro* under conditions in which most T 4 early genes are transcribed asymmetrically (Jayaraman, 1972). (e) A set of T 4 transcripts, the so-called "antimessenger," which is complementary to a large fraction of T 4 late RNA (Geiduschek and Grau, 1970; Jurale *et al.,* 1970), first appears 3 min after infection (Notani, 1973). The abundance of these transcripts is low (about 2% of 5-min RNA), but they are not particularly unstable relative to early T 4 RNA (Notani, 1973; di Porzio, 1973). They were first thought to originate by read-through from delayed early genes (i.e., by nontermination of RNA chains). However, their appearance is sufficiently synchronous in time and their complementarity to late RNA is sufficiently extensive that a glance at the genetic map of T 4 (Fig. 5) makes that guess rather implausible. However, antimessenger synthesis might be regulated by relief of RNA chain attenuation. This would imply the possible presence of short leader sequences for anti-late RNA among the earliest T 4 transcripts. As far as we know, short anti-late T 4 leader sequences have not been carefully sought. It has also been suggested that the synthesis of antimessenger reflects a change in initiation specificity of RNA synthesis (Notani, 1973). Antimessenger RNA chains start to grow at 2–3 min after infection, and they are not produced if protein synthesis is blocked with chloramphenicol. The *ts*G1 mutation also affects antimessenger synthesis (Snyder, 1975). The function of antimessenger RNA is not known.

The central problem of early transcriptional regulation is to understand this change of transcriptional pattern, which we call the "early regulatory event," and which occurs 1–2 min after infection. We are uncertain about whether all of the manifestations of the early regulatory event are due to a single effect or to several independent but concurrently acting regulatory effects. The rough coincidence in time of all the transcriptional changes could signify a unitary mechanism, but may mean nothing more than that 1–2 min is the minimum time span for the generation or mobilization of the regulatory components. Chloramphenicol-induced polarity and the delayed middle RNA chain initiation event *appear* to involve different mechanisms. Moreover, the phenotype of the gene *mot* mutation affects only the delayed middle RNA chain initiation event and this suggests that a unitary mechanism does not exist. (However, the persistence of chloramphenicol-induced polarity after infection with *mot* mutants has not yet been measured.) Nevertheless, plausible though complex arguments and evidence for a unitary basis of the early regulatory event have recently been presented (Thermes *et al.,* 1976).

By what means does chloramphenicol affect T 4 early transcription? It seems likely that the chloramphenicol effect is primarily due to induced polarity (Black and Gold, 1971; Young, 1975) which affects certain bacterial transcription units such as the *lac* and *trp* operons (Alpers and Tomkins, 1966; Morse and Yanofsky, 1969; Kuwano *et al.,* 1971; Morse, 1971; Hansen *et al.,* 1973). The mechanism of chloramphenicol-induced polarity is not known, but T 4 transcription becomes resistant to it concurrently with the early regulatory event. The T 4 early transcription units may be comparable with the *E. coli trp* and *gal* operons for which variations of polarity at internal loci are associated with transcription from different promoters (Imamoto, 1973*b;* Franklin, 1974; Adhya *et al.,* 1974; Segawa and Imamoto, 1974). In *E. coli,* uncoupling of transcription from concurrent translation makes some mRNA more labile (Morse and Yanofsky, 1969; Morse, 1970; Hansen *et al.,* 1973). The effect of chloramphenicol on T 4 mRNA also appears to be selective, varying for different messages and also perhaps depending on the time of chloramphenicol addition (Young, 1970; Young and van Houwe, 1970; Adesnik and Levinthal, 1970; Brody *et al.,* 1970*b;* Guthrie, 1971; Sakiyama and Buchanan, 1972). We would argue by analogy that a change of T 4 transcription susceptibility to chloramphenicol-induced polarity might be associated with the use of different T 4 promoters as the result of a change in transcription control. The application of the newly isolated *E. coli* polarity suppressor mutations (Section 1.1.1) to this problem will probably allow one to understand the chloramphenicol effect on early transcription much better.

The problem of unravelling the early regulatory event is made much more difficult by the special topography of the T 4 early genes. It is this topography which makes it especially difficult to distinguish between regulation of transcription at promoters and termination sites. The T 4 early genes are predominantly clustered in two regions of the genetic map (Fig. 5, 120 to 147 kb and 158 to 75 kb). In each of these regions there are genes classified as immediate early or delayed early on the basis of the order of appearance of their respective proteins (the immediate early proteins make their first appearance before the delayed early proteins). For the T 4 delayed early proteins, time of first appearance lacks the simple regulatory significance that it has for the corresponding phage T 7 class II and SPO 1 middle proteins. However, the T 4 immediate early genes are proximal to promoters, P_E, that are accessible to *E. coli* RNA polymerase, so that the corresponding messages and proteins are made first, *in vivo* and *in vitro*. Delayed early genes can be shown to be relatively distal to the P_E promoters (Salser *et al.*, 1967, 1970; Hosoda and Levinthal, 1968; Milanesi *et al.*, 1969, 1970; Brody *et al.*, 1970*a,b*, 1971; O'Farrell and Gold, 1973*a,b*; Hercules and Sauerbier, 1973, 1974); since the delayed early proteins also appear after a time lag, the structure of at least some transcription units read by unmodifed bacterial RNA polymerase must be

$$\cdots 3' \ P_E \ \text{(immediate early gene) (delayed early gene)}_n \ 5' \cdots$$

One real transcription unit with this organization includes the delayed early genes rIIA and rIIB and the contiguous delayed early region D1, which are transcribed *in vivo* as a single polycistronic message from a start signal located "upstream" from rIIA (Bautz *et al.*, 1969; Schmidt *et al.*, 1970; Brody *et al.*, 1970*a,b*; Fig. 8, part C).

In each such transcription unit, there must be at least one signal that is located between the P_E promoter and delayed early genes and is sensitive to whatever process constitutes chloramphenicol-induced polarity. In the same region there also are signals sensitive to ρ-induced termination of RNA chains *in vitro* at low ionic strength, since ρ restricts T 4 transcription by *E. coli* RNA polymerase to E-promoter-proximal segments that correspond, at least approximately, with immediate early RNA (Richardson, 1970*b*; Jayaraman, 1972). Neither type of site is necessarily relevant to normal *in vivo* regulation of T 4 transcription.

Other experiments on RNA synthesis *in vitro* with *E. coli* RNA polymerase holoenzyme also suggest the existence of signals or regions within which RNA synthesis stops at low ionic strength and, finally, intrinsic termination signals at which RNA chains ending with 3′U are released and RNA polymerase is also released (at high ionic strength) for reinitiation of RNA synthesis (Fuchs *et al.*, 1967; Millette *et al.*, 1970; Schäfer and Zillig, 1973*b*). The RNA chains terminated at these last-named signals are, by different estimates, between 7000 and 12,000 nucleotides long (Millette *et al.*, 1970; Schäfer and Zillig, 1973*b*). The longest of these RNA chains must contain transcripts of more than one immediate early gene and thus represent two or more contiguous transcription units of the types shown in Fig. 8. (Brody and Geiduschek, 1970; Milanesi *et al.*, 1970). If all T 4 early mRNA chains grow at the average rate of 1100 nucleotides per minute at 30°C (Section 4.2.6a), then no early gene can have its terminus further than 6000 to 7000 nucleotides from a promoter, since no early protein first appears later than 6 min after infection (O'Farrell and Gold, 1973a). The significance of the low ionic strength blocking signals for *in vivo* transcription is unclear, but the intrinsic T 4 early RNA chain-terminating sites have properties comparable with the termination signal of the early T 7 transcription unit (Section 2.3.3), and we anticipate that they will turn out to be significant for *in vivo* transcription. Even 7000 nucleotide-long T 4 transcripts are not normally seen *in vivo*: Either RNA chain termination *in vivo* employs an accessory chain termination factor and the *in vitro* synthesized RNA chains combine several transcription units, or early T 4 RNA is processed *in vivo*. It is not yet known whether *E. coli* RNase III deficiency affects T 4 messenger size (cf. Sections 1.1.1c and 2.3.2).

The early T 4-specified regulatory event intervenes in this framework of transcription units. The simplest interpretation of the far-from-complete evidence is that a new set of RNA chain-initiation specificities is introduced, utilizing a new set of initiation sites, P_M ("middle" promoters) some of which are located in the middle of transcription units served by P_E promoters ("early" promoters). According to this interpretation, many T 4 early genes can be transcribed in two overlapping reading frames at rates set by two different promoters. In the rII transcription unit a P_M promoter may be situated between rIIA and rIIB (Bautz *et al.*, 1969; Schmidt *et al.*, 1970; Gold *et al.*, 1973).* Separate P_M promoters may also be located near genes 43, 45, and 1 (O'Farrell and Gold, 1973a,b; Gold *et al.*, 1973; Hercules and Sauerbier, 1973, 1974; Cohen *et al.*, 1974). The gene 1 mRNA can be found in monocistronic and oligocistronic forms, and this might be due either to transcription in two reading frames or to a peculiarity of mRNA processing (Sakiyama and Buchanan, 1971, 1973). Only a small number of the early genes are transcribed exclusively from M promoters, and they yield only a small fraction of early RNA. Genes 1, 32, 40, and 41 may be in this group (Jayaraman, 1972; Gold *et al.*, 1973, Cohen *et al.*, 1974).† These must, by definition, be delayed early genes, but immediate early genes can also be read from

---

* Current experiments place this promoter *within* gene rIIA (Singer *et al.*, 1976). Another M-promoter appears to be located near the beginning of gene rIIA (Daegelen *et al.*, 1975).

† More precisely, the evidence regarding the gene 32 protein is that it is not synthesized in the cell-free system from uninfected bacteria programmed by T 4 DNA, but that the system can be programmed by *in vivo* messenger. The DNA-dependent *in vitro* system either synthesizes no gene 32 mRNA or produces it in a translationally inactive form. Translational inactivity might reflect the secondary structure of a very large transcript bearing the gene 32 message near its 3′ end.

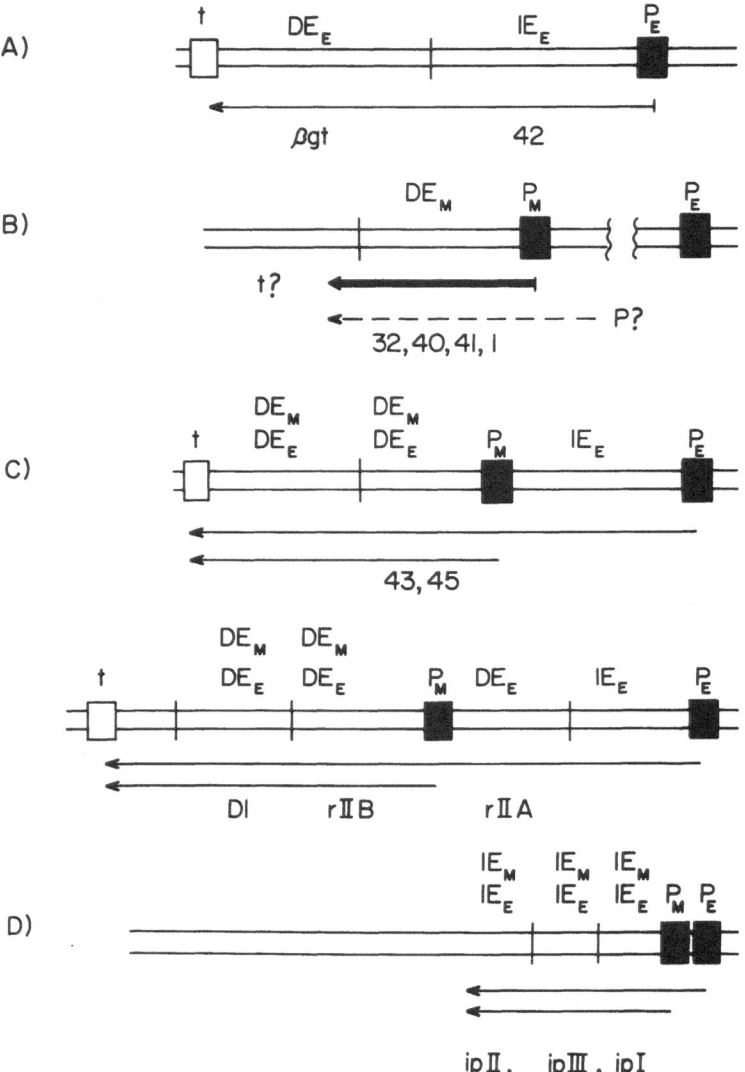

Fig. 8. Topography of T 4 prereplicative genes and transcription units (based on O'Farrell and Gold, 1973*a*). Different arrangements of the relation between genes and their conjugate transcription-regulating sites are shown. For the sake of being concise, in this illustration we assume that the regulation of middle transcription involves altered recognition of initiation rather than termination sites. Alternative possibilities are discussed in the text. P_E designates early promoters, P_M designates middle promoters, and t designates termini of transcription units, some of which are intrinsic termination sites while others may require auxiliary termination factors. Immediate early genes (IE) or delayed early genes (DE) are additionally characterized by a suffix, E or M, according to the promoter site at which their synthesis is initiated (e.g., DE_E designates a de-

additional M promoters. A model of these ideas (based on O'Farrell and Gold, 1973a) is presented in Fig. 8.

The evidence which comes closest to proving the existence of P_M comes from experiments in which the relative sensitivity of the expression of different genes to UV irradiation is measured. It has been suggested that transcription is sensitive to UV damage in DNA because transcription stops at pyrimidine dimers (Sauerbier *et al.*, 1970). In that case, the target determining the UV sensitivity of the activity of a gene is the entire DNA segment between its promoter and its terminus. Bräutigam and Sauerbier (1974) showed that if the entire T 7 genome is turned into a single transcription unit by the appropriate mutations (Studier, 1972; see Section 2.3.5), the UV sensitivity of the synthesis of T 7 proteins does indeed correlate with the order of their conjugate genes on the chromosome. But the UV sensitivity of protein synthesis of T 4 genes 43 and 45 relative to other proteins (α-glucosyl transferase and a protein, designated as U, whose gene has not yet been identified) changes during the course of infection, decreasing significantly several minutes after infection (Hercules and Sauerbier 1973, 1974). This is consistent with promoter proximities for these genes that change with time. However, alternative explanations are not excluded. For example, gene 43 protein, which autoregulates gene 43, participates in repairing that gene from UV damage. Also, general metabolic disturbances caused by the UV irradiation may influence these measurements.

---

layed early gene served by an early promoter, P_E). Arrows indicate transcription; line thickness indicates the relative abundance of transcription of a gene from different promoters. Numbers and symbols designate those genes to which models A–D probably apply. Type A: Tandem arrangement of IE and DE genes that are served by an E promoter. A site at which antibiotics induce polarity probably exists in the region between IE and DE. The adjacent genes 42 and βgt (Fig. 5) probably constitute a transcription unit of this kind. Type B: DE genes that are mainly or exclusively served by M promoters. The termini of these transcription units are not known. Genes 32, 40, 41, and 1 probably are parts of such transcription units. Limited transcription of genes 32 and 1 also occurs from distant E promoters (at least *in vitro*). Type C: DE genes that are served by E *and* M promoters. Proximity of such genes to conjugate promoters changes as a result of the early transcription-regulating event. Genes 43 and 45 probably are parts of such transcription units. A similarly organized but very long transcription unit contains DE genes rIIA and rIIB and the contiguous region D1. Type D: Like type C, but P_E and P_M are contiguous. Genes *ip*I, II, and III may fit into such a scheme. They are transcribed throughout infection. *ip*II and *ip*III are very close to E promoter(s) (Brody *et al.*, 1971). *ip*I is even closer to a different E promoter but it is not transcribed in the presence of chloramphenicol. A site at which chloramphenicol induces polarity must therefore be located between an E promoter and gene *ip*I (Black and Gold, 1971). The relative order of P_E and P_M is not known.

An alternative model of the early regulatory event is that it involves a change in specificity of termination, rather than initiation, of transcription. In this alternative model, early transcription units have exactly the structures shown in Fig. 8 but with a hypothetical closely spaced promoter-terminator pair, P_E t_E, substituting for the hypothetical P_M promoter. Genes with promoter regions P_M P_E and P_E t_E P_E would, for example, have formally equivalent regulatory properties. The transcription unit which is shown in Fig. 8, part C, as $P_E \cdots$ rIIa-P_M-rIIB-D1 $\cdots$ would, instead be represented as $P_E \cdots$ rIIA-P_E-t_E-rIIB-D1 $\cdots$. The T 4 regulatory substance would act as an antiterminator of transcription at t_E, as the phage λ gene N product presumably does (Thomas, 1971; Franklin, 1971, 1974; Georgopoulos, 1971; Dottin and Pearson, 1973; Adhya *et al.*, 1974; Segawa and Imamoto, 1974). The hypothetical termination signal t_E, like the termination signal at λ gene N, cannot be the same as the signal for intrinsic T 4 early chain termination, T_E, since it is not recognized by RNA polymerase *in vitro*.

Two major handicaps have, in the past, frustrated attempts to bring the analysis of the early regulatory event to a more decisive point. The first handicap involves the shortcomings of *in vitro* systems for RNA and protein synthesis which are exacerbated by the peculiar topography of the T 4 early genes. As mentioned elsewhere (Sections 2.4.4, 4.3, and 4.4) the *in vitro* systems—those consisting only of *E. coli* RNA polymerase and mature T 4 DNA, as well as the less purified cell-free systems for coupled transcription and translation—are, to some extent, defective in RNA chain termination. As a result, the *in vitro* systems exceed the transcriptional capacity of phage transcription *in vivo,* producing transcripts and enzymes that are not under the corresponding regulatory control *in vivo*. The extent of transcriptional read-through *in vitro* seems to vary considerably. For example, of the DE_M genes (Fig. 8, part B), gene 1 seems to be transcribed (and translated) quite well (Natale and Buchanan, 1972), gene 32 only to a small extent (O'Farrell and Gold, 1973*b*), and genes 40 and 41 not at all (Jayaraman, 1972). Even gene *e* (lysozyme), which appears as a late protein *in vivo,* can be transcribed and translated *in vitro* (Gold and Schweiger, 1970; Rabussay, 1970). This anomalous activity of the *E. coli* cell-free system is due to the fact that the lysozyme gene is transcribed with the same polarity as the early genes and is evidently transcribed *in vitro* as a read-through product from a strong early (P_E) promoter that is located several thousand nucleotides from the gene (Brody *et al.*, 1971).

A recently devised *in vitro* system is likely to be important for unraveling the early regulatory event. This system reproduces the chloramphenicol effect (of restricting transcription to promoter proximal regions), allows *in vitro* complementation experiments to be performed, and reproduces some aspects of the phenotype of the *mot* mutant (Thermes *et al.*, 1976). The first result of experiments with this system is that the early regulatory event involves a modification of the DNA template. It has been postulated that the modification is non-covalent and that it involves interaction with membranes of the host cell (Daegelen *et al.*, 1975; Thermes *et al.*, 1976).

The second major handicap in analyzing the early T 4 regulatory event has been the lack of mutations that affect it. In a general way, one can draw a functional analogy between this event and the control of phage SPO 1 middle transcription (Section 3). The association of at least one viral gene (gene 28) with strict control of the early SPO 1 regulatory event has greatly simplified the analysis of the otherwise much less studied *B. subtilis* phage. Mutations apparently affecting the early regulatory event have only recently been found. One of these mutations in a gene designated as *mot* has already been referred to (see p. 83) and has the following properties: Cells infected with this temperature-sensitive mutant (*ts*G1) appear to be delayed, though not completely defective, in the execution of the early regulatory event (Matson *et al.*, 1974). Other mutations, designated as *far* I, are also delayed in the execution of the early regulatory event but differ in one respect from the *mot* mutation: They lead to an overproduction of the rIIB protein, whereas the *mot* mutant is delayed in the synthesis of this protein. *mot* and *far* I mutations may be in the same cistron. Further biochemical and physiological analysis of these mutations and of certain mutations in rIIB with analogous phenotype (Nelson and Gold, 1975) might go far to clarify T 4 early transcription.* The simplest assumption is that the gene containing the *ts*G1 mutation might code for the effector that controls the early T 4 regulatory event. Recent *in vivo* experiments prove that the effector of the early regulatory event is a diffusible product (Daegelen *et al.*, 1975). The product of the *mot-far* I gene(s) is also diffusible in the sense that mutants are recessive to wild type. However, the observation that rIIB mutations can produce the same phenotype as *mot* mutations suggests alternative and less conventional explanations: that some or all of these mutations act pleiotropically on the early regulatory event through a primary defect of the cytoplasmic membrane. In this connection we want to reemphasize the observation that the early regulatory event occurs even if proteins are synthesized after infection with analogs for six amino acids (gln, lys, met, phe, pro, and try; Black and Gold, 1971; Gold, 1975) or in the

---

* See also Notes Added in Proof, p. 196, Note 3.

presence of the miscoding-inducing antibiotic, neamine (Brody, 1975). The early proteins that are synthesized in these circumstances even include gp 32 which can normally be read only from P_M promoters (Fig. 8), yet the viral proteins that are synthesized under these conditions, including any effectors of the early regulatory event, should not be biologically active. There are two types of possible interpretations of these experiments: the first class of explanations includes all excuses for ignoring the experiments (e.g., because the amino acid analogs do not totally abolish biological activity, etc.). Alternatively, these experiments may signify that the effector of the early regulatory event is *preformed* and is recruited to viral regulation. *Preformed* means that the phage or host product in question was synthesized before the infectious cycle in which it regulates viral gene expression. There still remains the apparent paradox that the recruitment process can occur even if nonfunctional proteins are synthesized after infection, although the recruitment process can be prevented by mutant (*mot* or *far* I) viral proteins. One can conceive of explanations which invoke the release of the preformed regulatory components from the plasma membrane. This process might not require newly synthesized phage-specified proteins but might be poisoned by mutant phage-coded membrane proteins. Such explanations do, of course, run counter to the experiments of Travers (1970a) which implied the existence of a conventional phage-coded transcription factor.

No T 4 promoters have yet been sequenced. The P_E promoters are now, in principle, accessible to analysis as regions protected by host RNA polymerase from nucleotide digestion (Giacomoni *et al.,* 1974; Niyogi and Underwood, 1975), locatable on DNA fragments that can be generated by restriction enzymes (Kaplan and Nierlich, 1975). Information about the initiation of RNA chains is also meager: They start predominantly with 5′ pppAp, but also with 5′ pppGp, *in vitro* and *in vivo* (Maitra and Hurwitz, 1965; Natale and Buchanan, 1974; Rabussay and Bieger, unpublished). RNA chain-initiation experiments with di- and trinucleotides have also been performed (Hoffman and Niyogi, 1973). However, with a template that contains several transcriptional units, their interpretation in terms of specific starting signals is complex.

To summarize the extensive discussion of this section, a transcription-regulatory event occurs in the first minutes after infection. It is subject to influences by mutations of the phage genome. This transcriptional regulation probably involves a change of the specificity of RNA chain initiation and possibly of RNA chain termination as well. Its manifestations are peculiarly difficult to analyze because of the particular topography of the T 4 early transcription units. Some recent genetic and biochemical experiments put the early regulatory event in a new light and hint at mechanisms that are not like those which we have discussed, for SPO 1, in Section 3.

4.3. Host RNA Polymerase Modifications

The search for the regulatory principles underlying the control of viral gene expression has led to very detailed analysis of RNA polymerase after T 4 infection. Such biochemical analysis must ultimately be the key to our understanding of the control of T 4 gene expression. That it has made almost no contributions so far is certainly due to the absence of suitable cell-free systems for analyzing all T 4 gene expression with purified components. The significance of this failure, or of any failure, is difficult to assess with assurance, except in retrospect. Nevertheless, we shall try to examine the roots of this problem.

Host RNA polymerase is involved in all viral transcription. That the β subunit is so involved is implied by the maintenance of rifamycin-response phenotype after T 4 infection (Haselkorn et al., 1969; Mizuno and Nitta, 1969; DiMauro et al., 1969; Zillig et al., 1970). This frequently quoted result naturally does not exclude the possible activity of virus-specified protein capable of polymerizing ribonucleotides on a DNA template in T 4-infected bacteria (Chamberlin, 1970). The conservation of all host RNA polymerase subunits after T 4 infection has been shown by isotope labeling experiments (Goff and Weber, 1970; P. Palm, quoted in Schachner and Zillig, 1971), but RNA polymerase undergoes a sequence of modifications (Table 5). In addition to these modifications, four newly synthesized peptides appear tightly bound to RNA polymerase core after infection (Stevens 1970, 1972, 1974). Two of these new peptides are the gene 33 product and the gene 55 product (Stevens, 1970, 1972; Horvitz, 1973; Ratner, 1974b). The products of both these genes are required for late gene expression (Epstein et al., 1963; Bolle et al., 1968a; Pulitzer, 1970; Pulitzer and Geiduschek, 1970). A third peptide is altered in a T 4 mutant that grows normally on laboratory strains of E. coli but for which nonpermissive hosts exist among a collection of "hospital strains" of E. coli (Horvitz et al., 1975). All the RNA polymerase modifications can be grouped into three events and ordered in time. These events create three different RNA polymerases which can be crudely classified as: (1) "chloramphenicol" or "altered," (2) "early" or "modified," and (3) "late." There is a rough correlation, in time and circumstance, between these

TABLE 5

Modifications of RNA Polymerase in T 4-Infected *E. coli* at Different Times after Infection

Host subunits	Changes after T 4 infection				References
	Within 30 sec and also in the presence of chloramphenicol	Early (2–5 min)	Middle (10 min)	Late (after 10 min)	
A. Subunits of bacterial RNA polymerase					
α	$\alpha_A = \alpha_{P,Ad}{}^a$	$\alpha_M = \alpha\text{-ADPR}^b$			Walter *et al.*, 1968; Seifert *et al.*, 1969; Goff and Weber, 1970; Seifert *et al.*, 1971; Schachner and Zillig, 1971; Zillig *et al.*, 1970; Rabussay *et al.*, 1972; Goff, 1974; Horvitz, 1974a,b; Rohrer *et al.*, 1975
β		$\beta_M{}^c$			Schachner and Zillig, 1971; Zillig *et al.*, 1970.
β'		$\beta'_M{}^c$	$\beta'_L{}^d$		Khesin, 1970, Travers, 1970b; Zillig *et al.*, 1970; Schachner and Zillig, 1971
σ	$\sigma_A = \sigma_P?^e$				Rabussay *et al.*, 1972; Horvitz, 1974a

Subunits		Presence after T 4 infection				References
Protein	Molecular weight	Within 30 sec and also in the presence of chloramphenicol	Early (2 5 min)	Middle (10 min)	Late (after 10 min)	
B. T 4-coded subunits						
gp 55	22,000	−	−	+	+	Stevens, 1970, 1972, 1974; Horvitz, 1973; Ratner, 1974b
gp alc (?)γ	14–15,000	−	−	+	+	Stevens, 1974; Horvitz et al., 1975; Snyder, 1976.
gp 33	12,000	−	−	+	+	Horvitz, 1973; Stevens, 1974
Unnamed g	10,000	−	−	+	+	Stevens, 1972, 1974

[a] The first change ("alteration," denoted by subscript A) involves only one-half or less of the cell's α-subunits. Alteration involves the addition of one adenine and two phosphates to α. The sites of "modification" (denoted by subscript M) and alteration of α may be identical. Alteration, like modification, might involve addition of ADP-ribose (ADPR), but this has not been proven unequivocally (Seifert et al., 1971; Horvitz, 1974a). Alteration is performed by an enzymatic activity which is present in the mature T 4 phage and is injected into the host cell upon T 4 infection (Rohrer et al., 1975). Alteration of RNA polymerase was also achieved in vitro. (See also Notes Added in Proof, p. 196, Note 4.)

[b] α_M is assumed to contain a covalently bound ADPR residue, most probably linked to an internal arginyl residue (Goff, 1974). However, it has not been unequivocally shown that the claimed modification in vitro (Goff, 1974) is identical with the modification in vivo rather than with alteration. (See also Notes Added in Proof, p. 196, Note 4.)

[c] Changes of antigenicity and tryptic peptide maps, not reflected in electrophoretic properties of the complete subunit (Zillig et al., 1970). The kinetics of these modifications were not measured. It is possible that the observed changes were due to contaminations by the subsequently discovered phage-specified low-molecular-weight subunits.

[d] It is not clear whether β_M and β'_L are different; β'_L was reported to differ in electrophoretic mobility from β' and β'_M (Travers, 1970), but there is disagreement on this point (D. Rabussay, unpublished).

[e] A fraction of σ (up to 25%) apparently becomes labeled with ^{32}P with the same kinetics as α_A [even in the presence of chloramphenicol (CAM)].

[f] This assignment is tentative (Snyder, 1976). (See also Notes Added in Proof, p. 196, Note 5.)

[g] See also Notes Added in Proof, p. 196, Note 6.

changes in RNA polymerase and the control and sequence of viral transcription (Schachner *et al.*, 1971; Rabussay *et al.*, 1972). However, we shall see that the first two of these events may not be essential for the control sequence of *viral* gene expression while the third change, to "late" RNA polymerase, is involved in the control of T 4 late RNA synthesis.

The *in vitro* transcription of mature T 4 DNA by these enzymes is not remarkably different. All three enzymes require σ to transcribe intact T 4 DNA and then read only the DNA *l*-strand to produce early RNA (Bautz and Dunn, 1969; Hager *et al.*, 1970; O'Farrell and Gold, 1973*b*). The enzymes differ with respect to certain detailed parameters of *in vitro* transcription whose significance for *in vivo* regulation is unclear: (1) *E. coli* σ has a lower affinity for all T 4-changed core enzymes (Seifert *et al.*, 1969; Bautz and Dunn, 1969; Travers, 1970*b*; Schachner *et al.*, 1971), yet this affinity is nevertheless great enough to allow full stimulation of the T 4-changed enzymes to the specific activity of host RNA polymerase by host σ with T 4 DNA as template (Bautz and Dunn, 1969; Schachner *et al.*, 1971); (2) termination factor ρ interacts with T 4 "altered" and "modified" enzyme at higher ionic strength ($\mu = 0.165$) (Schaefer and Zillig, 1973*b*; Rabussay, unpublished), whereas host RNA polymerase transcribing T 4 DNA is sensitive to ρ only at low ionic strength (Richardson, 1970*b*; Goldberg, 1970; Millette *et al.*, 1970); (3) "altered" and "modified" RNA polymerase differ from host RNA polymerase in the ionic strength dependence of their activity (Stevens, 1974; Kleppe, 1975). More important enzymatic parameters such as strength of binding to different initiation sites of the template and number of the template binding sites (Chamberlin, 1974*a*) have yet to be determined for the T 4-changed enzymes. The genetic evidence argues strongly that "alteration" and "modification" of RNA polymerase are not both required for gene expression that is essential for T 4 development. T 4 alt⁻ and T 4 mod⁻ grow well on several strains of *E. coli* (Horvitz, 1974*a,b*). However, it remains to be seen whether a double mutant T 4 alt⁻ mod⁻ can grow normally.

"Alteration" of RNA polymerase occurs rapidly and even in the presence of chloramphenicol (Seifert *et al.*, 1969; Rabussay *et al.*, 1972). It has recently been demonstrated that the enzyme which executes "alteration" of RNA polymerase is carried by the mature phage and gets injected into the host during T 4 infection (Rohrer *et al.*, 1975). The chemical nature of alteration involves covalent binding of a residue containing two phosphates and one adenyl group (probably ADP ribose) to one of the two α subunits present in an RNA polymerase protomer. A fraction of the σ, β, and β' subunits also becomes labeled with ^{32}P (Seifert *et al.*, 1971; Rabussay *et al.*, 1972; Horvitz, 1974a; Rohrer *et al.*, 1975). "Alteration" might be in part responsible for the shutoff of host transcription in normal infection (Nomura *et al.*, 1966; Terzi, 1967; Adesnik and Levinthal, 1970; Schachner *et al.*, 1971). A direct test of this possibility (see below) has not yet been reported. However, Snyder (1973) described an *E. coli* RNA polymerase mutation which leads to an ineffective shutoff of host transcription after T 4 infection, and which could involve a defective alteration process. "Alteration" also does not, by itself, account for chloramphenicol-restricted early transcription, since this enzyme, with σ, transcribes T 4 immediate and delayed early genes *in vitro*. "Alteration" is reversible and appears only transiently *in vivo*.

The second change of RNA polymerase ("modification") involves most probably ADP-ribosylation of all α subunits (Goff, 1974). This change also occurs very rapidly—it is one-half complete about 2 min after infection—but does require T 4-specific protein synthesis (Seifert *et al.*, 1969). Each α subunit covalently binds one ADP-ribosyl residue to a site which seems to differ from the site(s) that is(are) involved in "alteration" (Seifert *et al.*, 1971; Rabussay *et al.*, 1972; Goff, 1974). Most likely, the terminal ribose 1-carbon of ADPR forms a bond with the guanido nitrogen of an arginine residue. ADP-ribosylation of the mammalian EF-2 catalyzed by diphtheria toxin or *Pseudomonas aeruginosa* toxin is another example of enzymatic regulation via ADP-ribosylation (Honjo *et al.*, 1972; Iglewski and Kabat, 1975). NAD is also a necessary cofactor for the activation of adenylate cyclase by cholera toxin (Gill, 1975). The only potentially incisive attempt to demonstrate a changed T 4 transcription specificity of "modified" RNA polymerase has led to a negative answer: The test involves the *in vitro* synthesis of T 4 DNA unwinding protein (gene 32 protein) in cell-free systems from uninfected or T 4-infected *E. coli* partially dependent on exogenous RNA polymerase. Neither unmodified nor T 4-"modified" RNA polymerase yields this delayed early protein, whose message is thought to be read from an M promoter (Fig. 8; Gold, 1974, personal communication). The possible role of antitermination in early gene transcription control has already been mentioned and a changed termination activity of "modified" RNA polymerase is therefore of potential significance, but the observed change is in the wrong direction: toward more effective ρ-mediated termination of RNA chains *in vitro* rather than to antiterminating activity (see above).

Although attempts to relate "modification" to T 4 transcription specificity have been unsuccessful, a clear-cut depression of *E. coli* transcription in a cell-free system for DNA-directed RNA and protein synthesis has now been observed (Mailhammer *et al.*, 1975). "Modification" therefore might be involved in host shutoff although it occurs too slowly to be the primary shutoff mechanism. "Modification" is not essential for T 4 development, but it could be required for the expression of "nonessential" early genes. An examination of the T 4 transcription properties of modified RNA polymerase therefore remains of considerable interest.

4.4. Initiation Factors and Stimulating Factors

Selectivity of viral transcription over host transcription in the T-even phage-infected bacterium is almost complete. Yet the RNA polymerase isolated from T 4-infected bacteria is relatively inactive on T 4 DNA templates mainly because of the tendency, which we have already mentioned, of the "altered" and "modified" polymerase cores to bind σ less tightly and to lose much or all of their σ during purification, the extent of this loss depending on the isolation procedure and varying for "altered," "modified," or "late" core. σ is not degraded or inactivated, but it is partly altered very soon after T 4 infection (Table 5; Stevens, 1972, 1974; Rabussay *et al.*, 1972; Horvitz, 1974*a*). No significant differences in the stimulating properties of σ factors from T 4-infected and uninfected cells for *E. coli* RNA polymerase core have yet been shown to exist (Rabussay *et al.*, 1972; Stevens, 1974). The decreased affinity of cores for σ probably is not solely responsible for decreased host transcription but might be involved in changes of transcription specificity through an exchange of bacterial for phage-specified initiation factors (Travers, 1971). It is because of these relationships that the search for components that stimulate transcription of T 4 DNA by RNA polymerase has been pursued in several laboratories. Most of the work has been directed at the problem of "late" transcription (Section 4.6). There has, however, been one report claiming selective stimulation at what one would now call M promoters by a T 4 specific "factor": An activity was found in T 4-infected cells, but not in uninfected cells, which stimulated the transcription of delayed early RNA with core RNA polymerase from uninfected or infected *E. coli* (Travers, 1969; 1970*a,b*). Perhaps because of its instability, this factor has never been further purified. Accordingly, its binding to RNA polymerase core and specific role in initiation, i.e., in copolymerizing the first nucleotides, was not verified. Although the factor appeared to have different fractionation properties from host σ factor, the possibility that it might be a complex of host σ with some other protein was not eliminated (Losick, 1972). Notions on the control of early gene expression have been sufficiently refined that more precise tests for selective initiation of T 4 genes along the lines indicated in Fig. 8 and mentioned in the preceding section are now accessible. But there has been no subsequent or confirmatory report on this important effect.

Another way in which transcriptional specificity might be altered is by interaction of host σ factor with some antagonist. The experimental evidence here is also sparse, but some interesting observations on RNA polymerase-inhibiting activities in T 4-infected cells have been made. Phosphocellulose column chromatography of "altered" RNA polymerase yields a mixture of several nonadsorbed proteins (sometimes completely separated from σ) which inhibit host holoenzyme and core as well as T 4 "altered" core polymerase on T 4 and calf thymus DNA templates (Rabussay, 1972, unpublished). Khesin *et al.* (1972) reported a similar inhibitory activity in crude fractions of T 2-infected bacteria which was further investigated by Goldfarb (quoted by Bogdanova *et al.,* 1974); Stevens (1974) separated an inhibitory activity from "late" T 4 RNA polymerase which is probably the T 4-coded 10,000 molecular weight protein (see Table 5).* It has been proposed that the observed inhibition is due to an "anti-σ" activity. The identity of anti-σ and its mechanism of action remain to be determined.

4.5. Possible Translational Regulatory Elements

T 4 modifies components of the host translational machinery and also directs the synthesis of some new constituents. In the usual laboratory host strains, none of these modifications or T 4-specific products are essential for development: Nevertheless, at least some of these virus-induced processes are known to affect virus yield quantitatively. T 4 modifies host tRNA, a tRNA methylase, an aminoacyl tRNA synthetase, and the so-called *i* factors; T 4 also codes for new phage-specific tRNA. In the following paragraphs we discuss these effects and comment on possible translational shutoff mechanisms.

4.5.1. tRNA

The synthesis of phage-specific tRNAs is induced by several phages, including all the T-even phages, soon after infection. The existence of T 4-specific tRNA was first clearly proven several years

* See also Notes Added in Proof, p. 196, Note 6.

ago (Weiss *et al.*, 1968; Daniel *et al.*, 1968; Tillack and Smith, 1968).
Although the existence of T 4-induced RNA species in the size range of
tRNA had been known for a longer time (Nomura *et al.*, 1960; Baguley
et al., 1967), at least some of these subsequently proved to be mRNA
degradation products. T 4 tRNA, like bacterial tRNA, is stable relative
to T 4 mRNA (Adesnik and Levinthal, 1970; Lamfrom *et al.*, 1973).
Even some dimeric tRNA precursors show considerable stability
(Guthrie *et al.*, 1973; Lamfrom *et al.*, 1973), unlike the *E. coli* tyrosine-
tRNA precursor (Altman and Smith, 1971). Eight different T 4 tRNAs
are now known (McClain *et al.*, 1972; Paddock and Abelson, 1973);
their nucleotide sequences have been determined (Guthrie *et al.*, 1973;
Barrell *et al.*, 1973; Stahl *et al.*, 1973; Abelson *et al.*, 1975; Guthrie *et
al.*, 1975). The genes that code for these tRNAs and for two other stable RNA species (Paddock and Abelson, 1973) are clustered in a 1500-
nucleotide segment between genes *e* and 57. The genetic map of this
region is shown in expansion I of Fig. 5 (Abelson *et al.*, 1975, and personal communication).

T 4 tRNA^{ser} probably recognizes codons UCA and UCG (Abelson
et al., 1975). Ochre (pSu$_c^+$) and amber (pSu$_a^+$) suppressor mutants of
this tRNA gene have been isolated (McClain, 1970; Wilson and Kells,
1972; Comer *et al.*, 1974). Further genetic analysis (through the isolation of pSu$_c^-$ derivatives of pSu$_a^+$) has led to the identification of a
distant T 4 gene designated as *mb* and located near *r*II (Wilson, 1973).
The product of this gene is probably required for the processing of four
of the RNA molecules coded in this region, ser, pro, and ile tRNA and
"band 2" (Abelson *et al.*, 1975; Guthrie *et al.*, 1975). No other phage-
coded proteins are known to be required for the modification and
maturation of T 4 tRNA. Host proteins probably perform these functions; uridylic acid 5′ methylase (Björk and Neidhardt, 1973), RNase P
(Sakano *et al.*, 1974; Abelson *et al.*, 1975), and tRNA nucleotidyl
transferase (Guthrie *et al.*, 1975) have already been identified in these
roles. What is the biological function of phage tRNA? It can be
charged with arginine, glycine, isoleucine, leucine, proline, serine, glu-
tamine, and threonine (Daniel *et al.*, 1970; Scherberg and Weiss, 1972;
Pinkerton *et al.*, 1972; McClain *et al.*, 1972; Guthrie *et al.*, 1973). At
least some of this tRNA can function in protein synthesis *in vivo*
(McClain, 1970; Wilson and Kells, 1972) and *in vitro* (Scherberg and
Weiss, 1972). Although deletion of the entire tRNA region still yields
viable phage (McClain *et al.*, 1972; Wilson *et al.*, 1972; Wilson, 1973),
the entire tRNA region is highly conserved in the three T-even phages
(McClain, quoted in Guthrie and McClain, 1973; Kim and Davidson,
1974). T 2, T 4, and T 6 code for identical tRNA^{leu} and tRNA^{gly}. Even

"band 1" RNA, a 140-nucleotide-long, stable RNA molecule of unknown function, diverges at only six nucleotides among the T-even phage (Abelson *et al.*, 1975). Selective pressures must be at the root of this conservation. There is evidence that T 4 draws at least two advantages from its own tRNA production: (a) Phage tRNA provides an augmentation of code word recognition that is represented only by minor species of *E. coli* tRNA. Deletions of the entire tRNA region lower phage yield on common laboratory strains of *E. coli* (Wilson, 1973). (b) One of the T 4 tRNAs is essential for T 4 growth on an *E. coli* wild-type strain that differs from the common laboratory strains (Guthrie and McClain, 1973). The production of its own tRNAs therefore also widens the host range of T 4.

The tRNA region can be transcribed from an immediate early promoter: (1) Transcripts of the tRNA genes accumulate in the presence of chloramphenicol (Scherberg *et al.*, 1970). However, not all T 4 stable RNAs accumulate in mature form in the presence of chloramphenicol (Wu, personal communication). (2) The immediate early promoter from which the lysozyme (*e*) gene is transcribed during the first minutes of infection has been calculated to lie about 5–6000 nucleotides "upstream" in the direction of gene 57 (Black and Gold, 1971). This estimate would place the tRNAs in the same transcription unit with gene *e*. (3) T 4 tRNA can be transcribed and processed *in vitro* (Nierlich *et al.*, 1973; Lamfrom *et al.*, 1973). However, the existence of additional middle promoters (P_M) in this region is not excluded, and Abelson and co-workers (1975) have presented reasons for believing that such promoters might exist.

Bacterial tRNAs (Sueoka and Kano-Sueoka, 1964; Waters and Novelli, 1967, 1968; Kan *et al.*, 1968) and enzymes of tRNA modification (Wainfan *et al.*, 1965; Boezi *et al.*, 1967; Hsu *et al.*, 1967; Smith and Russell, 1970) are altered after T 4 infection. The specific cleavage of a host leucyl-tRNA within 1 min after infection has been studied in detail (Yudelevich, 1971). The biological role of these changes is not understood.

4.5.2. Alteration of tRNA Synthetase

A T 4-induced change in the valyl-tRNA synthetase of the host has been demonstrated by Neidhardt and Earhart (1966). The altered synthetase differs from the host product in its physical and enzymological properties. Certain temperature-sensitive *E. coli* valyl-tRNA synthetases become thermostable as a result of the T 4-specific alteration. The change of valyl-tRNA synthetase is caused by the addition of a T 4-specific protein, τ (M.W. $\simeq$ 10,000), to the host enzyme. The τ-modified tRNA synthetase binds at least one tRNA molecule rather tightly (Neidhardt *et al.*, 1969; Marchin *et al.*, 1972, 1974). The alteration begins 1–2 min after infection (30°) and continues until all of the valyl-tRNA synthetase has been converted into the modified form about 20 min after infection. The synthesis of τ-specific mRNA occurs even when *E. coli* is infected in the presence of chloramphenicol (Neidhardt *et al.*, 1969; Müller and Marchin, 1975). T 4 *vs.* mutants (cf. Table 4) grow normally in several host strains (Neidhardt *et al.*, 1969; McClain *et al.*, 1971). T 6 infection causes a similar or the same change, but other phage as λ, T 5, or S 13 do not (Neidhardt *et al.*, 1969). No other tRNA synthetase alterations have been found after T-even phage infection.

4.5.3. Alterations of Translation Factors and Ribosomes

Alterations of translation specificity in T 4-infected *E. coli* have been sought and detected under certain experimental conditions. Nevertheless, the significance of these findings for translational regulation *in vivo* is uncertain.

Translational discrimination is exhibited *in vitro* by different inhibitory *i*-factors from uninfected *E. coli* and, according to a widely held view, by two classes of the initiation factor IF3. Two *i*-factors, i_α and i_β, have been purified and relatively well studied. Both bind strongly to synthetic polynucleotides. Factor i_α is the 30 S ribosomal protein S1 and incorporates into Qβ RNA replicase as subunit I (Groner *et al.*, 1972b; Kamen *et al.*, 1972; Wahba *et al.*, 1974; for a review of phage Qβ replication, see Eoyang and August's article in Volume 2 of *Comprehensive Virology,* 1974). In the Introduction we briefly discussed the controversy surrounding this subject and considered the strong possibility that the *i*-factors may not play a translational regulatory role *in vivo*. On the other hand, S1 probably does interact with the 5′ end of messages, and this interaction may help to determine, even if it does not regulate, the diverse translational yields of different messages.

This is the background against which experiments on translational regulation must be considered. Translational control could be effected by selective alterations of IF3 or *i*-factors after infection. The unique IF3 of Wahba and collaborators has been purified unchanged and fully active from T 4 infected *E. coli,* although previous experiments had been interpreted to suggest that one of the IF3, $IF3_\alpha$, is selectively inactivated after infection (Lee-Huang and Ochoa, 1971; 1973; Schiff *et al.*, 1974). In contrast, electrophoretic mobility changes of two *i*-factors have been observed after T 4 infection (Revel *et al.*, 1973*b*). An altered S1 has now been purified from T 4-infected bacteria: $S1^{T4}$ retains the translation stimulating activity of S1 with S1-depleted ribosomes but lacks the translation-inhibiting (*i*-factor) property of excess S1 (Wahba, personal communication). Since S1 is thought to be an *i*-factor of T 4 early mRNA and RNA phage RNA (Section 1.1.3b), the loss of this property after T 4 infection is not directly related to early-late regulation. However, any change of S1 might affect the relative translational yield of T 4 messages. Translational control could also be effected by modifications of ribosomes after infection or by binding of phage-specific proteins to ribosomes (Smith and Haselkorn, 1969; Rahmsdorf *et al.*, 1972) or to messages. However, as we see next, the changes of translational discrimination that have been observed *in vitro* appear to be due to the factor fraction rather than the washed ribosomes.

Alterations of translational specificity after T 4 infection were first reported by Hsu and Weiss (1969) who could detect differentially decreased translation of RNA phage RNA relative to T 4 late mRNA. The differential effects could be traced to the ribosome wash fraction, which contains the initiation and *i*-factors, and were shown to affect initiation complex formation as well as mRNA-ribosome binding (Dube and Rudland, 1970; Steitz *et al.*, 1970; Klem *et al.*, 1970). These effects were variously supposed to arise from selective destruction of factor activity after T 4 infection or from the appearance of a selective inhibitor (Klem *et al.*, 1970; Schedl *et al.*, 1970). The production of factor(s) selectively binding to T 4 late initiation sites in single-stranded DNA was reported to occur after T 4 infection (Ihler and Nakada, 1971). These findings implied the possible existence of negative controls of T 4 early and host translation and positive controls of T 4 late translation.

However, in trying to understand the significance of these experiments, one encounters substantial problems with other observations that must be taken into account in thinking about translational control: (1) A number of late T 4 proteins have been synthesized *in vitro* in extracts from uninfected *E. coli* (Salser *et al.*, 1967; Wilhelm and Haselkorn, 1971*a,b*). IF3 from uninfected *E. coli* binds early and late T 4 RNA to ribosomes equally well (Yoshida, 1972). These observations argue against an absolute positive translational control over T 4 late protein synthesis. (2) T 4 early proteins can be synthesized in cell-free systems prepared late after T 4 infection. This argues against

absolute negative translational control over T 4 early protein synthesis.
(3) Relatively unpurified systems from uninfected and late T 4-infected
E. coli differ quantitatively, but not selectively, in their abilities to
synthesize a large number of *fmet* peptides (i.e., N-terminal segments
of proteins) with RNA phage, *E. coli,* T 7, T 4 early, and T 4 late RNA
(Goldman and Lodish, 1972). This result argues against positive or
negative translational control of the early-late switch. (4) The changes
of mRNA selectivity referred to in the preceding paragraph occur too
slowly to account entirely for the shutoff of host translation (Pollack *et
al.,* 1970).

We consider the evidence against general positive translational
control of late gene expression to be very convincing. The case regarding negative translational control of early gene expression is more
complicated. We have presented evidence that posttranscriptional regulation of early genes occurs (Section 4.2.2), and anticipate that it
should be possible to analyze this regulation biochemically. However,
we suspect that the prior *in vitro* work on negative translational control
of T 4 early protein synthesis after the T 4 early-to-late switch is not
relevant to this subsequently discovered posttranscriptional regulation.
The prior *in vitro* experiments do provide evidence for changes in
components of protein synthesis, including nucleic acid-binding proteins. However, only one of these changes (S1) has so far been precisely
identified, and the regulatory significance of such changes remains to
be discovered.

4.6. Postreplicative (Late) Regulation

The phase of phage development which starts shortly after the
onset of replication and is accompanied by assembly of progeny virus is
customarily called the "late" period. Two main classes of RNA and
proteins are synthesized during this period of phage development, the
"true late" and "quasi-late." The regulation of true late and quasi-late
transcription follows different rules. This dichotomy of postreplicative
transcription is one of the most interesting features of T 4 regulation
and probably involves a variety of mechanisms. The challenge of understanding these mechanisms at the biochemical level has yet to be met.

True late and quasi-late RNA synthesis differ in their template and
RNA polymerase requirements: True late transcription depends on
DNA replication and on several low-molecular-weight T 4-specific proteins which bind to RNA polymerase, but quasi-late transcription does
not have these strict requirements. The synthesis of true late proteins

also depends on the presence of hmC instead of C in the DNA template. The effect of C → hmC substitution on quasi-late gene expression is not known. There is no convincing evidence for posttranscriptional regulation of true late gene expression (as already discussed in Section 4.5).

4.6.1. True Late RNA and Proteins

Many viral proteins start to be synthesized within a narrow time span, several minutes after the onset of replication (Hosoda and Levinthal, 1968; Notani, 1973; O'Farrell and Gold, 1973a). Most of these proteins are structural components of the virion assembly. More than 30 true late proteins have been identified (Hosoda and Levinthal, 1968; O'Farrell et al., 1973; Hosoda and Cone, 1970; Laemmli, 1970; Dickson et al., 1970; Wood and Bishop, 1973; Vanderslice and Yegian, 1974; Fig. 6). The conjugate mRNAs first are detected at approximately the same time. They are transcribed predominantly, but not exclusively, from the DNA r-strand (Guha and Szybalski, 1968; Guha et al., 1971; Notani, 1973); l-strand true late RNA is under the same control as r-strand true late RNA (Guha et al., 1971; Notani, 1973).

4.6.2. Quasi-Late RNA and Proteins

A subclass of late RNA and proteins already appears during the prereplicative period. The RNA of this subclass is more abundant at the end of the viral eclipse period than before replication. The rate of synthesis of this subclass of RNA, which was originally termed "quasi-late" by Salser et al. (1970), increases relative to other prereplicative RNA between 6 and 10 min after infection (Notani, 1973; cf. Section 4.2.6). Confusion about "quasi-late" genes arose when others used the same term in different contexts. We shall use the term only in the original sense defined in this paragraph.

Bruner and Cape (1970) divided late genes into "true late," characterized by their absolute dependence on gp 55 function and newly replicated progeny DNA as template, and "quasi-late," which could be partly expressed without fulfilling these requirements. These criteria are consistent with quasi-late behavior in the sense of Salser et al. (1970). Complications arise only because of the diversity of quantitative criteria of gene expression that were employed in the work of Bruner,

Salser, and their collaborators. Thus, Bruner and Cape classified genes 12, 9, 34 and 35 as quasi-late because the corresponding proteins could be made at a low level from nonreplicating DNA. They investigated gene 12 in detail and concluded that gp 12 could be synthesized at greater rates in the absence than in the presence of functioning gp 55. Bruner and co-workers use potentially very sensitive but uncalibrated assays for gp 12 (Bruner and Cape, 1970; Berger and Bruner, 1973; Mason and Haselkorn, 1972). They did not show that gp 12 is synthesized in the prereplicative period and with increased rate later in infection. We suspect that gene 12 is not "quasi-late" in the original sense and that the regulatory features discovered by Bruner and Cape characterize certain true late genes. Analysis of T 4 proteins by two-dimensional gel electrophoresis would make a complete analysis of the regulation of gp 12 possible and would resolve this conflict. Gold and co-workers used the term "quasi-late" in yet another connection, applying it to those early genes whose expression is selectively blocked by rifampicin within the first 2 min of infection (O'Farrell and Gold, 1973a,b; Gold et al., 1973). In effect, Gold and collaborators substituted a specific and simple model of quasi-late gene regulation for the original empirical definition. We lack proof that this model accounts for all the properties of quasi-late regulation. We have therefore avoided the term "quasi-late" in discussing what we instead call T 4 "middle" promoters (Section 4.2.6). "Middle" promoters are named by analogy with SPO 1 transcription because they control transcription that starts between the onset of early and late transcription.

Genes that have "quasi-late" regulation in the above original sense should be continuously or almost continuously expressed throughout the eclipse period of phage infection according to the way in which their initiation is regulated. Since prereplicative transcription is either exclusively (Notani, 1973) or almost exclusively (Guha and Szybalski, 1968; Guha et al., 1971) from the l-strand of T 4 DNA, these genes should have the same polarity. Very few genes have been specifically identified as subject to quasi-late regulation. The most clearly identified is gene 32; gene 32 protein synthesis does start before replication but occurs at increasing rate after the onset of replication and is normally shut off toward the end of the eclipse period (Krisch et al., 1974). Since gene 32 mRNA is relatively stable (Russel et al., 1976), it may well constitute a large proportion of quasi-late RNA. However, as we have already discussed (Section 4.2.4), gp 32 is autogenously, and therefore individually, regulated. Other proteins which are synthesized both before and after replication include gp 57 (Wilhelm and Haselkorn, 1971b), which is required for tail-fiber maturation, and the three internal proteins of the phage head (Black and Gold, 1971; Howard *et al.*, 1972; O'Farrell and Gold, 1973*a*). The T-even tRNAs (Section 4.5) are synthesized both before and after replication, but the regulation of their transcription has not yet been investigated. One gene which is transcribed off the *r*-strand, gene 5, has been tentatively identified as subject to quasi-late regulation (Jayaraman and Goldberg, 1970). Gene *e*, codes for T 4 lysozyme and is regulated essentially according to the rules of late gene expression. Yet gene *e* must also be included in the quasi-late class for peculiar reasons: At least a large part of gene *e* is transcribed from the *l*-strand during the prereplicative phase although it is not translated. Gene *e* RNA is synthesized in increasing amounts during the postreplicative period (see Sections 4.2.1. and 4.2.6b).

Upon examining the results of a hybridization-competition analysis of T 4 RNA and separated strands of DNA, Guha and co-workers (1971) concluded that quasi-late genes would turn out to be diversely, rather than uniformly, regulated. If gene 32 is a member of this set, then this conjecture must be correct.

4.6.3. Regulation of True Late Gene Expression

T 4 true late gene expression is remarkable in requiring modification both of the transcription system of the host and of the viral DNA template. The modifications of the host transcription system involve RNA polymerase which binds several T 4 proteins, including the transcription control proteins gp 33, gp 55, and, gp 45; the last one is also an essential protein of viral DNA replication. We shall describe these transcription control elements in Section 4.6.3e. The modification of the *E. coli* RNA polymerase core subunits precedes late gene expression in time but is not absolutely required for true late transcription (Table 5 and Section 4.3). The precise chemical nature of the DNA template suitable for late transcription is not fully understood. It is clear, however, that the phage DNA must contain hmC in place of C. C-containing DNA severely impairs late gene expression (Hosoda and Levinthal, 1968; Kutter and Wiberg, 1969; Bolle *et al.*, 1968*b*; Kutter *et al.*, 1975; Wu and Geiduschek, 1975). Modifications ("processing") of the template also seem to be essential, but the nature of these modifications is not entirely clear.

Several years ago, Bruner and Cape (1970) focused a discussion of late gene expression around two distinct models—a "cell state" and a "DNA state" model. The "cell state" model involves control of gene action through small molecular effectors, regulatory macromolecules, special sites of DNA binding for transcription, and changes of the host's transcription machinery. The "DNA state" model involves control of gene action through DNA "processing." The "cell state" changes that regulate true late transcription will be discussed in Section 4.6.3e. The experimental evidence regarding the "DNA state" model and DNA "processing" is less complete and will be discussed in Sections 4.6.3c and 4.6.3d.

4.6.3a. The Timing of True Late Transcription

When does true late transcription start? We estimate that it begins at 8.5–9 min, approximately 2.5–3.5 min after the start of replication. It has been estimated to begin somewhat later, on the basis of hybridization-competition experiments (Bolle *et al.*, 1968a), but low levels of true late transcription could be masked by antimessenger RNA (Geiduschek and Grau, 1970; Jurale *et al.*, 1970). Our estimate is more consistent with measurements of the time at which synthesis of some late proteins becomes rifamycin-resistant (Sheldon, 1971) and with calculations based on the time of first appearance of several large late proteins (O'Farrell and Gold, 1973a) and on the average elongation rate of T 4 RNA chains (1100 nucleotides per minute; Section 4.2.6a).

The causes of the short lag between the onset of replication and the start of late transcription are not known. The lag might reflect the time required for the accumulation, arrangement, and action of several interacting components, or it might reflect the incompetence of the DNA template for late transcription during the first rounds of replication. The replication mechanisms of these first and subsequent rounds of replication probably differ. (For a provocative and up-to-date review of this subject, see Broker and Doermann, 1975.) An answer to this problem of competence of first-phase DNA could bear strongly on the kind of DNA processing which is involved in creating a true late transcription-competent structure.

4.6.3b. T 4 DNA Replication

At this point, it is appropriate to summarize the available information regarding T 4 DNA replication (for recent reviews, see Doermann, 1973; Kornberg, 1974). The genes directly or indirectly involved in this process are listed in Table 4. Some of the corresponding gene functions are known; others are completely obscure. Since DNA replication is mechanistically complex and spatially organized (Kornberg, 1974), it is anticipated that the replication proteins should be part of a complex. The interaction between the components of this replication complex must be weak since it never has been isolated (in contrast with the multicomponent complex of RNA polymerase). There is, however, evidence for the interaction of several T 4 replication proteins: (a) gp 32, the DNA binding protein, interacts with DNA and with gp 43, the T 4 DNA polymerase (Alberts and Frey, 1970; Huberman et al., 1971); (b) gp 44 and gp 62 form a relatively stable complex (Barry and Alberts, 1972); (c) gp 45 stimulates the DNA-dependent ATPase activity of the gp 44–gp 62 complex, and all three of these proteins interact with DNA and with gp 43 (Alberts et al., 1973, 1975). In addition, it has been suggested that the enzymes dCMP hydroxymethylase (gp 42) and deoxynucleotide kinase (gp 1) have structural roles in replication in addition to their catalytic roles in nucleotide metabolism (Wovcha et al., 1973; Collinsworth and Mathews, 1974; Chiu et al., 1976). The participation of host proteins in T 4 replication in vivo is not excluded, although a cell-free system consisting of gp 43, gp 32, gp 41, gp 44, gp 62, gp 45, the four ribo- and four deoxynucleoside triphosphates, and appropriate salts seems to be capable of initiating and synthesizing T 4 DNA pieces of considerable size by a semiconservative mechanism with almost the in vivo rate (Alberts et al., 1975). Although there is now encouraging progress in analyzing T 4 DNA polymerization, the organization of replicons within the T 4 chromosome is still quite unclear. Experimental and interpretational difficulties arise from the enormous recombination frequency, the asynchronous initiation of rounds of replication, and the circular permutation of the T 4 genome. The model for T 4 DNA replication with the best experimental support involves multiple origins and bidirectional, semiconservative DNA synthesis (Doermann, 1973; Miller et al., 1970; Miller and Kozinski, 1970; Delius et al., 1971; Carlson, 1973; Emanuel, 1973; Howe et al., 1973; Kozinski and Doermann, 1975). A different model with a single origin near gene 43 (Fig. 5) and clockwise unidirectional replication (Werner, 1968; Mosig and Werner, 1969; Mosig, 1970; Marsh et al., 1971) now seems less likely to be valid. Large concatemeric molecules of T 4 DNA which exist in the infected cell (Frankel, 1968) probably result from recombination rather than by a rolling circlelike mechanism (Miller et al., 1970; Broker and Lehman, 1971; Broker, 1973; Broker and Doermann, 1975). In fact, all of the replicating T 4 DNA of a cell has been shown to be linked in a single complexly interconnected structure (Huberman, 1968). This complex structure is membrane-associated (Frankel, 1968; Miller et al., 1970; Altman and Lerman, 1970; Boikov and Gumanov, 1971). Membrane association of T 4 DNA is an early event that does not require viral gene expression (Miller, 1972; Earhart *et al.*, 1973), and DNA replication predominantly occurs close to the sites of membrane association (Huberman, 1968; reviewed by Siegel and Schaechter, 1973). Furthermore, certain viral mutations which lead to a premature arrest of replication (the so-called DA phenotype; Table 4) also involve membrane detachment of the DNA (Shalitin and Naot, 1971; Shah and Berger, 1971; Naot and Shalitin, 1973; Wu and Yeh, 1974).

4.6.3c. DNA Nucleotide Modification and True Late Gene Expression

T 4 true late transcription exhibits a specific requirement for nucleotide modifications in DNA. These modifications also influence restriction of T 4 DNA by bacterial and phage-specified nucleases. However, with appropriate attention to the use of proper phage and host strains (Kutter and Wiberg, 1969; Georgopoulos and Revel, 1971; Bruner *et al.*, 1972), transcription and restriction effects can be separated; we concern ourselves here with the former. As we have already mentioned, substitution of C for the normally present hmC in T 4 DNA blocks true late gene expression. The primary effect is surely on transcription, but there is, in addition, a residual production of true late transcripts which are not translated, for reasons that are not yet understood (Kutter *et al.*, 1975; Wu and Geiduschek, 1975). Early gene expression remains largely unaffected, and quasi-late gene expression has not been studied in detail but is certainly not abolished since gene 32 protein is overproduced (Kutter *et al.*, 1975; Wu and Geiduschek, 1975). These defects clearly arise from the C content of T 4 DNA and not from DNA degradation (Bolle *et al.*, 1968a; Kutter and Wiberg, 1969; Kutter *et al.*, 1975; Wu and Geiduschek, 1975). However, the requirement for hmC is not absolute. Several percent of C in the DNA are tolerated (Kutter and Wiberg, 1969). Recently, the tolerance for C has been greatly enhanced. T 4 triple mutants (56[-], *den*B[-], plus a third mutation) whose DNA contains a high amount of C instead of hmC make normal amounts of true late RNA and proteins. They grow well on *E. coli* hosts (Snyder *et al.*, 1976). The characterization of the third mutation (tentatively designated as *alc* for *a*llows *l*ate gene expression in *c*ytosine-containing DNA) suggests that the *alc* gene may specify the 15,000 dalton T 4-specified RNA polymerase binding protein (Snyder, 1976; Table 5).*

---

* See also Notes Added in Proof, p. 196, Note 5.

The hmC in T 4 DNA normally is fully glucosylated (Lehman and Pratt, 1960). Glucosylation protects T-even DNA against restriction (Revel, 1967; Georgopoulos and Revel, 1971). Glucosylation can also affect early and late transcription under special circumstances. However, there is no absolute dependence on glucosylation since T-phage multiplication can occur in the almost total absence of glucosylation (Luria and Human, 1952; Revel, 1967). Unglucosylated T 4 DNA is a better template for early protein and RNA synthesis *in vitro*. The enhanced template activity may be differential among different early genes (Cox and Conway, 1973). No detailed comparison of early gene action on unglucosylated and normally (α- and β-glucosylated) viral genomes *in vivo* has been undertaken.

A direct effect of β-glucosylation on T 4 late gene expression has recently been found in certain bacterial mutants which carry RNA polymerase mutations (probably in the β-subunit) and are cold sensitive for late gene expression. The RNA polymerase mutation also affects host shutoff and, remarkably enough, makes phage DNA replication initially cold sensitive. T 4 DNA lacking β-glucosyl groups suppresses the transcription and replication defects. α-Glucosylation appears not to affect either process. Mutations in gene 45 can compensate for the RNA polymerase defect and allow replication and late transcription. A persuasive argument has been made that these gene 45 compensatory mutations provide less gp 45 to the infected cell (Snyder, 1972; Montgomery and Snyder, 1973; Snyder and Montgomery, 1974). Additional compensatory mutants, which have yet to be characterized, map between α-gt and gene 55 and presumably define a new cistron (Snyder and Montgomery, 1974). The biochemical analysis of these mutations should be of great help in understanding T 4 replication and regulation.

4.6.3d Structure of the DNA Template for True Late Gene Expression

True late RNA synthesis *in vivo* appears not to involve intact linear duplex DNA but to occur on T 4 DNA whose structure has been further modified or "processed." The evidence on this subject includes the following observations: (1) Effective late transcription normally is coupled to concurrent T 4 DNA replication. (2) In several quantitative, but not qualitative, respects inferior late transcription occurs in the absence of replication. This replication-independent late transcription, which occurs under special conditions, can be influenced by manipulating genes that affect T 4 DNA stucture.

It is known that true late transcription is normally coupled to concurrent replication (Lembach *et al.*, 1969; Riva *et al.*, 1970*a*). Is that coupling direct, in the sense that competence is directly conferred upon a DNA segment by its replication? We do not know the answer to this question. Unfortunately, unequivocal experiments which show transcription of a particular segment of the chromosome correlated with the replication of that piece of DNA are probably not feasible. An examination of the sequence of first appearance of late proteins also would yield little information about the correlation between replication and true late transcription because of the multiple origins and bidirectionality of T 4 replication. By the time that late transcription starts, T 4 replication probably is irretrievably asynchronous.

The replication-created or replication-coupled competent state of the DNA seems to have a limited lifetime *in vivo*. When replication is stopped (e.g., by shifting a T 4 DNA polymerase *ts* mutant to the nonpermissive temperature), transcription of true late genes is depressed by a factor of 10 within some minutes (Riva *et al.*, 1970*a*). A residual low level of late transcription is maintained in the absence of any DNA replication. This low level could be due to a second mode of true late transcription which does not require DNA replication because of its small contribution to normal total late transcription. We shall discuss this question in another connection below.

True late transcription is not simply proportional either to the total amount of DNA or to the total rate of DNA replication. Several mutations which severely restrict replication permit relatively high rates of true late gene expression. These include mutations in genes 30, 41, 46, and 47 (Bolle *et al.*, 1968*b*; Hosoda and Levinthal, 1968; Wu *et al.*, 1975). All of these mutations lead to the formation or stabilization of interruptions in the primary or secondary structure of DNA. Indeed, the T 4 DNA that has been replicated in the absence of gene 30 and 46 function retains its competence for late transcription upon cessation of replication relatively well. It occurred to Riva and co-workers (1970*b*) that this might be due to the presence of unligated but stable progeny DNA. They proposed that breaks in the DNA primary structure and their attendant disruptions of secondary structure might be required for replication-uncoupled and, by analogy, for normal T 4 true late transcription.

It appears that some true late transcription can occur in the absence of replication. There had been indications that some lysozyme could be synthesized *in vivo* in the absence of replication (Kutter and Wiberg, 1969; Mark 1970). (Lysozyme shows some regulatory peculiarities and can be synthesized *in vitro* in cell-free coupled systems that make early proteins and no other late proteins, so that lysozyme is often regarded with suspicion as a marker for true-late proteins. Nevertheless, in these experiments, lysozyme synthesis was an indicator of the replication-independent synthesis of true late proteins.) A systematic investigation of bacteria infected with replication-defective (DO) mutants in many genes showed that true late genes are transcribed asymmetrically to some extent in the absence of replication and that all identifiable (unprocessed) late proteins can be detected (Wu *et al.*, 1973, 1975; Wu and Geiduschek, 1975). This result means that even parental DNA can become competent for late transcription. Late transcription under these conditions, however, is characteristically temperature sensitive (in DO amber infection), occurs with a delay, and never reaches wild-type levels; its abundance relative to total transcription depends on the number of infecting phage. Also, in this case, introduction of mutations into the T 4 DNA ligase produce (modest) increases in late transcription and make the true late transcription less temperature-sensitive. A more striking inverse correlation of the proportion of true late transcription with DNA ligase activity has been shown in experiments in which the activities of both phage and bacterial DNA ligases were manipulated (Wu and Geiduschek, 1974). The correlation between enhanced late transcription and defective DNA ligase is again consistent with the hypothesis that single-strand breaks or gaps may play a role in late transcription. Unfortunately, the examination of single-strand breaks has not so far been informative. Only the distribution of breaks between the *l* and *r* DNA strands has thus far been examined, and no significant differences were found in a study of all the DO mutants, including gene 45 (Wu *et al.*, 1975). This represents a major gap in the chain of evidence concerning the template for true late transcription. The major problem of these experiments is that no mutants affecting endonucleolytic cleavage parallel to T 4 late transcription are yet known. There is no identified T 4 gene that is comparable with T 5 gene D15 which is essential for late T 5 transcription and codes for a nuclease activity (Section 5.3.3). Neither is there any assurance that the observed breaks and gaps in unreplicated DNA have exclusively transcriptional significance. In fact, it seems unlikely that this should be so. Nevertheless, we believe that at least replication-independent true late transcription depends on the availability of nicks or small gaps in T 4 DNA. Such nicks and gaps also exist during replication, at least of DNA ligase-exonuclease (genes 30, 46) deficient phage (Riva *et al.*, 1970*a,b*). But it is not established that precisely the same DNA structures control true late transcription of normally replicating wild-type phage. Indeed, the proteins which create the competent DNA template for true late transcription in the normal development of T 4 need not be identical with those that make unreplicated DNA competent for a low level of true late gene activity.

Only very circumstantial evidence connects most of the replication proteins with true late transcription. It has been observed that introduction of additional amber mutations in replication genes introduces a peculiar temperature sensitivity into replication-independent (*pol⁻ lig⁻ exonuclease⁻*) true late transcription (Wu *et al.*, 1975). This is consistent with the possibility that many of these replication proteins interact weakly with the true late transcription system. A speculative model of how replication proteins might work as a template-processing machine in the absence of replication has been proposed (Wu *et al.*, 1975).

Gene product 45 is the exceptional replication protein. The clearly demonstrated double function of gp 45 in DNA replication and late transcription (Wu *et al.*, 1973, 1975; Wu and Geiduschek, 1975) leads us back to a cardinal problem of the normal true late transcription process: its spatial correlation with DNA replication. One simple interpretation would be that late transcription occurs at, or very close to, the replication fork. Gene product 45 would act as a link between DNA replication and late transcription since it can bind to RNA polymerase (Section 4.6.3e) and is a part of the T 4 replication complex (Section 4.6.3b).

The significance of the special template structural requirements for true late transcription must be either that T 4 DNA has no true late promoters that can be recognized in unnicked (unprocessed) linear duplex DNA or that a premature chain-termination mechanism for true late T 4 RNA operates in linear duplex DNA. We consider the former of these possibilities more likely. Our discussion therefore next turns to two aspects of RNA chain initiation: (1) effects of DNA coiling on RNA chain initiation and (2) conceivable unconventional mechanisms of RNA chain initiation.

RNA chain initiation by *E. coli* holopolymerase involves cooperative alterations of the DNA helix in the promoter region, which slightly unwind the DNA (Saucier and Wang, 1972). As a result, negative DNA supercoiling affects promoter strength (Hayashi and Hayashi, 1971; Puga and Tessman, 1973*a*; Zimmer and Millette, 1975), probably because it is accompanied by some unwinding of the primary DNA helix (Dean and Lebowitz, 1971). Such unwound regions of DNA are stabilized by gp 32 (Alberts and Frey, 1970). Supercoiling of DNA can also occur in DNA that is not circular or closed if chain ends can be otherwise prevented from rotating (Worcel and Burgi, 1972; Pettijohn and Hecht, 1973). We judge that these specific considerations are not relevant to true late transcription. Evidence of extensive supercoiling in intact vegetative T 4 DNA has not been found (Pettijohn, personal communication). DNA ligation decreases rather than promotes true late transcription, as we have already discussed in detail. Gene product 32, which is continuously required for T 4 replication and which stimulates T 4 replication *in vitro,* is not absolutely required for replication-independent T 4 true late transcription. For these reasons, we think that DNA supercoiling is not used as a part of the recognition-initiation mechanism of true late transcription.

The search for special features of true late transcription is in part motivated by the failure of conventional *in vitro* systems of purified and even unpurified components to give clear evidence for *de novo* initiation of true late T 4 RNA chains (Section 4.6.3f). The template requirements for late transcription which have just been discussed may account for a part of that failure. Another possibility is that the mechanism of true late RNA chain initiation is unconventional, requiring a primer of polyribo- or polydeoxyribonucleotide. Such primers have thus far been sought only in connection with DNA replication. Ribonucleotide pieces of different size have been found in early replicating DNA *in vivo* (Buckley *et al.,* 1972), in DNA replicating *in vitro* (Dicou, 1975), and in mature T 4 DNA (Speyer *et al.,* 1972). T 4 DNA replication is greatly stimulated by the four ribonucleoside triphosphates *in vitro* in the absence of RNA polymerase (Alberts *et al.,* 1975). If there exists a mechanism for primer RNA synthesis in T 4 (Dicou, 1975; Okazaki *et al.,* 1975) which does not use the host RNA polymerase, this mechanism could also be involved in initiation of late transcription. Freshly synthesized DNA pieces which have not been joined yet might also act as primers. RNA polymerase holoenzyme is able to use short primers for initiating RNA synthesis on double-stranded DNA (Niyogi, 1972). The same reaction has not thus far been investigated with single-stranded templates or with T 4 modified RNA polymerase cores. No pertinent experiments on rifamycin sensitivity of primer-mediated initiation of RNA synthesis *in vitro* have been reported. True late transcription *in vivo* remains subject to the rifamycin response of the β-subunit of RNA polymerase (Haselkorn *et al.,* 1969; Mizuno and Nitta, 1969; Sheldon, 1971).

4.6.3e. Transcription Control Elements and RNA Polymerase—the Gene 33, 45, and 55 Products

True late transcription involves both T 4 DNA strands, although *l*-strand specific transcripts comprise only a very small fraction of true late RNA. Gene product 33, gp 45, and gp 55 are directly involved in, and the latter two proteins are continuously required for, replication-coupled as well as replication-independent true late transcription (Bolle *et al.*, 1968*b*; Pulitzer, 1970; Wu *et al.*, 1973, 1975; Coppo *et al.*, 1975*a,b*). The involvement of gp 45 in true late transcription is direct as a transcription control protein, and indirect as a replication protein (Epstein *et al.*, 1963; Wu *et al.*, 1973, 1975). The direct transcriptional role is seen in two ways. First, gp 45 is required for replication-independent true late transcription (Wu *et al.*, 1973, 1975). Second, T 4 mutations (*com* or *gor*) which suppress bacterial mutations affecting T 4 true late transcription (*tab*-D and certain *rif*ʳ mutations) are located in gene 45 as well as in gene 55 and in an as-yet uncharacterized locus (Snyder and Montgomery, 1974; Coppo *et al.*, 1975*a*). Since all other early proteins that have so far been resolved are synthesized in the absence of gp 45, the effect of gp 45 on true late transcription appears to be direct (Wu *et al.*, 1973, 1975).

Gene product 33, gp 45, gp 55, and two other T 4-specified proteins interact with RNA polymerase (cf. Table 5). With the exception of gp 45, these T 4-specified proteins bind sufficiently tightly to the modified core polymerase that they remain associated with it through all the usual RNA polymerase purification steps,* including chromatography on T 4 DNA cellulose (Horvitz, 1973) and precipitation with antibodies to host RNA polymerase core (Goff and Weber, 1970). However, their binding intensities vary considerably. The 10,000 and 15,000 molecular weight proteins are each present in about one copy per RNA polymerase protomer (Stevens, 1972). Gene product 55 and σ, the latter partially in a modified form, are present in substoichiometric amounts (Bautz, 1968; Seifert *et al.*, 1969; Stevens 1972; Rabussay *et al.*, 1972; Horvitz, 1974*a*). Gene product 33 can be removed from the polymerase complex by phosphocellulose chromatography (Horvitz, 1973).

The gene 45 product (M.W. ∼ 29,000) does not copurify with modified RNA polymerase core, but there is direct biochemical evidence, in addition to the already mentioned genetic evidence, that it does interact with the modified core enzyme (Snyder and Montgomery, 1974; Ratner, 1974*a*; Coppo *et al.*, 1974, 1975*a,b*). All T 4-specific proteins, except gp 45, can bind sufficiently tightly to the unmodified host RNA polymerase core to allow their separation from crude extracts by an affinity-column procedure in which RNA polymerase is covalently linked to Sepharose and used as adsorbent. However, gp 45 is only retained to a significant extent on carrier-fixed, T 4-modified RNA

---

* Gene product 33 is removed by phosphocellulose chromatography; some disagreement exists on the behavior of the 10,000 dalton protein. (See, however, Notes Added in Proof, p. 196, Note 6.)

polymerase core (Ratner, 1974a). The only certain difference between the two enzymes is the α-modification of the T 4 core (Table 5), since all the T 4-specific proteins are present in the cell extract and bind to the RNA polymerase on the column. Thus, this experiment may imply a functional difference between the α-modified and unmodified RNA polymerase. Since the modification of the α-subunits does not seem to be essential for T 4 development (Horvitz, 1974b; Section 4.3), it appears likely that a relatively weak interaction with unmodified RNA polymerase suffices for gp 45 function in true late transcription.

The 15,000-dalton protein may be coded by the *alc* gene (Section 4.6.3c) and probably is responsible for the hmC requirement for true late transcription (Snyder, 1976). The fourth protein (M.W. between 7000 and 10,000) seems to bind less tightly than the others (compare Horvitz, 1973; Stevens, 1972); its structural gene is unknown. The kinetics of synthesis of the small, T 4-coded proteins as well as other criteria classify them as early proteins. They are synthesized 3–15 min after infection at 30°C (Stevens, 1972; Horvitz, 1973; Pulitzer, 1970).

As mentioned, purified late T 4 RNA polymerase is inhomogeneous. These inhomogeneities involve the phage-specific subunits and also α and σ. Depending on the host and phage strains used, up to 95% of the α-subunits are modified by ADP-ribosylation, but at least a trace of unmodified α is always detectable. The inhomogeneities with respect to the phage-specific subunits probably result from an enzyme purification procedure that is not ideally suitable for relatively weakly bound proteins. However, we guess that the weakness of these interactions is itself significant. The model that we have in mind is of one RNA polymerase core interacting with different sets of host- and T 4-specified proteins. We guess that one set of components, which is responsible for quasi-late transcription, initiates RNA chains very readily; the other set, which is responsible for true late transcription, does not. As a consequence, the transcription capacity of quasi-late and true late transcription units differs considerably and the former operate *in vivo* in template excess, whereas the latter are template-limited even when replication occurs. This interpretation is consistent with *in vitro* experiments on complementation between RNA polymerase and crude cell extracts (Rabussay and Geiduschek, unpublished).

Some ideas about what combinations of proteins might serve which functions can be derived from the mode of interaction between the core RNA polymerase and its accessory proteins and between the accessory proteins themselves. The gp 33 content of RNA polymerase, for example, is decreased if gp 55 is not functional (Horvitz, 1973). Whether gp 55 has a direct or indirect effect on the binding of gp 33 is not known. The interaction of the host σ-subunit with the T 4-modified core polymerase is certainly weakened relative to the unmodified core enzyme but nevertheless clearly demonstrable (compare Section 4.3). Gene product 33 and σ seem to compete with each other for the same binding site at the core RNA polymerase (Ratner, 1974a). Gene product 33, like σ, can be selectively removed from the enzyme complex by phosphocellulose chromatography (Horvitz, 1973). Although these observations could be taken to indicate that gp 33 functions as a σ analog, no such function has been observed *in vitro*: RNA polymerase $[(\beta\beta'\alpha_2)_M$ gp 55 gp 33 (15K)] behaves like host core when tested on native T 4 or calf thymus DNA (compare Section 4.3). In fact, we think it unlikely that a specific σ analog for T 4 late transcription exists. σ is specifically adapted for preventing RNA polymerase core-DNA binding and for preventing RNA chain initiation at single strand breaks (see Chamberlin, 1974a). It may also be specifically adapted for promoter recognition in double-helical DNA. These functions would be gratuitous to true late transcription on competent DNA as we have pictured it.

Considering the known facts about initiation of RNA synthesis (compare Section 1.2 and Chamberlin, 1974a) and late T 4-modified RNA polymerase structure, we can present a rather speculative and schematic model for postreplicative transcription which is not excluded by the currently available information: Postreplicative transcription of quasi-late genes is achieved by a RNA polymerase complex consisting of $\beta\beta'$ $(\alpha_2)_M$, the 15,000- and 10,000-dalton proteins. In addition, a factor, which could be the host σ, is needed for efficient initiation. The template is "unprocessed" parental and/or progeny DNA. Most or all of the true late genes are transcribed from a "processed" DNA template by a transcription complex containing $\beta\beta'$ $(\alpha_2)_M$, gp 55, gp 33, gp 45, the 15,000- and 10,000-dalton protein, and probably one or more loosely bound transcription proteins. The requirement for an initiation factor of the σ type is probably dispensable if initiation of late transcription occurs at single strand breaks or regions or involves priming. Gene product 33, gp 55, and/or other as yet unidentified proteins may promote chain initiation but probably are not σ factors. Dausse *et al.* (1972b) have shown that single strand breaks, not gaps or frayed ends, are the preferred initiation sites for *E. coli* RNA polymerase core. σ restricts initiation from these sites. Therefore σ could, in principle, control quasi-late transcription positively while controlling late transcripton negatively. Gene product 33 could then be assumed to function as a σ competitor, which does not inhibit initition at single-strand breaks. The relative amounts of gp 33 and σ would be crucial.

4.6.3f. Transcription of True Late RNA *in Vitro*

The analysis of T 4 postreplicative transcription suffers from the lack of suitable *in vitro* systems. The failures are of two kinds: (1) Purified RNA polymerase from T 4-infected cells either transcribes early RNA, in the presence of σ, or transcribes intact T 4 DNA extremely poorly, in the absence of σ (compare Section 4.3). Substitution of vegetative for mature DNA also fails to yield true late transcription. Transcription-stimulatory factors have been sought *in vitro* but, with one exception, they do not promote T 4 late transcription. There is one report of a T 4 *r*-strand transcription-stimulating factor active with late RNA polymerase and mature T 4 DNA (Travers, 1970*b*). No further or confirming analysis of this factor has been made (see previous discussion by Wu *et al.*, 1973). (2) Crude systems from T 4-infected cells that synthesize late RNA *in vitro* off their endogenous templates can be prepared (Hall and Crouch, 1968; Snyder and Geiduschek, 1968; Rabussay and Geiduschek, 1973; Jörstad *et al.*, 1973; Maor and Shalitin, 1974). Analysis of these systems is hampered by uncertainties regarding the *de novo* initiation of RNA chains *in vitro* and by the failure of many kinds of complementation experiments.

4.7. What Is the Significance of the Complexity of T 4 Regulation?

The regulation of gene expression during phage T 4 development clearly is elaborate. Since many aspects of this regulation are not understood at the molecular level, it is therefore appropriate to close the section on T 4 with a consideration of the following two questions. Do all the phenomena that appear to be regulatory actually reflect regulatory events? And, are all these apparently complex regulatory patterns necessary for T 4 multiplication and do they affect the yield of progeny phage quantitatively or are they rather tolerated secondary consequences of properties that are immanent in this particular biological system? The existence of a regulatory relationship in phage development frequently is inferred from the observation that changes in gene A (and, generally, in gp A) affect the synthesis of gp W. Does this necessarily mean that gp A regulates gene W directly or indirectly? The answer, of course, is: "No, not necessarily." However, the exceptions are not numerous. Gene product A can validly be said to exert regulatory functions either if it is a *bona fide* regulator of transcription or translation of gene W in the sense described for the *lac* repressor, the *ara* regulatory protein, T 4 gp 32, and Qβ coat protein, or if, during the course of normal infection, gp A interacts with, or alters, components of translation or transcription that set the constituitive level of function of gene W. Such interaction may be direct, for example through ribosomal binding or modification, or it may be indirect, for example through membrane interactions which release or immobilize an effector.

In contrast, gp A should not be said to regulate in the following hypothetical circumstance: gp A forms a complex with gp B, C, and D. In the normal course of infection, the {gp A, B, C, D} complex does not exert regulatory activity in either of the terms listed above. However, in the absence of intact gp A, the complex {gp B, C, D} directly or indirectly blocks some aspect of gene expression.

Which are the *bona fide* regulatory proteins and components of T 4 development? (a) Gene products 33, 55, and 45 certainly are regulatory proteins. The wild-type proteins bind to RNA polymerase; mutants in these genes quantitatively affect late T 4 gene transcription; gp 55 and 45 are continuously required for late RNA synthesis. The genetic suppression pattern associated with gene 45 and 55 mutations and the phenotypes of gene 45, 55, and 33 mutants provide strong assurance that these gene products regulate late transcription. Gene 33 is somewhat less studied physiologically and genetically, and the phenotype of gene 33 mutants is also less simple than that of the other mutants in this group. Specifically, there is a nonnegligible level of T 4 late transcription in gene 33 *am* mutant-infected bacteria, even in *str*^{r} antisuppressing bacteria infected with a double mutant in gene 33 (Wu and Geiduschek, unpublished) in which the yield of progeny phage is very low indeed. Defects of gp 33 may not produce exactly the same phenotype as defects of gp 55, and it is therefore possible that either gp 33 does not participate equally in all late transcription or that gp 33 only *shifts* the early/late mRNA ratio by its RNA polymerase binding function. (b) Late gene expression is regulated by DNA replication. Shutting off replication by means of inhibitors or by inactivating replication proteins produces comparable effects on late gene expression. The basically unresolved questions of the mechanism by which replication is coupled to gene expression are discussed at length in Section 4.6.3. However, we do not doubt the involvement of DNA replication in the regulation of late gene expression. (c) The shutoff of host RNA synthesis is regulated by T 4 gene products. These genes probably include *mod* and *alt* but are otherwise not identified. A host RNA polymerase mutation affects the shutoff (Snyder, 1973). There also are posttranscriptional events in shutoff. (This subject was discussed in detail by Mathews in Volume 7 of *Comprehensive Virology*.)

If viral gene products regulate these posttranscriptional events, they are not yet identified. (d) T 4 middle gene expression is regulated at the transcriptional level. The regulatory components are not identified, and it is not yet established whether they are preformed or synthesized in the infectious cycle in which they regulate. The function of the *mot-far* I gene (or genes) has recently been associated with the onset of middle gene expression. The analysis of these mutants should provide a real impetus to further work on the problem of middle gene expression. However, further analysis is still required to decide whether gp (*mot-far* I) regulates middle transcription or whether certain mutants of this kind poison the onset of middle transcription by indirect means. (e) The complexity of the shutoff of prereplicative protein synthesis has been discussed at length in Sections 4.2.2 to 4.2.5 together with the evidence that diverse mechanisms, including posttranscriptional mechanisms, are at work. Mutants in gene *reg*A affect that shutoff particularly strongly in the absence of DNA replication. It is our prejudgment of the gp (*reg*A) that it will turn out to be a posttranscriptional regulator. At the present time, however, the same caveat must be applied to the *reg*A as to the *mot* and *far* I gene products. (f) Certain other mutants, such as the *far* II mutants (Chace and Hall, 1975), have pleiotropic effects on the synthesis of early proteins, which differ from those of the *far* I, *mot,* and *reg*A mutations. It has been suggested that the effect of the *far* II deletions is on posttranscriptional events and relates to the selective, very early shutoff of certain viral proteins. The comments that we have made about *mot* and *far* I also apply to these mutants. (g) The DNA polymerase and the DNA unwinding protein are negative regulators of their own synthesis. The chain of evidence that these proteins are indeed the effectors of their autoregulation is strong. It includes the demonstration of reversible changes in autoregulation, involves nonsense as well as temperature-sensitive mutants, and, in the case of the nucleic acid unwinding protein, involves some biochemical evidence about the mechanism of action.

What part of this T 4 regulation is necessary for viral multiplication? Certainly, the proteins that control positively regulated middle and true late gene expression are. The early regulatory event must also be required to the extent that the expression of at least one essential gene is almost completely dependent on its occurrence. In wild-type phage, the hydroxymethylation of cytosine is required for true late gene expression, but this requirement is not absolute, and we are just learning that it can be substantially circumvented. On the other hand, the regulation of early gene shutoff, presently exemplified by gp *reg*A and gp *far* II, may not substantially affect phage yield on standard laboratory host strains. The status of the other regulatory relationships is less clear. The viral proteins that are so far identified as being involved only in host shutoff appear to be individually nonessential. However, host shutoff occurs at several levels and involves many viral activities. It is only possible to guess that a total failure of host shutoff would substantially affect T 4 development in all bacterial hosts. The significance of the autoregulatory properties of gp 43 and 32 is uncertain. It is particularly attractive to speculate on all the dramatic consequences that might flow from a substantial overabundance of functional nucleic acid binding protein. Unfortunately there exist as yet no experimental data by which to test such speculation.

5. BACTERIOPHAGE T 5

5.1. Introduction

We have chosen bacteriophage T 5 as the fourth of our prototypes because it incorporates one unique regulatory feature in its development: The activity of T 5 genes in the infected bacterium is limited in time by the stepwise transfer of T 5 DNA to the cell interior. In addition, T 5 represents a link between the relatively simply regulated T 3 and T 7 phage and the T-even phage with their rather complex regulation pattern. For example, we shall relate that T 5, like T 4, probably transcribes viral late genes from processed DNA, but that the T 5 processing mechanisms are not obscured by coupling to concurrent replication.

Bacteriophage T 5 of *E. coli* is another member of the collection of lytic phages that was isolated by Demerec and Delbrück. T 5 encases its DNA in an icosahedral head to which a long, hollow, noncontractile tail is attached. The tail bears distal tail fibers.

Bacteriophage BF 23 is a very close relative of T 5. These two phages undergo phenotypic mixing, substitution of gene products, and genetic recombination, and their genetic maps have been found to be colinear (Mizobuchi *et al.*, 1971). Advantage has been taken of this relationship in determining gene order (Beckman *et al.*, 1973). Three other phages, PB, BG 3, and 29 α, are also closely related to T 5; however, T 5 has been studied more extensively, and we shall concentrate on it. Certain aspects of T 5 development have been reviewed by Lanni (1968) and by Calendar (1970); a comprehensive review by McCorquodale (1975) has only recently been completed.

5.2.   DNA Structure, Genetic Map, and Transcription Units

Each complete T 5 virion contains approximately 120 kb of double-stranded DNA (molecular weight 8×10^7; Lang, 1970) which includes a unique segment of 100 kb flanked by two 11 kb (9%) terminally redundant segments (Rhoades and Rhoades, 1972). One of the DNA strands, the "light" strand, contains five breaks, n_1 to n_5, which separate contiguous segments of approximately 4, 4, 9, 15, 33, and 36% of the length of the DNA (Davison *et al.*, 1964; Burgi *et al.*, 1966; Abelson and Thomas, 1966; Bujard, 1969; Jacquemin-Sablon and Richardson 1970; Bujard and Hendrickson, 1973; Hayward and Smith, 1972a,b; Hayward, 1974) (Fig. 9). The position of the nick separating the two 4% pieces (n_1) varies between individual genomes and is entirely missing from a fraction of T 5 DNA molecules (Hayward, 1974). The other four nicks are present in all T 5$^+$ DNA molecules, and their positions are constant. An additional pattern involving a larger number of specific nicks appears in only a small fraction of all DNA molecules (Hayward and Smith, 1972a; Hayward, 1974). The nick n_1, which is located in the terminally redundant segment of T 5 DNA, appears to be introduced only at one end of the molecule (Fig. 9).

The genetic map of T 5 is presented in Fig. 10. It had been observed that conditional lethal T 5 mutants fell into four linkage groups; two factor crosses between mutants in different linkage groups tended to give maximal or almost maximal proportions of recombinants. The segmentation of the recombination map into three short and one long linkage group seemed plausibly relatable with the partition of the genome into separate regions by the single-strand breaks (Hendrickson and McCorquodale, 1971; Beckman *et al.*, 1973;

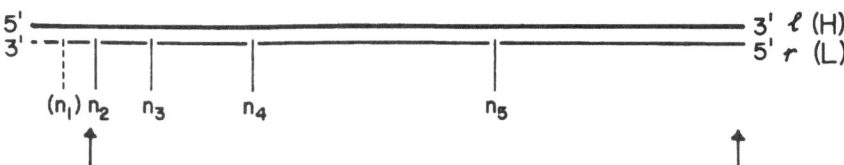

Fig. 9.   T 5 DNA, its interruptions, and its terminal redundancy. n_1 to n_5 are breaks in the *r* strand. n_1 is present in only a portion of T 5 DNA molecules. The vertical arrows point to two more recently detected breaks in the *r* strand (see Fig. 11).

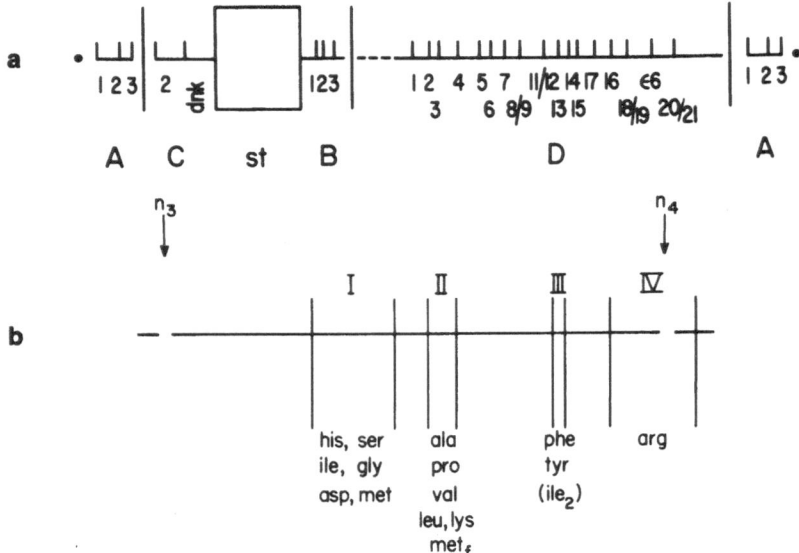

Fig. 10. (a) The general genetic map of T 5, (after McCorquodale, 1975). Linkage groups A, B, C, and D and the *st* region have been placed as discussed in the text. The numbers refer to genes identified in the prior genetic map of Hendrickson and Mc-Corquodale (1971). The corresponding gene products are identified in Table 7. The *st* region is deleted in T 5 *st* mutants and contains tRNA genes. Within each linkage group, the horizontal (linkage) line is extended to the most distal of the more recently mapped mutants (McCorquodale, 1975). Within entirely separate linkage groups, the polarity of the genes could not be established by recombination analysis alone. However, there is a residual gradient of recombination frequency between mutations in groups B and D (Hendrickson and McCorquodale, 1971) and this has been used to orient D on the total map. (b) Expansion of the *st* region of the T 5 map, showing the location of the T 5 tRNA genes (after M.-J. Chen *et al.*, 1976). This is a physical map, derived from an electron-microscopic mapping of heteroduplexes of T 5 deletion mutants and from hybridization studies with T 5 tRNA. Regions I, II, III, and IV which contain tRNA genes and the intervening segments are defined by a set of overlapping deletions. The segment between single-strand breaks n_3 and n_4 is taken as 14.9% of the total length of T 5 wild-type DNA. (Ile_2) is a tentatively identified second ile tRNA gene. (See also Notes Added in Proof, p. 196, Note 7.)

Hendrickson and Bujard, 1973). However, the number and size of break-segregated DNA segments does not correspond with the number and size of linkage groups. In addition, the breaks are actively repaired early in infection (Herman and Moyer, 1974). The newer genetic map of Fig. 10 supports entirely different explanations for the separate linkage groups: (1) Linkage group A corresponds with the terminally redundant segment of the genome. Since no genes outside A can be very close to both termini, the probability of their recombination with genes in one A segment are always high. (2) A deletable region of the genome which contains no essential genes is interposed between linkage groups C and B. The size of this region of nonessential genes accounts for the lack of linkage between essential genes in its contiguous segments C and B. (3) Since the other failures of linkage can be "explained" in conventional ways, the remaining large unlinked segment, D, may simply reflect a paucity of markers in the region between it and B.

5.3. Regulation

The T 5 genome, like the other phage genomes that we have analyzed, is expressed in a defined temporal sequence (Table 6). Three classes of gene products are distinguished: Pre-early or class I proteins are synthesized from the first until about the eighth minute after infection (37°C) (McCorquodale and Buchanan, 1968). The corresponding RNA synthesis is shut off around 4 min after infection (Moyer and Buchanan, 1969; Sirbasku and Buchanan, 1970a,b; Moyer and Buchanan, 1970b). Class II or early proteins and RNA are synthesized from about the fifth minute after infection (at 37°C). Two subclasses of early proteins can be distinguished: Many class II proteins are shut off between the 20th and 25th minute after infection (class IIa), but some proteins continue to be synthesized at significant rates until lysis (Hendrickson and Bujard, 1973; Hayward and Smith, 1973). The terms

TABLE 6

The Program of T 5 Protein Synthesis

	Timing of protein synthesis (min at 37°C)		Control of protein synthesis	
Class of genes	On	Off	On	Off
I. Pre-early	0	8	—	A1[b]
IIa. Early	4	20–25 (not coordinate?)	SST[a]	D5
IIb. Early; quasi-late	4	lysis	SST	—
III. Late	13	lysis	SST, C2, D15	—

[a] Second-step transfer.
[b] A1, C2, D5, and D15 are genes whose products regulate T 5 protein synthesis.

"pseudo-late" or "quasi-late" have been applied to these transcripts because they appear to correspond with the appropriate regulatory class of T 4 genes (Sections 4.2 and 4.6.2). At least some class II RNA synthesis is also shut down later in infection (Sirbasku and Buchanan 1970a). Phage DNA synthesis starts about 8 min after infection (Crawford, 1959; Moyer and Buchanan, 1970b; Pispa et al., 1971; Hendrickson and McCorquodale, 1971). The transcription of class III or late RNA follows so that late proteins appear from about 13 min on and are synthesized until lysis occurs at 45 to 60 min after infection (McCorquodale and Buchanan, 1968; Moyer and Buchanan 1969; Sirbasku and Buchanan, 1970a; Pispa et al., 1971).

5.3.1   First-Step Transfer (FST)

Viral gene expression during the first minutes of infection is confined to the T 5 class I genes by a remarkable restriction—the incomplete injection of the T 5 genome. Only 8% of the viral DNA, the FST or first-step-transfer piece, is initially injected. Transfer of the remaining 92% requires the activity of T 5 proteins that must be synthesized in the infected cell. Consequently, when T 5 infection occurs in starvation medium or in the presence of chloramphenicol, only the FST piece enters the cytoplasm (Lanni, 1960a, b, 1965). FST DNA has been isolated (Lanni and McCorquodale, 1963; McCorquodale and Lanni, 1964; Bujard and Hendrickson, 1973; Labedan et al., 1973). Is FST DNA always transferred from the same redundant end of T 5 DNA molecules? That is the conclusion that Labedan et al (1973) and Herman and Moyer (1975) have drawn, but McCorquodale (1975) presents persuasive arguments for regarding the issue as yet unresolved. The role of break n_2 in limiting FST is inextricably bound up with the polarity of injection—n_2 cannot be significant unless FST to the cytoplasm is exclusively from the left end (Fig. 9) of T 5 DNA.

Under normal, permissive conditions the first-step transfer is followed after 3–4 min by a second-step transfer (SST) of the remaining 92% of T 5 DNA (Lanni, 1965, 1968; Moyer and Buchanan, 1970a, b). The occurrence of SST is dependent on the function of the products of genes A1 and A2 (McCorquodale and Lanni, 1970; Lanni, 1969). gp A1 and gp A2 are proteins of molecular weights 57,000 and 15,000, respectively, which can form a DNA-binding complex (Beckman et al., 1971). SST seems to work by pulling or releasing the DNA into the host cell with the help of gp A1 and gp A2 (and perhaps other phage and/or host components) rather than by phage-mediated injection. This is the conclusion drawn from an experiment in which phage coats were stripped off after FST and DNA was nevertheless shown to be completely transferred into the host cell (Labedan and Legault-Demare, 1973). The other phage functions residing in the FST segment can be analyzed by radical surgery, that is, by shearing off the infecting phage particles containing SST DNA after FST has been completed. Such FST-infected cells break down host DNA rapidly and completely and stop host RNA and protein synthesis (Lanni and McCorquodale, 1963; Beckman et al., 1971). The genes on FST DNA yield Class I RNA and proteins which are synthesized and shut off (McCorquodale and Buchanan, 1968; Sirbasku and Buchanan, 1970a). The shutoff of class I RNA synthesis requires protein synthesis after FST. Therefore, the class I genes must be capable of regulating their own shutoff. However, it is not certain that the timing of class I autoregulation and of normal class I shutoff are identical and the participation of class II genes in normal class I regulation is not excluded. There is, of course, no synthesis of class II and III proteins, and neither is DNA replicated nor is new phage produced.

If FST-infected cells are formed as above and, after some time, are superinfected with T 5 in the presence of chloramphenicol, the entire superinfecting T 5 genome is transferred without interruption (Lanni, 1965). This again supports the view that the residual SST DNA can be pulled into the host once the necessary phage-specified proteins have been synthesized within the infected cell. If superinfection is delayed until class I proteins have already been shut off, the superinfecting intact genome induces almost no synthesis of class I proteins or RNA, but class II proteins and RNA are made without the usual lag (McCorquodale and Buchanan, 1968; Moyer and Buchanan, 1969; Sirbasku and Buchanan, 1970a). Therefore, the shutoff of class I genes that is produced during the FST period also suppresses the class I gene expression of the superinfecting DNA and must be capable of keeping the terminally redundant homolog of the FST genes silent. In summary, when only the first-step transfer is allowed to occur, T 5 class I proteins and RNA are synthesized and shut off and host DNA is broken down.

Only three genes have thus far been located in the FST segment and linkage group A. The function of gene A2 is, as already mentioned, essential for the SST (Lanni, 1968, 1969; McCorquodale and Lanni, 1970), whereas gene A3 is responsible for the arrest of growth of T 5^+ in cells carrying the colicinogenic factor col Ib (Moyer et al., 1972; Beckman et al., 1973). Mutants in gene A1 are pleiotropically defective: They fail to break down the host DNA, fail to arrest host RNA and protein synthesis, express class I proteins incompletely, do not execute SST, and fail to shut off class I protein synthesis (Lanni, 1969; McCorquodale and Lanni, 1970). Beckman and co-workers (1971) suggested that the gene A1 product is directly involved in all these processes through its participation in different multisubunit complexes. They assumed that a complex of gp A1 and gp A2, which has been found in cell extracts, acts in SST (see above). They also suggested that an oligomeric form of gp A1 alone, which has also been detected, could be active in the shutoff of class I proteins* whereas the monomer might degrade host DNA. According to this suggestion, gp A1 would be an extraordinarily versatile and interesting protein, executing or participating in three different DNA-related functions: cleaving host DNA, transporting T 5 DNA, and repressing certain regions of T 5 DNA.

Individual class I RNA species have been separated on polyacrylamide gels (Sirbasku and Buchanan, 1970a; Hayward and Smith, 1973); their molecular weights range from 1×10^6 to 6×10^4 daltons. The average chemical half-life of this RNA has been measured to be 4–5 min, whereas functional mRNA half-life is only 2–4 min at 37°C (Sirbasku and Buchanan, 1971). These RNA chemical lifetime measurements were made with rifamycin and pulse-chase methods, and the decay of protein-synthesizing capacity was measured after rifampicin addition. Under the conditions of the experiment, it is conceivable that the physiological shutoff mechanisms for class I transcription were at least partly functional. We mention these details because: (1) it is, as we have already stated (Section 4.2.2), difficult to assess the physiological significance of mRNA lifetime measurements made with antibiotics; (2) the shutoff of class I genes might affect messenger lifetime (cf. Section 4.2.2). However, these experiments do indicate that functional inactivation of class I RNA detectably precedes its chemical degradation to nucleotides.

The smallest (6×10^4 daltons) pre-early transcript is chemically stable. Degradation products of the larger species which are the 4 S size of tRNA (see below) are formed during degradation of class I RNA. However, it appears that the pre-early segment of T 5 does not code for tRNA (M.-J. Chen et al., 1976).

---

* The formation of an oligomer of gp A1 would effectively require a finite, minimal concentration of monomer. The time required to accumulate this critical concentration would be the time allotted to group I RNA synthesis (initiation). Such a model of negative autoregulation is plausible and not unique. (For another example, see Section 4.)

5.3.2. Class II (Early ; Pseudo-Late) Gene Expression

The activity of the class II genes starts immediately upon their (second step) transfer into the cytoplasm. It is likely (see Section 5.3.4) that no modification of bacterial transcription specificity is required for class II gene expression. Many of the characterized class II functions are directly or indirectly involved in T 5 DNA replication (Table 7) and include many enzymes of nucleotide metabolism. T 5 completely breaks down host DNA to nucleotides (Lanni, 1968). However (unlike T 7), these nucleotides are not incorporated into viral DNA but are instead degraded further by a T 5-specific 5′ nucleotidase and excreted (Zweig *et al.,* 1972; Warner *et al.,* 1975).* Consequently, large quantities of DNA precursors must be synthesized *de novo.* Several T 5 gene products are absolutely essential for DNA synthesis; one of these is a T 5-specific DNA polymerase (Hendrickson and McCorquodale, 1972). Mutations in other genes result in premature arrest (DA) or delayed onset (DD) of DNA replication. The product of gene D6 is essential for T 5 DNA stability (Hendrickson and McCorquodale, 1972).

The products of three class II genes, D5, C2, and D15, are associated with the regulation of gene expression. We shall discuss the latter two in connection with late genes (Section 5.3.3), but consider D5 here. We have already stated that, while the psuedo-late proteins (IIb) are synthesized throughout the remainder of the phage development and increase in abundance at later times, the class IIa proteins are gradually shut off starting 20–25 min after infection. The product of gene D5 is associated with this shutoff. This abundant nucleic-acid-binding protein is a member of class IIa and is also required for DNA replication (Chinnadurai and McCorquodale, 1974*b*). The shutoff of class IIa proteins by a class IIa protein is remarkable in that the duration of gene activity is relatively long and that the presumptive regulatory protein is accumulating throughout this time. While some class II RNA synthesis declines at later times (Sirbasku and Buchanan, 1970*a*) the regulation of the class IIa proteins could be either at the transcriptional or posttranscriptional level. The similarities between T 5 gp D5 and T 4 gp 32 are therefore particularly suggestive. However, the appropriate analysis of the T 5 protein has not yet been carried out and, specifically, no analysis of T 5 class II and late protein synthesis in cell-

---

* The strategic significance of nucleotide secretion is not clear. We doubt that there is such a thing as wanton destruction in the world of viruses. Perhaps excretion is a way of keeping modified nucleotides out of the DNA precursor pool.

TABLE 7

T 5 Genes, Functions, and Proteins (Adapted from McCorquodale, 1975)

Gene	Function or Product
A1	SST; host DNA degradation; regulation
A2	SST
A3	(compensate ColIb nonpermissiveness)[a]
Unknown	RNA-polymerase-binding protein
Unknown	5′ nucleotidase
C2	Regulation of late gene expression
Unknown	RNA polymerase-binding protein
dnk	5′ deoxyribonucleotide kinase
(st)[b]	tRNA
B1	Replication (DS)[c]
B2	Replication (DS)
B3	Dihydrofolate reductase
B4	Replication (DS)
Unknown	Thioredoxin
Unknown	Thymidilate synthetase
Unknown	Ribonucleotide reductase
D1, 2	Replication (DD and DS)
D3	Replication (DS)
D4	Replication (DA; degraded)
D5	Nucleic acid binding protein; Replication and regulation (DS)
D6	Replication (DA; degraded)
D7	Replication (DO)
D819	DNA polymerase (DO)
D11/12	Replication
D13	Replication (DA)
D14	Replication (DA)
D15	5′ exonuclease; DNA processing and regulation
ε6	(compensate defective T 5 tail assembly of gro E E. coli)[a]
D16–D19	Tail proteins[d]
D20/21	Head protein[d] (DA)
Unknown	endolysin

[a] These are phenotypes, not functions. *E. coli* carrying the ColIb episome are nonpermissive for T 5. A3 (*h*) mutants overcome the nonpermissiveness. *E. coli gro* E mutants are nonpermissive for T 4 and T 5. These bacteria are defective in T 4 head and T 5 tail assembly. The ε6 T 5 mutants compensate for this defect as do certain T 4 head mutants. (Georgopoulos *et al.*, 1972; Coppo *et al.*, 1973; Zweig and Cummings, 1973a; Sternberg, 1973a, b; Takano and Kakefuda, 1973).

[b] This deletable region contains all the tRNA genes (M.-J. Chen *et al.*, 1976) and may also contain other genes.

[c] This designation means that the rate of replication is affected. Conceivably, some of these gene products might affect replication only indirectly. Designation of replication phenotypes as in Section 4: DO, no replication; DA, replication arrests prematurely; DS, DNA synthesis is at an abnormally low rate; DD, the start of DNA synthesis is delayed.

[d] D19 and 20/21 mutations affect replication: gene D19 (DS); gene D20/21 (DA).

free systems has yet been undertaken. Class II transcription also produces a series of small RNA molecules with molecular weights of 2.6 to 8.0 $\times$ 10^4 which are synthesized until lysis (class IIb; Sirbasku and Buchanan, 1970b; Ikemura and Ozeki, 1975). At least some of these species are T 5 tRNAs and their precursors; The T 5 genome codes for an astonishing 15 or more different tRNAs (Scherberg and Weiss, 1970; M.-J. Chen et al., 1974, 1976). Transfer RNA is generated from primary transcripts by the RNase processing activities of several nucleases (Section 1.1.1c; Ozeki et al., 1975). At least two of these transcripts, with molecular weights of 2.6 and 5.3 $\times$ 10^4 appear to be cleavage products of larger class II precursor RNA molecules. All tRNA genes are located in the deletable segment of the genome, between linkage groups C and B (M.-J. Chen et al., 1974, 1976). The physical map of the T 5 tRNA genes, which has recently been completed, is shown in Fig. 10b. The general questions that relate to the presence of tRNA genes in a viral genome have already been discussed in Section 4. There is very little that can be added to the subject here,

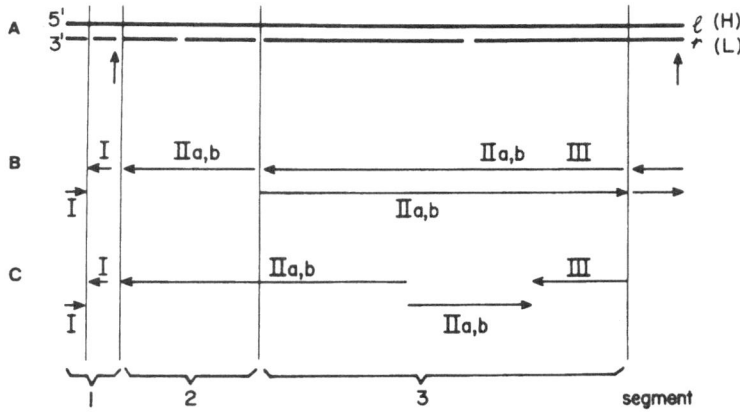

Fig. 11. A tentative transcription map of phage T 5 adapted from C. W.-S. Chen (1975) and McCorquodale (1975). Line A repeats the fragmentation map of Fig. 9, labels the DNA strands l(H) and r(L) and shows their polarity. The gaps in the r strand are the breaks n_1–n_5 (left to right). The vertical arrows point to two additional breaks postulated by C. W.-S. Chen and observed by Delius [quoted C. W.-S. Chen (1975) and McCorquodale (1975)]. Line B is a presentation of the results of hybridization experiments with isolated strand fragments of T 5 DNA. The vertical lines separate chromosome segments that are distinguished in the analysis. Leftward pointing arrows represent transcripts of the l(H) strand. The roman numerals refer to classes of RNA (Section 5.3). Line C is a model of transcription which places class III transcripts in the region of the map occupied by the late genes, places class I transcripts in only one of the terminally redundant segments (but see Section 5.3.1), and arranges polarities of transcription so as to minimize polarity switches along the chromosome.

because so little is known about the specific roles of the set of T 5 tRNA genes.*

Class I and II gene expression differs in its dependence on divalent cations and their concentrations. Kay (1952) had found that T 5 development was dependent on relatively high Ca^{2+} concentrations although higher concentrations of Mg^{2+} or Sr^{2+} could substitute for Ca^{2+}. Subsequently, Moyer and Buchanan (1970a,b) demonstrated that $10^{-4}M$ Ca^{2+} allowed normal class I but only very little class II and no class III RNA synthesis. Class II proteins could not be detected under these conditions. It is not clear whether Ca^{2+} acts directly or indirectly on transcription. All viral DNA is delivered from the phage to the bacterium so that the block is not at the level of SST. Nevertheless, it has been suggested that the low Ca^{2+} defect may involve failure to allow phage DNA to penetrate into the cytoplasm (Herman and Moyer, 1974, 1975).

5.3.2a. Aborted Early Gene Activity in a Nonpermissive Host

E. coli carrying the wild-type *col* Ib plasmid are nonpermissive for T 5 and BF 23 phage (Strobel and Nomura, 1966; Nisioka and Ozeki, 1968; Moyer *et al.*, 1972). The nonpermissive infection progresses as follows: (1) The entire phage genome is injected into the bacterial cytoplasm, class I RNA and proteins are synthesized, and host DNA is degraded as in the productive infection of *col* Ib-less bacteria. (2) All RNA and protein synthesis ceases 6–8 min after infection (at 37°C), but infected phage DNA is not degraded. (3) Cells lyse prematurely, about 15 min after infection (Nisioka and Ozeki, 1968; Moyer *et al.*, 1972). The restriction of T 5 and BF 23 phage results from the action of viral and plasmid genes. Host range (*h*) mutants of these phages have been isolated. The *h* mutations, which overcome the *col* Ib⁺-specific block to viral multiplication, are located in gene A3 (Table 7; Mizobuchi *et al.*, 1971; Beckman *et al.*, 1972). *h* mutants fail to synthesize a small, pre-early protein (Szabo *et al.*, 1975).

There is a clear resemblance between F plasmid-dependent restriction of T 7 and *col* Ib-dependent restriction of T 5 and BF 23. In both cases, there is a complete cessation of macromolecular synthesis soon after infection, within a few minutes of the switch from one to another class of viral gene expression. T 5 and BF 23 class II genes are either expressed very transiently or not at all (Moyer *et al.*, 1972; Mizobuchi and McCorquodale, 1974; Herman and Moyer, 1975). It is therefore possible that the nonpermissiveness might reflect a transcriptional

---

* See also Notes Added in Proof, p. 196, Note 7.

defect. RNA polymerase is not inactivated in the nonpermissive host (Szabo and Moyer, 1975), but a transcriptional model which ascribes abortive development to a failure of DNA ligation has been offered (Herman and Moyer, 1975). With one eye on the T 7–F$^+$ incompatibility, which we have already analyzed (Section 2.6.1), one might object that the observations which motivate this theorizing only concern temporal coincidences and that the latter can be misleading. Indeed, if the T 5–BF 23 defect is transcriptional, then viral protein synthesis should continue for a few minutes after the cessation of transcription, and this appears not to be the case (Moyer *et al.*, 1972). Instead, it seems entirely possible that the *col* Ib$^+$–T 5 incompatibility involves a general, nonspecific shutdown of macromolecular synthesis that results from a membrane defect. McCorquodale (1975) refers to unpublished observations that T 5-infected *col* Ib$^+$ bacteria, like T 7-infected F$^+$ bacteria, continuously leak low-molecular-weight substances.

5.3.3. Class III (Late) Gene Expression

The late genes code for many of the T 5 proteins. At least 17 late proteins can be resolved by one-dimensional gel electrophoresis. These proteins range in molecular weight from 140,000 to 15,000 and account for an appreciable fraction of the coding capacity of the genome; they include an endolysin and at least 10 proteins which are a part of the virion or are required for phage morphogenesis (Zweig and Cummings, 1973*b*; Chinnadurai and McCorquodale, 1974*a*). Some of these proteins are also required for the continuation of DNA replication, and T 5 replication is, in fact, continuously dependent on protein synthesis (Crawford, 1959; Hendrickson and McCorquodale, 1972) like SPO 1 DNA synthesis (Levner and Cozzarelli, 1972) but unlike T 4 or T 7 DNA synthesis. The late genes are activated approximately 9–10 min after infection, after the start of T 5 DNA replication. Their transcription is entirely from one strand (Hendrickson and Bujard, 1973; Hayward and Smith, 1973; C. Chen, 1975; see Section 5.3.4) and includes very large transcripts (molecular weights 2.3 and 2 × 10^6; Hayward and Smith, 1973). The induction of late RNA synthesis depends on the activity of two class II genes—C2 and D15—but is not absolutely dependent on replication, although less class III protein is made when DNA replication does not occur at all (Pispa *et al.*, 1971; Hendrickson and McCorquodale, 1971). If the gene C2 protein is inactive, class III proteins and RNA are not made, and all other (class II) protein synthesis ceases (Chinnadurai and McCorquodale, 1974*a* and personal communication). Surprisingly, some RNA synthesis and a considerably diminished DNA synthesis are sustained. There is no phage DNA breakdown, and cells do not leak their nucleotide pools. The gene C2 protein (molecular weight 90,000) binds to RNA polymerase (Szabo *et al.*, 1975), and it is therefore natural to suppose that the regulation of class III gene expression is at the transcriptional level. Yet, transcriptional control does not account for the total phenotype of C2 mutants. The general shutoff of protein synthesis may be a secondary consequence of the failure to make late proteins and late RNA. Perhaps (the analogy is, of course, with T 4 gene 32 regulation, Section 4.2.4) the shutoff of protein synthesis involves the overproduction of a nucleic acid binding protein which affects translation.

Although T 5 late gene expression is not absolutely dependent on initiation or concurrent continuation of DNA replication, it is dependent on the activity of a protein which probably "processes" DNA for transcriptional competence (Chinnadurai and McCorquodale, 1973). The early gene D15 codes for a nuclease which plays a central role in the regulation of class III transcription and also, perhaps indirectly, affects DNA replication. The replication phenotype of a gene D15 mutant is the following: DNA synthesis starts at the normal time at about 8 min after infection at 37°C, proceeds with almost normal rate for 10 min, then slows down and stops prematurely ("DNA arrest" phenotype; Hendrickson and McCorquodale, 1972). At the starting time for normal class III transcription, D15⁻ DNA is replicated at the wild-type rate. Normal T 5 DNA replication involves the synthesis of concatemeric T 5 DNA (form A). Approximately mature-size DNA intermediates (form B) are cleaved from this precursor and subsequently converted to mature virion DNA. Form A and B DNA contain single-stranded regions as judged by their binding to nitrocellulose. In mature T 5 DNA these gaps are filled mainly with DNA but perhaps partly with RNA, which has been found in mature T 5 DNA (Rosenkranz, 1973; compare T 4, Section 4.6.3d). A defect in gene D15 blocks the conversion of type A DNA to type B and mature DNA (Frenkel and Richardson, 1971*b*; Carrington and Lunt, 1973).

The gene D15 product is a well-characterized 5′ exonuclease, which degrades single- or double-stranded DNA to deoxyribonucleoside 5′ triphosphates and acid-soluble 5′ phosphoryl-terminated oligonucleotides of various lengths (Paul and Lehman, 1966; Frenkel and Richardson, 1971*a,b*). Gene product D15 action on replicating T 5 DNA probably involves the introduction of specific gaps into concatemeric DNA (form A) at specific sites as a prerequisite for cleavage to form B. Form A DNA probably also contains other single-stranded regions, for example, at replicative forks and recombination sites (Carrington and Lunt, 1973).

The regulatory phenotype of gene D15 mutants is that class III protein synthesis is severely defective but that class IIb proteins continue to be synthesized and that some class IIa proteins fail to shut off (see below). It seems likely that this phenotype, at least with respect to class III transcription, reflects a defect of transcriptional regulation; it has been discussed that way by others, and we do likewise (Chinnadurai and McCorquodale, 1973, 1974*a*; Szabo *et al.*, 1975). However, the appropriate direct analysis of class III transcripts has not yet been reported, and other models could be constructed.

What template processing for late RNA synthesis does gp D15 provide? Bearing in mind that it is a nuclease and recalling the information and theorizing about T 4 true late gene expression (Section 4.6), the processing should involve nicks or breaks in T 5 DNA. The genetic and transcriptional maps of T 5 (Figs. 10 and 11 suggest that this processing might occur at interior locations of T 5 DNA, not at the ends. Since gp D15 is also required for cleavage of concatemeric DNA, it either is associated in its activity with one or more endonucleases or possesses its own specific and as-yet uncharacterized endonuclease activity. A large number of interesting experiments on the function of gp D15 in class III transcription seem accessible both *in vivo* and *in vitro* (but, see below) since a special template, a processing enzyme, and a modified RNA polymerase have all been identified. Such experiments will hopefully widen our understanding of template-processing-dependent transcription.

Two other effects caused by D15⁻ infection should be mentioned: (1) Three of the class IIa proteins are not shut down after 25 min but are continuously made until lysis (Chinnadurai and McCorquodale, 1974*a*). One of these proteins, the gene D5 product, is particularly interesting because it is the one responsible for the shut-down of the other class IIa proteins (Chinnadurai and McCorquodale, 1974*b*). It may be that all the class IIa proteins are not coordinately repressed and that these three proteins require a class III function for their shut-down. (2) One late protein which plays a role in phage head assembly is also made in D15⁻ infection. Its synthesis, however, is delayed and reduced, resembling the synthesis of late T 4 proteins under conditions that uncouple late transcription from replication (Section 4.6.3c).

5.3.4. The T 5 Transcriptional Map

The single-strand breaks of T 5 DNA permit the separation of discrete segments of the genome. These segments, which are nature's modest substitute for restriction enzyme fragments, have been used to map the transcription of the T 5 genome (Hendrickson and Bujard, 1973; Hayward and Smith, 1973; C. Chen, 1975). The map presented in Fig. 11 has the following major features: (1) Both DNA strands contain template segments for transcription, and (2) it is not possible to construct a transcription map without several strand "switches."

The FST, class I genes are read from both strands but predominantly from the intact H strand. [The designations H and L derive from experiments with completely ligated T 5 DNA (Bujard and Hendrickson, 1973). The naturally intact strand of this DNA complexes poly(U,G) or polyG and hence is denser, or "heavy" (H), in poly(U,G)-CsCl fractionation than the ligated complementary strand which is designated as L. An alternative nomenclature has strand H designated as l and its complement as r]. Almost 90% of the class I RNA hybridizes to the terminally redundant portion of the H(l) strand. The remainder hybridizes to the complementary L(r) stand and, more specifically, to the DNA fragment bounded by breaks n_1 and n_2 as well as to the large fragment bounded at its 3' end by break n_5 (Hendrickson and Bujard, 1973; C. Chen, 1975). Class II RNA hybridizes to both strands. It hybridizes only to the H(l) strand of segment 2 but to both strands of segment 3. Two-thirds to three-quarters of class II RNA hybridization is to the H(l) strand. A separate analysis of class IIa and class IIb RNA has not been done. Indeed, the transcriptional control of the shutoff of class IIa proteins is barely analyzed, as we have already mentioned in Section 5.3.2. Class III RNA hybridizes only to strand H(l) in segment 3. The location of the late genes (Fig. 10) places class III transcripts at the right side of the genome. The class II transcripts are arranged to minimize strand switching in transcription. This consideration is arbitrary, as we have explained, and may also be gratuitous since the number of T 5 promoters probably is large (Schäfer *et al.*, 1973*b*; Schäfer and Zillig, 1973*b*).

5.4. RNA Polymerase and Transcriptional Regulation

Throughout its development, T 5 RNA synthesis shares the host's sensitivity to inhibition by rifamycin. Consequently, at least the β-subunit and, in all probability, all of the host RNA polymerase core are used for the transcription of all essential viral genes (Beckman *et al.*, 1972). In view of the relatively elaborate program of T 5 gene expression, it is reassuring to know that at least three T 5-induced proteins bind to RNA polymerase. One of these is a class I protein (molecular weight 11,000) and two are class II proteins (molecular weights 15,000 and 90,000). The 90,000 molecular weight protein is tentatively identified as the gene C2 product, and gene C2 regulates class III gene expression (Section 5.3.3). Of these three proteins only the 15,000 molecular weight subunit binds tightly; the others separate during enzyme purification. *E. coli* σ is not inactivated after T 5 infection and copurifies with RNA polymerase from T 5-infected cells (Szabo *et al.*, 1975; Szabo and Moyer, 1975).

Since gene C2 positively regulates class III gene expression, the 90,000 molecular weight protein has been called a T 5 σ factor. There is as yet no justification for such a designation: (1) *E. coli* σ copurifies with RNA polymerase after T 5 infection, and the C2 protein is lost; (2) the C2 protein has not been shown to affect RNA polymerase function in general or the formation of rapidly initiating complexes of polymerase with DNA in particular; (3) class III transcription might not require a σ-like initiation factor. The analysis of T 5 transcription *in vitro* with these materials is only just beginning, and no decisive results are at hand: It is known that RNA polymerase from infected cells with or without the 15,000 molecular weight subunit, but containing σ, has the same T 5 transcription specificity as *E. coli* holoenzyme (Szabo and Moyer, 1975).

As a matter of fact, the *in vitro* transcription of mature T 5 DNA by unmodified *E. coli* RNA polymerase poses a practical problem for the analysis of T 5 transcriptional regulation *in vitro*, which must be dealt with before progress can be made in the analysis of T 5 regulation. T 5 DNA is an excellent template for RNA polymerase holoenzyme. There are numerous initiation sites for RNA synthesis that are recognized by the enzyme. The number of these is between 35 (Bautz and Bautz, 1970) and 14 (Schäfer *et al.*, 1973*a*). [The conditions used for counting γ-^{32}P 5′ mRNA ends in the presence of rifampicin (Bautz and Bautz, 1970) may come close to the true number of initiation sites, whereas the method employed by Schäfer *et al.* may underestimate it.] These initiation sites probably do not include the single strand breaks; randomly placed single strand breaks are binding but not initiation sites for holopolymerase on T 7 DNA (Dausse *et al.*, 1972; Hinkle *et al.*, 1972). Of course, the problem of assessing the role of T 5 breaks is that they are not randomly placed. We do not know what the effects of specifically placed breaks at specific nucleotides in the initiation site would be. In Section 4.6.3, we speculated that they might promote initiation and might change the dependence on σ. On the other hand, specifically placed breaks near, but not at, the initiation site might immobilize RNA polymerase at locations that would block initiation by free enzyme. *In vitro* transcription and formation of stable initiation complexes with ligated T 5 DNA has not shown significant differences as compared with natural T 5 DNA (Knopf and Bujard, 1975; Schäfer *et al.*, 1973*b*).

The major problem of this transcription is that it yields class I, II, and III RNA *in vitro* (Bautz *et al.*, 1970; Pispa and Buchanan, 1971; Szabo and Moyer, 1975). It is true that the major component is class II RNA and that class III RNA constitutes only 10–20% of the whole. But the fraction of the T 5 genome that codes for late genes is probably not much greater than 20% (Fig. 11). When the promoters for T 5 *in vitro* RNA synthesis are located, and when T 5 protein synthesis in cell-free coupled systems is analyzed, the picture may become clearer and simpler. However, at this preliminary stage, T 5 (class III) late transcription resembles T 4 middle transcription in this practical respect: The minimal *in vitro* system of RNA polymerase and DNA makes T 5 RNA *in vitro* that is under positive control *in vivo*. In Section 4.2.6 we argued that the T 4 problem is related to the topography of the early and middle genes, which are interspersed, and to failures of RNA chain termination. That explanation may not fit T 5: (1) The class III genes are clustered on the (relatively incomplete) T 5 map; (2) T 5 *in vitro* transcripts are only about 4000 nucleotides long, and none are longer than 7000 nucleotides (Schäfer, 1972).

Finally, we wonder about the strategic significance of the stepwise transfer of the T 5 genome into the bacterial cytoplasm. Is the "aim" to create a two-step program of prereplicative gene expression without the kind of transcriptional regulation that SPO 1 or T 4 employ? In spite of the reservations expressed in the preceding paragraph, it seems likely that class II genes can, in fact, be transcribed by unmodified bacterial RNA polymerase. The stepwise transfer probably does therefore create a time sequence which would not otherwise exist. However, McCorquodale (1975) has instead argued that the primary strategic aim of the two-step transfer is to allow time for the elimination of one or more host restriction mechanisms before the major part of the viral genome is introduced: in a sense, the FST portion of T 5 is a helper (sub)phage of the whole. We have already considered that T 3 and T 7 also include antirestriction functions in their early genes.

6. TRANSCRIPTIONAL AUTARCHY

6.1. Introduction

The last two bacteriophages that we consider, N 4 and PBS 2, show distinctive and novel regulatory features that are not encompassed by the four preceding prototypes. N 4 is an *E. coli* phage. Its virions contain approximately 60 kb of double-stranded DNA enclosed in an isometric head to which a short, noncontractile tail is attached (Schito *et al.*, 1966*a*,*b*). N 4 is a virulent phage that kills its host. However, it can enjoy an extended period of intracellular accumulation during its multiplication cycle because of the lysis inhibition of its host (Pesce *et al.*, 1969; Schito, 1974). The physiological genetic analysis of this phage has recently been started with the isolation of a collection of amber mutants (Schito, 1973). PBS 2 is a *B. subtilis* phage with one of the largest bacterial viral genomes. Its approximately 280 kb of double-stranded DNA contain uracil in substitution for thymine (Hunter *et al.*, 1967). PBS 2 is remarkable in being a clear plaque variant of a temperate phage, PBS 1. No other temperate phage with an unusual nucleotide is known, and the nucleotide constitution of proviral PBS 1 DNA has not yet been determined. It is conceivable that a T-containing state of PBS 1 genomes might exist and that differential PBS 1 gene expression might depend on the T vs. U content of viral and proviral DNA. Such a situation would resemble, at least formally, the specific requirement of T 4 late gene expression for hmC-containing DNA (Section 4.6.3).

6.2. Selective Effect of N 4 Infection on *E. coli* Catabolite-Sensitive Genes

The synthesis of *E. coli* catabolic enzymes is generally repressed when the bacteria utilize glucose as their carbon source. This regulation is mediated by cAMP, which controls the transcription of these catabolite-sensitive genes through its interaction with a receptor protein (CAP). Transcription of the catabolite-sensitive genes requires cAMP-CAP (Section 1.1.2a). When phage N 4 infect *E. coli* K12, host DNA is not degraded and most host RNA and protein synthesis continues until the postreplicative or late phase of viral development. However, catabolite-sensitive operons, such as the lactose operon, are selectively shut off very soon after infection. Promoter mutations which remove the lactose operon from normal, cAMP-CAP-mediated positive control also make β-galactosidase synthesis resistant to rapid, phage N 4-mediated shutoff after infection. Addition of sufficient cAMP to the culture medium blocks the rapid shutting off of β-galactosidase synthesis. Evidently a function associated with phage N 4 infection reduces the concentration of cAMP and/or the activity of cAMP-CAP and in this way achieves a selective regulation of host gene expression

(Rothman-Denes *et al.,* 1972; Rothman-Denes, 1975). The detailed mechanism of this selective host shutoff remains to be worked out. Neither is the significance of phage-mediated regulation of catabolite-sensitive bacterial genes clear. The selective advantages that bacteria and phage derive from shutting off catabolite-sensitive genes must differ. For the bacterial host cell, such regulation spares a fraction of protein synthesis under conditions that allow very rapid multiplication. One suspects that this cannot be the sole advantage that the developing phage derives during its lytic cycle; the primary target of this regulatory phenomenon perhaps remains to be discovered. A somewhat similar repression of catabolite-sensitive genes is also known to occur during the lytic development of phage λ (Echols, 1974).

6.3. Novel RNA Polymerase Activities of N 4- and PBS 2-Infected Bacteria

The transcription systems that initially act on all the viruses that we have so far discussed involve the bacterial host's major RNA polymerase, whose β-subunit normally binds, and is inhibited by, rifampicin. Phage PBS 2 and subsequently N 4 have been shown to follow different and novel strategies of viral transcription (Price and Frabotta, 1972; Rothman-Denes and Schito, 1974). All the phages that we have considered thus far initially utilize the major bacterial RNA polymerase for their own early gene expression. In fact, the hmU phages of *B. subtilis,* the T-even and T 5 phages of *E. coli,* and many others utilize the major bacterial RNA polymerase, with modifications, throughout infection. Phage PBS 2 transcription is rifampicin resistant (rif^r) in rifampicin-sensitive (rif^s) bacteria throughout infection. Phage PBS 2 must therefore execute the transcription of all its essential genes without the bacterial RNA polymerase β-subunit. As an *in vitro* corollary, no PBS 2 transcription units appear to be well read by *B. subtilis* RNA polymerase *in vitro* (Clark *et al.,* 1974). Selective incompatibility of a genome with the host's transcription system is, of course, basic to virus-mediated positive control of gene action; what is special about PBS 2 DNA is the degree of its incompatibility. PBS 2 gene expression must surely rely on one of two solutions to this problem of genetic incompatibility. Both solutions involve the utilization of preformed transcribing activities that are distinct from bacterial RNA polymerase: (1) The phage might recruit RNA polymerizing activities of the host, which are not normally used for mRNA production

(Bouché *et al.*, 1975), for making at least some viral mRNA. (2) The phage might introduce its own polymerase(s) or its own substitute(s) for the β-subunit of the host RNA polymerase.

A partly comparable situation exists in phage N 4-infected *E. coli* and has been analyzed in somewhat more detail (Rothman-Denes and Schito, 1974). N 4-infected bacteria manifest rifampicin- and strep-tolydigin-resistant transcription activity *in vivo* (in sensitive bacteria) which is selective for N 4 genomes and is evident in cells that are prevented from synthesizing any protein after infection. This transcrib-ing activity is therefore preformed. Experiments with a *ts* mutant of N 4 strongly favor the notion that the virion carries a component of the preformed activity and injects it upon infection (the second of the above alternatives; Falco and Rothman-Denes, 1975*b*).*

The program of transcription and protein synthesis during N 4 development has been partly worked out. At the transcriptional level two classes of RNA can be distinguished (van der Laan and Rothman-Denes, 1975): N 4 Class I RNA synthesis is rifampicin resistant and does not depend on protein synthesis after infection, i.e., it is chloram-phenicol resistant. This RNA synthesis probably is dependent on a virion protein, since it is continuously temperature-sensitive, in the absence of protein synthesis, in bacteria infected with a phage *ts* mutant (*ts*213) in one of the viral "early" genes (Falco and Rothman-Denes, 1975*b*). Class II RNA synthesis is dependent on protein synthesis after infection and on the activity of two "early" phage genes. No viral transcripts that are made exclusively with bacterial RNA polymerase have been detected by hybridization-competition analysis. The transcribing activity for class II viral transcripts appears to be located on the *E. coli* inner membrane (Falco and Rothman-Denes, 1975*a*). N 4 DNA replication is not inhibited by rifampicin (Rothman-Denes and Schito, 1974).

However, the relationship of RNA to protein synthesis appears to involve some complicating factors. Three classes of N 4 proteins can be distinguished (Rothman-Denes and Schito, 1974). The earliest-appear-ing class of proteins (I) is probably made off class I RNA as template; the middle and late classes of proteins (II and III) are probably made from class II mRNA. Yet the synthesis of class III proteins (which also depends on N 4 DNA replication) is blocked by rifampicin and is sensi-tive to the thermal stability of RNA polymerase β- and β'-subunits (Rothman-Denes and Schito, 1974; Schito, 1975; Rothman-Denes, 1975). One can speculate on a variety of ways in which this anomalous

* See also Notes Added in Proof, p. 196, Note 8.

relationship between transcription and protein synthesis for certain viral genes might arise. The experiments to resolve this quandary remain to be done, but it appears possible that some N 4 genes may be regulated posttranscriptionally.

A rifampicin-resistant DNA-dependent RNA polymerase from phage PBS 2-infected *B. subtilis* (Price *et al.*, 1974) has been purified (Clark *et al.*, 1974). It is a multisubunit enzyme, but none of its subunits are taken from the bacterial RNA polymerase. The enzyme, which has been isolated late after virus infection, preferentially transcribes PBS 2 DNA and poly(dA-dT). It transcribes other *B. subtilis* phage DNA and bacterial DNA relatively poorly. The timing of synthesis of this enzyme suggests that it is not the preformed rifampicin-resistant RNA polymerase activity (Clark, 1975). The nature of the initial rifampicin-resistant transcribing activity therefore remains to be determined.

The peculiarities of RNA synthesis during infection by the lytic DNA phages N 4 and PBS 2 raise questions that have been appreciated for some time, particularly in animal virology (Kates and McAuslan, 1967). Several kinds of viruses must overcome a basic genetic incompatibility with their hosts: an inability to generate viral mRNA. The origins of this incompatibility are diverse and include the packaging of anti-mRNA or double-stranded RNA in the virion, as in the myxo- and reoviruses, replication of an RNA virus through a DNA intermediate, as in the RNA tumor viruses, and cytoplasmic multiplication of a DNA virus, as in the pox viruses (for review, see Baltimore, 1971). The solution in all instances is to package in the virion and to introduce into the cell upon infection those enzymes, be they transcriptases or reverse transcriptases, that repair the genetic incompatibility. Without these accessory protein components, the viral nucleic acid is not infectious. The recently discovered *pseudomonas* phage, $\phi6$, which contains double-stranded RNA, apparently fits into this framework since it contains an RNA polymerase activity in the virion (van Etten *et al.*, 1973).

On the other hand, N 4 and PBS 2 are double-stranded DNA phages. For both of these phages, the incompatibility between virus and host may reside in promoters of viral transcription units that are all unrecognized by bacterial RNA polymerase. In the face of this total or almost total incompatibility, these DNA phages establish transcriptional autarchy. In the case of N 4, the viral and host transcriptional systems work side by side during at least a substantial part of the phage developmental cycle.

7. CONCLUDING COMMENTS

In these final pages, we reemphasize certain common features of the development of the five groups of bacterial viruses that have been the objects of this analysis of regulated gene action. Our comments are organized under five headings.

7.1. Lytic Phages Establish Diverse Regulatory and Metabolic Relationships with Their Hosts

The lytic phages lead their hosts to the common end of lysis and progeny phage liberation along diverse and characteristic paths. On the way to that common end, different phages shut off host macromolecular synthesis to varying degrees and at different rates and degrade the host genome to different extents. Host components involved in transcription, translation, and membrane functions are more or less extensively altered or replaced. The phages kill their hosts in many different ways individually as well as collectively. For example, T 7 encodes several functions (e.g., the gp 0.7 protein kinase, endolysin, and nucleases), each of which ensures cell death. The T-even phages also encode numerous cell-killing functions. They shut off host-DNA synthesis very quickly and affect the cell membrane after attachment.* Other phages, like the hmU-containing *B. subtilis* phages and N 4, shut down host-DNA synthesis more slowly.

Degradation of DNA is not a ubiquitous concomitant of lytic phage development; neither is it a prerequisite for inactivation of genes. For example, the *B. subtilis* DNA from hmU phage-infected bacteria retains transforming ability, although most bacterial genes are shut off rapidly after normal phage infection. In T-phage infection, the breakdown of bacterial DNA to nucleotides does not start until several minutes after the shutting off of gene expression, and much of the deoxyribonucleotides generated by the breakdown is incorporated into viral progeny DNA. The nucleotides generated by the complete digestion of *E. coli* DNA after T 7 and T 3 infection constitute the major

---

* T-even phage ghosts are themselves capable of shutting off *E. coli* DNA replication, in part because they cause a leakage of nucleotides from the cytoplasm. After infection by intact phage these leaks are repaired, but other viral functions intervene to sustain the shut-off or viral replication. The cessation of replication after attachment of phage ghosts may also result from dissipation of the membrane-associated high-energy state ("∼"). This is *separately* required for some step, possibly at the substrate level, of *E. coli* DNA replication (Mevel-Ninio and Valentine, 1975).

source of precursors of viral DNA. In striking contrast, the nucleotides generated by the T 5-induced degradation of *E. coli* DNA are excreted and are not used as precursors of T 5 progeny DNA.

Host messenger RNA and protein synthesis is shut off rapidly after infection by most of the phages that we have discussed. The shutoff mechanisms certainly involve inhibition of transcription and, in at least some instances, translation. In the case of the T-even phages, ADP-ribosylation of the α-subunits of RNA polymerase probably shuts off host RNA synthesis selectively. An ADPR transferase activity is injected into the T 4-infected cell so that polymerase alteration can begin before the synthesis of any viral protein chains has been completed. Infection by phage N 4 provides an example of the selective shutoff of a specific set of genes, the catabolite-sensitive genes. In the case of the hmU phages, we observe the selective sparing of the few bacterial transcription units that code for stable RNA. The mechanism of this sparing is unknown but probably has important implications for the mechanism of regulation of stable RNA transcription.

The following pattern regarding RNA polymerase modification emerges from the already accumulated information and may turn out to be significant for our understanding of how RNA polymerase interacts with promoters: Host transcription shutoff can be mediated by chemical modification of RNA polymerase core, or by phage-specified RNA polymerase subunits; positive regulation of phage transcription requires phage-specified RNA polymerases or polymerase subunits and is not effected by chemical modification alone.

Many bacteriophages change properties of the host-cell membrane, presumably by introducing virus-specific proteins either as peripheral or as integral components of the membrane. A frequent effect of these membrane changes is the exclusion of superinfecting phage, although a variety of other superinfection-exclusion mechanisms exist. Interactions of virus-specific and host proteins also must be responsible for membrane defects that appear to be the cause of nonpermissiveness in several *E. coli* phage systems. The abortive infection of T 7 in F$^+$ *E. coli* (Section 2.6.1) and the nonpermissiveness of Col Ib$^+$ *E. coli* for T 5 (Section 5.3.2a) apparently involve membrane failure caused by interactions of host and viral proteins. Another example is provided by *E. coli* λ lysogens carrying the *rex* gene which are only permissive for T 4 phage carrying intact rIIA and B genes. In the absence of functional rII proteins, T 4-infected λ lysogens stop respiring several minutes after the onset of replication, cease to replicate DNA, and make almost no RNA or protein. The rII proteins interact with the cytoplasmic membrane (Takacz and Rosenbusch, 1975; Huang, 1975).

Host–virus interactions are so diverse and in many instances sufficiently complex that genetic analytical methods dealing specifically with these interactions are potentially of great scope. Isolated genetic facts about nonpermissiveness have been known for a long time. For example, Benzer's fine-structure genetic analysis of the T 4 rII genes, which forms a cornerstone of molecular genetics, relied on the existence of permissive and nonpermissive hosts. Only more recently have the genetics of nonpermissiveness been pursued systematically. One consequence has been the further analysis of the "nonessential" genes (which generally improve the exploitation of host resources for phage multiplication). Another consequence has been the identification of nonlethal host alterations which either interfere with specific steps of viral development or compensate for specific genetic defects of the phage (Section 4.6.3e). These interferences and compensations affect regulation, replication, and assembly of the phages and attest to the complex adaptation of virus to host at the molecular level.

7.2. Some Phages Introduce Enzymes or Subunits of Multicomponent Enzyme Complexes into the Infected Cell

In their blender experiments on the role of DNA as the genetic material, Hershey and Chase (1952) showed that some protein was co-transferred to the infected cell with viral, T-even phage DNA. It was not until long after the genetic role of DNA was securely understood that the diversity of viruses was fully appreciated. Viruses in general, including DNA phages, introduce proteins together with nucleic acids into the cytoplasm of the infected cell. Many kinds of viruses sequester in the virion genetic material that, upon reintroduction into a cell, requires accompanying accessory enzymes for expression. It is conceptually important to have DNA phages (N 4, PBS 2) now included among such viruses, for it seems likely that the incompatibility of their genetic material with the cellular apparatus for gene expression includes incompatibilities of regulatory sites. Undoubtedly, the injected proteins of the DNA bacteriophages do not all have such easily dramatized roles. Injection of the internal proteins of the T-even phages, for example, probably is of little or no consequences for phage development. However, it is interesting that one protein injected along with T 4 DNA is an enzyme capable of modifying the bacterial RNA polymerase α-subunit (Section 4.3). The RNA polymerase modification effected by this enzyme probably is part of the process of host transcriptional shutoff.

7.3. The Propagation of Phages Involves Time-Ordered Sequences of Gene Expression

The time sequences of gene expression in phage development generally follow the blueprint of T 4, which was the first phage to be analyzed in these terms (Epstein *et al.,* 1963). In all the cases that we have discussed (Sections 2–6), two or three stages of gene expression— early, (middle), and late—can be distinguished. What might be the objective of this common strategy? As mentioned above, the first stage of viral development includes many kinds of host-cell inactivation. The targets of inactivation include host-cell activities that would otherwise prevent viral multiplication. These activities include restriction enzymes (T 7; Section 2.3.4) and other nucleases that might inactivate the invading genome (T 5; Section 5). T 5 exemplifies the rare strategy of delaying the entry of the major part of the viral genome until this preliminary stage of host-cell inactivation is completed. Most phages inject all of their DNA at once but activate few viral genes during the first stage of development. Two purposes might be served by the inactivity of large segments of these genomes during the very first minutes of development: (1) The yield of the earliest viral proteins would be increased by the silence of viral late genes, since active genes are a source of competition for the transcriptional and translational machinery of the cell; the inactivity of viral late genes serves somewhat the same purpose as the inactivation of host genome expression. (2) Genetic inactivity may, in some instances, be the protective device for, or a by-product of, keeping segments of viral DNA from attack by nucleases. Several viruses activate a second class of genes, the middle genes, a few minutes after infection. The products of these genes are prominently involved in DNA replication and recombination. The mechanism of activating middle genes in phage SPO 1 infection now seems to be well on the way to detailed biochemical analysis. It is mediated by one or more phage genome-coded proteins. At least one of these changes the DNA binding and RNA chain initiation specificity of the host RNA polymerase core (Section 3.3). In the case of T 4 and in other phages, however, the switch to middle gene expression probably is more complex (Sections 4.2.6b, 4.3 and 4.4).

Delayed assembly of phage is commonly (but not universally; see

Section 6.3) associated with transcriptional regulation of a third class of genes, the late genes. These genes code for many virion proteins, endolysins, and assembly functions. They are activated after the onset of viral DNA replication. Regulatory linkages of the synthesis of late proteins to DNA replication are common (Sections 3.2, 4.6, and 5.3.3) but are especially striking in the case of the T-even phages. The advantage of delayed virus assembly is generally surmised to be the accumulation of sufficient DNA for abundant gene expression and encapsidation, and the prevention of extremely precocious lysis.

Although the arrangement of gene expression in two or three phases is ubiquitous in these phages, we have emphasized (Section 4) that the regulation of T-even phage gene expression is much more elaborate than the simple early-middle-late stereotype. The quantity and variety of genetic regulation that is executed by this bacteriophage is striking. We have commented on the significance of this imposing complexity in Section 4.7.

7.4.   Regulation of Viral Gene Activity Is Predominantly at the Transcriptional Level

There is a striking predominance of transcriptional regulatory mechanisms in generating the large-scale developmental sequences of viral proliferation, such as early, middle, and late gene expression. With some variations that we have noted (Section 2.3.1 and 4.2.4) this regulated transcription generates relatively short-lived messages. The general nature of some of these regulatory mechanisms is now understood.

Posttranscriptional regulation of some viral gene expression in the development of DNA phages has just recently been clearly identified and analyzed (section 4.2.4); other potentially posttranscriptional features of regulation have been identified in Sections 2.6.2, 4.2.2, and 6.3.) The biochemical mechanisms, of course, remain to be worked out. At least a part of the time lag in identifying posttranscriptional mechanisms comes from their having been sought in an inappropriate context—that of the major "switches" of phage development. Instead, the clearly established case of posttranscriptional regulation in T 4 development involves a single gene and a stable message which codes for T 4 gp 32, the single-stranded DNA-binding protein (O'Farrell *et al.*, 1976). In this instance it is primarily message activity rather than lifetime that is regulated.

However, if translational regulation is relegated to minor roles in

DNA phage development, translational discrimination is not. The constitutive protein yields of genes are partly set at the translational level through different rates of translation. In at least one instance, discrimination does not involve differences of messenger lifetime (Section 4.2.4). The best studied instances of translational discrimination involve viruses that are not included among our prototypes (Section 1.2.3), but there is evidence that translational discrimination also produces widely varying yields of different T 7 proteins (Section 2.4.2). More instances of important translational discriminatory effects will doubtless be discovered and analyzed. We suspect that translational discrimination is a general feature of phage protein synthesis and that discriminatory diversity in translation is as important as the regulatory structure of transcription in determining the fitness of a virus to its host.

Transcriptional discrimination is also a feature of positively regulated viral gene expression, as it is for gene expression of the host. The clearest example of such discrimination that we have considered occurs in the case of T 7: The class II and class III genes of T 7 are regulated by the T 7 RNA polymerase. The class II genes yield fewer messages (and less protein) because they are transcribed less frequently, *in vitro* and *in vivo*. Evidently T 7 RNA polymerase discriminates at least two different kinds of promoters in the late region of the T 7 chromosome in the absence of any other protein factors.

Transcriptional regulation probably is a concomitant of certain features of genome structure that are common to the bacteriophages. Phage genomes and viral genomes generally are arranged as blocks of genes or transcription units with common regulatory features: early and late genes are clustered, for example (Figs. 4, 5, 10, and 11). Moreover, genes participating in common biochemical pathways (e.g., DNA replication) frequently are closely linked within these blocks. Indeed, it has been pointed out (Botstein and Suskind, 1974) that organization of the genomes of temperate bacteriophages of *Escherichia coli* and *Salmonella typhimurium* is extremely highly conserved. Regulatory coherence and transcriptional regulation may help to ensure the stability of genome organization.

7.5. The More-or-Less Common Time-Ordered Sequences of Transcription Are Generated by a Remarkable Variety of Control Mechanisms

The common features of early, middle, and late stages of viral development and of viral gene expression are generated in the prototypes that we have discussed by a remarkably wide range of mechanisms. In T 3, T 7, and PBS 2, new RNA polymerases are synthesized. The hmU phages make several proteins which bind to RNA polymerase. One or two of these proteins act precisely as template- and transcription unit-specific determinants of DNA binding and RNA chain initiation. In the T-even phages, late transcription also involves positive control by RNA polymerase-binding proteins, but we have presented arguments that these proteins most probably act by mechanisms that differ from those of the hmU phage prototype. Phages also distinguish their own transcription from that of the host by incorporating peculiar bases (hmU, hmC) into their DNA. These special templates are selectively transcribed by phage-modified polymerases (Sections 3.3 and 4.6.3c). The nature of this selectivity may differ in different phage systems. Recent genetic analysis of T 4 promises to lead us to a better understanding of the role of the modified nucleotide, hmC, in genetic regulation. The T-even phages further complicate regulation by making their true late transcription dependent on "processing" of the template, which usually occurs during DNA replication. It is already evident that there is no unique or exclusive biochemical solution to the biological problem of generating a sequence of viral transcription. We suspect that the catalog of positive regulatory mechanisms in phage development is still incomplete. Indeed, it will be surprising if the phages do not collectively exploit all the mechanisms of positive gene regulation of the uninfected bacterial cell. We anticipate that an example of positive regulation of viral genes by DNA-binding proteins is likely to be found in lytic phages; examples are already at hand for temperate phages (Section 1.1.2a). It also is likely that positive regulation by antitermination, comparable with regulation by gp N of phage λ, will also be found to occur in the regulation of some lytic DNA phage. It had previously been thought that the middle gene regulation of T 4 would be the prototype of antitermination transcriptional control. In Section 4 we have presented the reasons for considering this possibility plausible in principle but less likely to be actually in effect.

In one important respect, the regulatory events of phage development differ from the regulatory events of vegetative bacteria, and this might be thought to imply a divergence of regulatory mechanisms. Regulation of gene expression in vegetative bacteria is (irregularly) reversible: Catabolite gene expression, for example, must be turned on and off as the nutrient state of the cell demands. Reversible regulation involves small molecules as effectors of positive and negative transcriptional regulation (Section 1.1.2). At this time, one knows of no positive regulation of gene expression in phage development which is mediated by a small molecular effector. However, phage λ utilizes a repressor for regulating viral gene expression in a situation that is not strictly reversible. The λ repressor (and perhaps the P 22 repressor also) are probably not capable of cyclic reuse except in special laboratory-created situations. The λ repressor may not have a small molecular effector (Sussman and Zeev, 1975). Instead, operator activation is accompanied by proteolytic cleavage of the repressor into inactive fragments (Roberts and Roberts, 1975). One can regard the λ repressor as a prototype of a DNA-binding protein which is a negative and nonreversible regulator of transcription.

If this emphasis on the equivalence of cellular and viral transcriptional regulation is valid, then there is no reason why the regulatory mechanisms that are considered unique to the phages should not also be found operating in procaryotic differentiation and in vegetative regulation. There is still no *a priori* reason why ν^{28}-like initiation factors or a transcription unit-specific RNA polymerase, among others, should not be found as positive regulatory elements of transcription in uninfected bacteria. If either the transcription unit-specific initiation factor or the transcription unit-specific RNA polymerase is to be used for positive regulation concurrently with general transcription by one or more general polymerases, the following conditions probably must be met: (1) The differential affinity of these regulatory components for specific sites on the template must be sufficient to prevent sequestering of the regulatory components by nonspecific interactions. (2) The transcription unit-specific factors must be prevented from acting as general repressors of transcription.

In discussing hmU and T-even phage regulation, we inferred that different and competing determinants of RNA chain-initiation specificity coexist, and that they compete with each other or function concurrently for varying periods of time at different stages of different phage developments. Perhaps comparable, but more complex, situations exist in the cells of higher eucaryotes whose purified RNA polymerases have a remarkably large numer of small subunits (Chamberlin and Losick, 1976). Perhaps these enzymes represent families of RNA polymerases with identical basic subunits and varying additional subunits exhibiting different transcription specificities.

8. REFERENCES

Abelson, J., and Thomas, C. A., Jr., 1966, The anatomy of the T 5 bacteriophage DNA molecule, *J. Mol. Biol.* **18,** 262.
Abelson, J., Fukada, K., Johnson, P., Lamfrom, H., Nierlich, D. P., Otsuka, A., Paddock, G. V., Pinkerton, T. C., Sarabhai, A., Stahl, S., Wilson, J. H., and Yesian, H., 1975, Bacteriophage T 4 tRNAs: Structure, genetics, and biosynthesis, *in* "Process-

ing of RNA" (J. J. Dunn, ed.), pp. 77–88, Brookhaven National Laboratory, New York.

Adams, M. H., 1959, "Bacteriophages," Interscience, New York.

Adesnik, M., and Levinthal, C., 1970, RNA metabolism in T 4-infected *Escherichia coli, J. Mol. Biol.* **48,** 187.

Adhya, S., Gottesman, M., and de Crombrugghe, B., 1974, Release of polarity in *Escherichia coli* by gene *N* of phage λ: Termination and antitermination of transcription, *Proc. Nat. Acad. Sci. USA* **71,** 2534.

Adhya, S., Gottesman, M., de Crombrugghe, B., and Court, D., 1975, λ *N* function, transcription termination and polarity. Abstract, Meeting on "RNA Polymerases," Cold Spring Harbor, August 1975.

Alberts, B. M., and Frey, L., 1970, T 4 bacteriophage gene 32: A structural protein in the replication and recombination of DNA, *Nature (London)* **227,** 1313.

Alberts, B. M., Barry, J., Hama-Inaba, H., Moran, L., and Mace, D., 1973, Some proteins of the T 4 bacteriophage DNA replication apparatus. Abstracts, Ninth International Congress of Biochemistry, Stockholm, p. 126.

Alberts, B. M., Morris, C. F., Mace, D., Sinha, N., Bittner, M., and Moran, L., 1975, Reconstruction of the T 4 bacteriophage DNA replication apparatus from purified components, *in* "DNA Synthesis and Its Regulation" (M. M. Goulian, P. C. Hanawalt, and C. F. Fox, eds.), pp. 241–269, W. A. Benjamin, Inc., Menlo Park.

Allet, B., and Solem, R., 1974, Separation and analysis of promoter sites in bacteriophage lambda DNA by specific endonucleases, *J. Mol. Biol.* **85,** 475.

Alpers, D. H., and Tomkins, G. M., 1966, Sequential transcription of the genes of the lactose operon and its regulation by protein synthesis, *J. Biol. Chem.* **241,** 4434.

Altman, S., and Lerman, L. S., 1970, Kinetics and intermediates in the intracellular synthesis of bacteriophage T 4 deoxyribonucleic acid, *J. Mol. Biol.* **50,** 235.

Altman, S., and Meselson, M., 1970, A T 4-induced endonuclease which attacks T 4 DNA, *Proc. Nat. Acad. Sci. USA* **66,** 716.

Altman, S., and Smith, J. D., 1971, Tyrosine tRNA precursor molecule polynucleotide sequence, *Nature (London) New Biol.* **233,** 35.

Anderson, D., 1973, personal communication.

Anderson, W. B., Gottesman, M. E., Pastan, I., 1974, Studies with cyclic adenosine monophosphate receptor and stimulation of *in vitro* transcription of the *gal* operon, *J. Biol. Chem.* **249,** 3592.

Añon, M. C., 1974, Desarollo del Bacteriofage SPO 1: Su Control. Ph.D. Thesis, Universidad Nacional de La Plata (Argentina).

Añon, M. C., and Grau, O., 1974, personal communication.

Apirion, D., 1973, Degradation of RNA in *Escherichia coli*: A hypothesis, *Mol. Gen. Genet.* **122,** 313.

Apirion, D., 1975, The fate of mRNA and rRNA in *Escherichia coli, in* "Processing of RNA" (J. J. Dunn, ed.), pp. 286–306, Brookhaven National Laboratory, New York.

Armelin, M. C. S., 1973, Cell-free model system of SPO 1 phage transcription—further characterization, M.S. Thesis, University of California at San Diego.

Avila, J., Hermoso, J. M., Viñuela, E., and Salas, M., 1970, Subunit composition of *B. subtilis* RNA polymerase, *Nature (London)* **226,** 1244.

Avila, J., Hermoso, J. M., Viñuela, E., and Salas, M., 1971, Purification and properties of DNA-dependent RNA polymerase from *Bacillus subtilis* vegetative cells, *Eur. J. Biochem.* **21,** 526.

Baguley, B. C., Berquist, P. L., and Ralph, R. K., 1967, Low-molecular-weight T 4 phage-specific RNA, *Biochim. Biophys. Acta* **138,** 51.

Baker, R., and Yanofsky, C., 1972, Transcription initiation frequency and translational yield for the tryptophan operon of *E. coli*, *J. Mol. Biol.* **69**, 89.

Baltimore, D., 1971, Expression of animal virus genomes, *Bact. Rev.* **35**, 235.

Barrell, B. G., Coulson, A. R., and McClain, H., 1973, Nucleotide sequence of a glycine transfer RNA coded by bacteriophage T 4, *FEBS (Fed. Eur. Biochem. Soc.) Lett.* **37**, 64.

Barrell, B. G., Seidman, J. G., Guthrie, C., and McClain, W. H., 1974, Transfer RNA biosynthesis: The nucleotide sequence of a precursor to serine and proline transfer RNAs, *Proc. Nat. Acad. Sci. USA* **71**, 413.

Barry, J., and Alberts, B., 1972, *In vitro* complementation as an assay for new proteins required for bacteriophage T 4 DNA replication: Purification of the complex specified by T 4 genes 44 and 62, *Proc. Nat. Acad. Sci. USA* **69**, 2717.

Barry, J., Hama-Inaba, H., Moran, L., Alberts, B., and Wiberg, J., 1973, Proteins of the T 4 bacteriophage replication apparatus, *in* "DNA Synthesis *in Vitro*" (R. D. Wells and R. B. Inman, eds.), pp. 195–214, University Park Press, Baltimore.

Bautz, E. K. F., and Bautz, F. A., 1970, Initiation of RNA synthesis: The function of σ in the binding of RNA polymerase to promoter sites, *Nature (London)* **226**, 1219.

Bautz, E. K. F., and Dunn, J. J., 1969, DNA-dependent RNA polymerase from phage T 4-infected *E. coli*: An enzyme missing a factor required for transcription of T 4 DNA, *Biochem. Biophys. Res. Commun.* **34**, 230.

Bautz, E. K. F., and Hall, B. D., 1962, The isolation of T 4-specific RNA on a DNA cellulose column, *Proc. Nat. Acad. Sci. USA* **48**, 401.

Bautz, E. K. F., Kasai, T., Reilly, E., and Bautz, F. A., 1966, Gene-specific mRNA. II. Regulation of mRNA synthesis in *E. coli* after infection with bacteriophage T 4, *Proc. Nat. Acad. Sci. USA* **55**, 1081.

Bautz, E. K. F., Bautz, F. A., and Dunn, J. J., 1969, *E. coli* σ factor: A positive control element in phage T 4 development, *Nature (London)* **223**, 1022.

Bautz, E. K. F., Dunn, J. J., Bautz, F. A., Schmidt, D. A., and Mazaitis, A. J., 1970, Initiation and regulation of transcription of RNA polymerase, *in* Lepetit Symposium, 1969, on "RNA Polymerase and Transcription" (L. Silvestri, ed.), pp. 90–109, North Holland Publishing Company, Amsterdam, London.

Beckendorf, S. K., Kim, J. S., and Lielausis, I., 1973, Structure of bacteriophage T 4 genes 37 and 38, *J. Mol. Biol.* **73**, 17.

Beckman, L. D., Hoffman, M. S., and McCorquodale, D. J., 1971, Pre-early proteins of bacteriophage T 5: Structure and function. *J. Mol. Biol.* **62**, 551.

Beckman, L. D., Witonsky, P., and McCorquodale, D. J., 1972, Effect of rifampicin on the growth of bacteriophage T 5, *J. Virol.* **10**, 179.

Beckman, L. D., Anderson, G. C., and McCorquodale, D. J., 1973, Arrangement on the chromosome of the known pre-early genes of bacteriophages T 5 and BF 23, *J. Virol.* **12**, 1191.

Behme, M. T., and Ebisuzaki, K., 1975, Characterization of a bacteriophage T 4 mutant lacking DNA-dependent ATPase, *J. Virol.* **15**, 50.

Beier, H., and Hausmann, R., 1973, Genetic map of bacteriophage T 3, *J. Virol.* **12**, 417.

Bendis, I. K., and Shapiro, L., 1973, Deoxyribonucleic acid-dependent ribonucleic acid polymerase of *Caulobacter crescentus*, *J. Bacteriol.* **115**, 848.

Benzer, S., 1953, Induced synthesis of enzymes in bacteria analyzed at the cellular level, *Biochim. Biophys. Acta.* **11**, 383.

Berger, H., Warren, A. J., and Fry, K. E., 1969, Variations in genetic recombination due to *amber* mutations in T 4D bacteriophage, *J. Virol.* **3**, 171.

Berger, T., and Bruner, R., 1973, Late gene function in bacteriophage T 4 in the absence of phage DNA replication, *J. Mol. Biol.* **74,** 743.

Bernardi, A., and Spahr, P. F., 1972, Nucleotide sequence at binding site for coat protein of R 17, *Proc. Nat. Acad. Sci. USA* **69,** 3033.

Bernstein, H., 1968, Repair and recombination in phage T 4. I. Genes affecting recombination, *Cold Spring Harbor Symp. Quant. Biol.* **23,** 325.

Bertrand, K., Korn, L., Lee, F., Platt, T., Squires, C. L., Squires, C., and Yanofsky, C., 1975, New features of the structure and regulation of the tryptophan operon of *Escherichia coli, Science,* **189,** 22.

Beyreuther, K., Adler, K., Geisler, N., and Klemm, A., 1973, The amino acid sequence of lac repressor, *Proc. Nat. Acad. Sci. USA* **70,** 3576.

Björk, G. R., and Neidhardt, F. C., 1973, Evidence for the utilization of host tRNA (m⁵U) methylase to modify tRNA coded by phage T 4, *Virology* **52,** 507.

Black, L. W., 1974, Bacteriophage T 4 internal protein mutants: Isolation and properties. *Virology* **60,** 166.

Black, L. W., and Abremski, K., 1974, Restriction of phage T 4 internal protein I mutants by a strain of *Escherichia coli, Virology* **60,** 180.

Black, L. W., and Gold, L. M., 1971, Pre-replicative development of the bacteriophage T 4: RNA and protein synthesis *in vivo* and *in vitro, J. Mol. Biol.* **60,** 365.

Block, R., and Haseltine, W. A., 1974, *In vitro* synthesis of ppGpp and pppGpp, *in* "Ribosomes" (M. Nomura, A. Tissieres, and P. Lengyel, eds.), pp. 747–761, Cold Spring Harbor Laboratory, New York.

Blumberg, D. D., and Malamy, M. H., 1974, Evidence for the presence of nontranslated T 7 late mRNA in infected F′ (PIF⁺) episome-containing cells, *J. Virol.* **13,** 378.

Blumberg, D. D., Mabie, C. T., and Malamy, M. H., 1976, T 7 protein synthesis in F factor containing cells: Evidence for an episomally induced impairment of translation and its relation to an alteration in membrane permeability, *J. Virol.,* **7,** 94.

Blumenthal, T., Landers, T. A., and Weber, K., 1972, Bacteriophage Qβ replicase contains the protein biosynthesis elongation factors EF Tu and EF Ts, *Proc. Nat. Acad. Sci. USA* **69,** 1313.

Blundell, M., Craig, E., and Kennell, D., 1972, Decay rates of different mRNA in *E. coli* and models of decay, *Nature (London) New Biol.* **238,** 46.

Boezi, J. A., Armstrong, R. L., and DeBacker, M., 1967, Methylation of RNA in bacteriophage T 4 infected *Escherichia coli, Biochem. Biophys. Res. Commun.* **29,** 281.

Bogdanova, E. S., Gorlenko, Zh. M., and Khourgess, E. M., 1974, On the effect of bacteriophage infection on host RNA polymerase, *Mol. Gen. Genet.* **133,** 261.

Boikov, P. Ya., and Gumanov, L. L., 1971, Interaction between phage T 4B DNA and membrane structures of the cell, *Mol. Biol.* **5**(3), 409.

Bolin, R. W., and Cummings, D. J., 1974, Structural aberrations in T-even bacteriophage. IV. Parameters of induction and formation of lollipops, *J. Virol.* **13,** 1368.

Bolle, A., Epstein, R. H., Salser, W., and Geiduschek, E. P., 1968*a*, Transcription during bacteriophage T 4 development: Synthesis and relative stability of early and late RNA, *J. Mol. Biol.* **31,** 325.

Bolle, A., Epstein, R. H., Salser, W., and Geiduschek, E. P., 1968*b*, Transcription during bacteriophage T 4 development: Requirements for late messenger synthesis, *J. Mol. Biol.* **33,** 339.

Bolund, C., 1973, Influence of gene 55 on the regulation of synthesis of some early enzymes in bacteriophage T 4-infected *Escherichia coli, J. Virol.* **12,** 49.

Bordier, C., 1974, Inhibition of rifampicin-resistant RNA synthesis by rifampicin-RNA polymerase complexes, *FEBS (Fed. Eur. Biochem. Soc.) Lett.* **45**, 259.

Bordier, C., and Dubochet, J., 1974, Electron microscopic localization of the binding sites of *E. coli* RNA polymerase in the early promoter region of T 7 DNA, *Eur. J. Biochem.* **44**, 617.

Bose, S. K., and Warren, R. J., 1967, On the stability of phage messenger RNA, *Biochem. Biophys. Res. Commun.* **26**, 385.

Bose, S. K., and Warren, R. J., 1969, Bacteriophage-induced inhibition of host functions. II. Evidence for multiple, sequential bacteriophage-induced deoxyribonucleases responsible for degradation of cellular deoxyribonucleic acid, *J. Virol.* **3**, 549.

Botchan, P., Wang, J. C., and Echols, H., 1973, Effect of circularity and superhelicity on transcription from bacteriophage λ DNA, *Proc. Nat. Acad. Sci. USA* **70**, 3077.

Both, G. W., Banerjee, A. K., and Shatkin, A. J., 1975, Methylation-dependent translation of viral messenger RNAs *in vitro*, *Proc. Nat. Acad. Sci. USA* **72**, 1189.

Botstein, D., and Suskind, M., 1974, Regulation of lysogeny and the evolution of temperate bacterial viruses, in "Mechanisms of Virus Disease" (W. S. Robinson, and C. F. Fox, eds.), pp. 363–384, W. A. Benjamin, Menlo Park.

Bouché, J.-P., Zechel, K., and Kornberg, A., 1975, *dna*G gene product, a rifampicin-resistant RNA polymerase, initiates the conversion of a single-stranded coliphage DNA to its duplex replicative form, *J. Biol. Chem.*, **250**, 5995.

Boyer, H. W., 1974, DNA restriction and modification mechanisms in bacteria, *Fed. Proc.* **33**, 1125.

Brack, C., and Pirrotta, V., 1975, Electron microscopic study of the repressor of bacteriophage λ and its interaction with operator DNA, *J. Mol. Biol.* **96**, 139.

Bräutigam, A. R., and Sauerbier, W., 1973, Transcription unit mapping in bacteriophage T 7. I. *In vivo* transcription by *Escherichia coli* RNA polymerase, *J. Virol.* **12**, 882.

Bräutigam, A. R., and Sauerbier, W., 1974, Transcription unit mapping in bacteriophage T 7. II. Proportionality of number of gene copies, mRNA and gene product, *J. Virol.* **13**, 1110.

Bremer, H., and Yuan, D., 1968, Chain growth rate of messenger RNA in *Escherichia coli* infected with bacteriophage T 4, *J. Mol. Biol.* **34**, 527.

Bremer, H., Konrad, M. W., Gaines, K., and Stent, G. S., 1965, Direction of chain growth in enzymic RNA synthesis, *J. Mol. Biol.* **13**, 540.

Brevet, J., 1974, Direct assay for sigma factor activity and demonstration of the loss of this activity during sporulation in *Bacillus subtilis*, *Mol. Gen. Genet.* **128**, 223.

Brevet, J., and Sonenshein, A. L., 1972, Template specificity of ribonucleic acid polymerase in asporogenous mutants of *Bacillus subtilis*, *J. Bacteriol.* **112**, 1270.

Britton, J. R., and Haselkorn, R., 1975, Permeability lesions in male *E. coli* infected with bacteriophage T 7, *Proc. Nat. Acad. Sci. USA* **72**, 2222.

Brody, E. N., 1975, personal communication.

Brody, E. N., and Geiduschek, E. P., 1970, Transcription of the bacteriophage T 4 template. Detailed comparison of *in vitro* and *in vivo* transcripts, *Biochemistry* **9**, 1300.

Brody, E. N., Diggelmann, H., and Geiduschek, E. P., 1970a, Transcription of the bacteriophage T 4 template. Obligate synthesis of T 4 prereplicative RNA *in vitro*, *Biochemistry* **9**, 1289.

Brody, E., Sederoff, R., Bolle, A., and Epstein, R. H., 1970b, Early transcription in T 4-infected cells, *Cold Spring Harbor Symp. Quant. Biol.* **35**, 203.

Brody, E. N., Gold, L. M., and Black, L. W., 1971, Transcription and translation of sheared bacteriophage T 4 DNA *in vitro, J. Mol. Biol.* **60**, 389.

Broker, T. R., 1973, An electron microscopic analysis of pathways for bacteriophage T 4 DNA recombination, *J. Mol. Biol.* **81**, 1.

Broker, T. R., and Doermann, A. H., 1975, Molecular and genetic recombination of bacteriophage T 4, *Ann. Rev. Genet.* **9**, 213.

Broker, T. R., and Lehman, I. R., 1971, Branched DNA molecules: Intermediates in T 4 recombination, *J. Mol. Biol.* **60**, 131.

Bruner, R., and Cape, R. E., 1970, The expression of two classes of late genes of bacteriophage T 4, *J. Mol. Biol.* **53**, 69.

Bruner, R., Souther, A., and Suggs, S., 1972, Stability of cytosine-containing deoxyribonucleic acid after infection by certain T 4 rII-D deletion mutants, *J. Virol.* **10**, 88.

Brunovskis, I., and Summers, W. C., 1971, The process of infection with coliphage T 7. V. Shutoff of host RNA synthesis by an early phage function, *Virology* **45**, 224.

Brunovskis, I., and Summers, W. C., 1972, The process of infection with coliphage T 7. VI. A phage gene controlling shutoff of host RNA synthesis, *Virology* **50**, 322.

Brunovskis, I., Hyman, R. W., and Summers, W. C., 1973, Pasteurella bacteriophage H and *Escherichia coli* bacteriophage ϕII are nearly identical, *J. Virol.* **11**, 306.

Buckley, P. J., Kosturko, L. D., and Kozinski, A. W., 1972, *In vivo* production of an RNA–DNA copolymer after infection of *Escherichia coli* by bacteriophage T 4, *Proc. Nat. Acad. Sci. USA* **69**, 3165.

Bujard, H., 1969, Location of single-strand interruptions in the DNA of bacteriophage T 5⁺, *Proc. Nat. Acad. Sci. USA* **62**, 1167.

Bujard, H., and Hendrickson, H. E., 1973, Structure and function of the genome of coliphage T 5. 1. The physical structure of the chromosome of T 5⁺, *Eur. J. Biochem.* **33**, 517.

Burgess, R. P., Travers, A. A., Dunn, J. J., and Bautz, E. K. G., 1969, Factor stimulating transcription by RNA polymerase, *Nature* **221**, 43.

Burgi, E., Hershey, A. D. and Ingraham, L., 1966, Preferred breakage points in T 5 DNA molecules subjected to shear, *Virology* **28**, 11.

Cairns, J., Stent, G. S., and Watson, J. D., 1966, "Phage and the Origins of Molecular Biology," Cold Spring Harbor Laboratory, New York.

Calef, E., and Neubauer, Z., 1968, Active and inactive states of the C_I gene in some λ defective phages, *Cold Spring Harbor Symp. Quant. Biol.* **33**, 765.

Calendar, R., 1970, The regulation of phage development, *Ann. Rev. Microbiol.* **24**, 241.

Capco, G. R., and Mathews, C. K., 1973, Bacteriophage-coded thymidylate synthetase. Evidence that the T 4 enzyme is a capsid protein, *Arch. Biochem.* **158**, 736.

Carlson, K., 1973, Multiple initiation of bacteriophage T 4 DNA replication: Delaying effect of bromodeoxyuridine, *J. Virol.* **12**, 349.

Carmichael, G. G., Weber, K., Niveleau, A., and Wahba, A. J., 1975, The host factor required for RNA phage Qβ RNA replication *in vitro*. Intracellular location, quantitation, and purification by poly(A)-cellulose chromatography, *J. Biol. Chem.*, **250**, 3607.

Carrascosa, J. L., Viñuela, E., and Salas, M., 1973, Proteins induced in *Bacillus subtilis* infected with bacteriophage ϕ29, *Virology* **56**, 291.

Carrington, J. M., and Lunt, M., 1973, Studies on the replication of bacteriophage T 5. *J. Gen Virol.* **18**, 91.

Cascino, A., Riva, S., and Geiduschek, E. P., 1970, DNA ligation and the coupling of late T 4 transcription to replication, *Cold Spring Harbor Symp. Quant. Biol.* **35,** 213.

Cascino, A., Riva, S., and Geiduschek, E. P., 1971a, Host DNA synthesis after infection of *Escherichia coli* with mutants of bacteriophage T 4, *Virology* **46,** 437.

Cascino, A., Geiduschek, E. P., Cafferata, R. L., and Haselkorn, R., 1971b, T 4 DNA replication and viral gene expression, *J. Mol. Biol.* **61,** 357.

Cashel, M., and Gallant, J., 1974, Cellular regulation of guanosine tetraphosphate and guanosine pentaphosphate, *in* "Ribosomes" (M. Normura, A. Tissières, and P. Lengyel, eds.) pp. 793–745, Cold Spring Harbor Laboratory, New York.

Chace K. V., and Hall, D. H., 1973, Isolation of mutants of bacteriophage T 4 unable to induce thymidine kinase activity, *J. Virol.* **12,** 343.

Chace, K. V., and Hall, D. H., 1975, Characterization of new regulatory mutants of bacteriophage T 4. II. New class of mutants, *J. Virol.* **15,** 929.

Chadwick, P., Pirotta, V., Steinberg, R., Hopkins, N., and Ptashne, M., 1970, The Λ and 434 phage repressors, *Cold Spring Harbor Symp. Quant. Biol.* **35,** 283.

Chakraborty, P. R., Sarkar, P., Huang, H. H., and Maitra, U., 1973, Studies on T 3-induced ribonucleic acid polymerase. III. Purification and characterization of the T 3-induced ribonucleic acid polymerase from bacteriophage T 3-infected *Escherichia coli* cells, *J. Biol. Chem.* **248,** 6637.

Chakraborty, P. R., Bandyopadhyay, P., Huang, H. H., and Maitra, U., 1974, Fidelity of *in vitro* transcription of T 3 DNA by bacteriophage T 3-induced RNA polymerase and by *E. coli* RNA polymerase, *J. Biol. Chem.* **249,** 6901.

Chamberlin, M., 1970, Transcription 1970: A summary, *Cold Spring Harbor Symp. Quant. Biol.* **35,** 851.

Chamberlin, M., 1974a, The selectivity of transcription, *Annu. Rev. Biochem.* **43,** 721.

Chamberlin, M., 1974b, Isolation and characterization of prototrophic mutants of *Escherichia coli* unable to support the intracellular growth of T 7, *J. Virol.* **14,** 509.

Chamberlin, M., and Berg, P., 1964, Mechanism of RNA polymerase action: Formation of DNA-RNA hybrids with single-stranded templates, *J. Mol. Biol.* **8,** 297.

Chamberlin, M., and Losick, R. (eds.), 1976, "RNA Polymerases," Cold Spring Harbor, New York.

Chamberlin, M., and Ring, J., 1972, Characterization of T 7-specific RNA polymerase. III. Inhibition by derivatives of rifamycin SV, *Biochem. Biophys. Res. Commun.* **49,** 1129.

Chamberlin, M., and Ring, J., 1973a, Characterization of T 7-specific ribonucleic acid polymerase. I. General properties of the enzymatic reaction and the template specificity of the enzyme, *J. Biol. Chem.* **248,** 2235.

Chamberlin, M., and Ring, J., 1973b, Characterization of T 7-specific ribonucleic acid polymerase. II. Inhibitors of the enzyme and their application to the study of the enzymatic reaction, *J. Biol. Chem.* **248,** 2245.

Chamberlin, M., McGrath, J., and Waskell, L., 1970, New RNA polymerase from *Escherichia coli* infected with bacteriophage T 7, *Nature (London)* **228,** 227.

Chan, V. L., and Ebisuzaki, K., 1973, Intergenic suppression of amber polynucleotide ligase mutation in bacteriophage T 4. II. *Virology* **53,** 60.

Chaney, S. G., and Boyer, P. D., 1972, Incorporation of water oxygens into intracellular nucleotides and RNA. II. Predominantly hydrolytic RNA turnover in *Escherichia coli,* *J. Mol. Biol.* **64,** 581.

Chao, L., and Speyer, J. F., 1973, A new form of RNA polymerase isolated from *Escherichia coli,* *Biochem. Biophys. Res. Commun.* **51,** 399.

Chen, C. W.-S., 1975, The transcriptional map of bacteriophage T 5, Ph.D. Thesis, University of Texas at Dallas.

Chen, M.-J., Locker, J., and Weiss, S. B., 1974, Structure and physical mapping of T 5 transfer RNA genes, *Fed. Proc.* **33**, 1280.

Chen, M.-J., Locker, J., and Weiss, S. B., 1976, The physical mapping of bacteriophage T 5 transfer tRNAs, *J. Biol. Chem.* **251**, 536.

Chinnadurai, G., and McCorquodale, D. J., 1975, personal communication.

Chinnadurai, G., and McCorquodale, D. J., 1973, Requirement of a phage-induced 5'-exonuclease for the expression of late genes of bacteriophage T 5, *Proc. Nat. Acad. Sci. USA* **70**, 3502.

Chinnadurai, G., and McCorquodale, D. J., 1974a, Regulation of expression of late genes of bacteriophage T 5, *J. Virol.* **13**, 85.

Chinnadurai, G., and McCorquodale, D. J., 1974b, Dual role of gene D5 in the development of bacteriophage T 5. *Nature (London)* **247**, 554.

Chiu, C.-S., Tomich, P. K., Wovcha, M., and Greenberg, G. R., 1974, Function of T 4 phage-induced early enzymes *in vivo*, *Fed. Proc.* **33**, 1420.

Chiu, C.-S., Tomich, P. K., and Greenberg, G. R., 1976, Simultaneous initiation of synthesis of bacteriophage T 4 DNA and of deoxyribonucleotides, *Proc. Nat. Acad. Sci. USA* **73**, 757.

Clark, S., 1975, Novel multi-subunit RNA polymerase induced by *Bacillus subtilis* phage PBS 2. Abstract. Meeting on "RNA polymerases" Cold Spring Harbor, August 1975.

Clark, S., Losick, R., and Pero, J., 1974, New RNA polymerase from *Bacillus subtilis* infected with phage PBS 2, *Nature (London)* **252**, 21.

Cocito, C., 1974, Origin and metabolic properties of the RNA species formed during the replication cycle of virus 2C, *J. Virol.* **14**, 1482.

Cohen, P. S., 1970, Regulation of cessation of early enzyme synthesis in bacteriophage T 4-infected *Escherichia coli*: Separate mechanisms for different enzymes, *Virology* **41**, 453.

Cohen, P. S., 1972, Translational regulation of deoxycytidylate hydroxymethylase and deoxynucleotide kinase synthesis in T 4-infected *Escherichia coli, Virology* **47**, 780.

Cohen, P. S., Zetter, B. R., and Walsh, M. L., 1972, Evidence that more deoxynucleotide kinase mRNA is transcribed than translated during T 4 infection of *Escherichia coli, Virology* **49**, 808.

Cohen, P. S., Natale, P. J., and Buchanan, J. M., 1974, Transcriptional regulation of T 4 bacteriophage-specific enzymes synthesized *in vitro, J. Virol.* **14**, 292.

Cohen, S. S., 1968, "Virus-Induced Enzymes," Columbia University Press, New York.

Coleman, J. E., 1974, The role of Zn(II) in transcription by T 7 RNA polymerase, *Biochem. Biophys. Res. Commun.* **60**, 641.

Collinsworth, W. L., and Mathews, C. K., 1974, Biochemistry of DNA-defective *amber* mutants of bacteriophage T 4: IV. DNA synthesis in plasmolyzed cells, *J. Virol.* **13**, 908.

Comer, M. M., Guthrie, C., and McClain, W. H., 1974, An *ochre* suppressor of bacteriophage T 4 that is associated with a transfer RNA, *J. Mol. Biol.* **90**, 665.

Condit, R. C., 1975, F-factor mediated inhibition of bacteriophage T 7 growth: Increased membrane permeability and decreased ATP levels following T 7 infection of male *E. coli, J. Mol. Biol.* **98**, 45.

Condit, R. C., and Steitz, J. A., 1975, F-factor mediated inhibition of bacteriophage T 7 growth: Analysis of T 7 RNA and protein synthesis *in vivo* and *in vitro* using male and female *E. coli, J. Mol. Biol.,* **98**, 31.

Coppo, A., Manzi, A., Pulitzer, J. F., and Takahashi, H., 1973, Abortive bacteriophage T 4 head assembly in mutants of *E. coli, J. Mol. Biol.* **76**, 61.

Coppo, A., Manzi, A., Martire, G., Pulitzer, J. F., and Takahashi, H., 1974, *Atti. Ass. Genet. Ital.* **19**, 47.

Coppo, A., Manzi, A., Pulitzer, J. F., and Takahashi, H., 1975*a*, Host mutant (*tab*D) induced inhibition of bacteriophage T 4 late transcription: I. Isolation and phenotypic characterization of mutants, *J. Mol. Biol.* **96**, 579.

Coppo, A., Manzi, A., Pulitzer, J. F., and Takahashi, H., 1975*b*, Host mutant (*tab*D) induced inhibition of bacteriophage T 4 late transcription: II. Genetic characterization of mutants, *J. Mol. Biol.* **96**, 601.

Cox, G. S., and Conway, T. W., 1973, Template properties of glucose-deficient T-even bacteriophage DNA, *J. Virol.* **12**, 1279.

Craig, E., Cremer, K., and Schlessinger, D., 1972, Metabolism of T 4 messenger RNA, host messenger RNA and ribosomal RNA in T 4-infected *Escherichia coli* B, *J. Mol. Biol.* **71**, 701.

Crawford, L. V., 1959, Nucleic acid metabolism in *Escherichia coli* infected with phage T 5, *Virology* **7**, 359.

Crepin, M., Cukier-Kahn, R., and Gros, F., 1975, Effect of a low-molecular-weight DNA-binding protein, H_1 factor, on the *in vitro* transcription of the lactose operon in *Escherichia coli, Proc. Nat. Acad. Sci. USA* **72**, 333.

Cukier-Kahn, R., Jacquet, M., and Gros, F., 1972, Two heat-resistant low molecular weight proteins from *Escherichia coli* that stimulate DNA-directed RNA synthesis, *Proc. Nat. Acad. Sci. USA* **69**, 3643.

Daegelen, P., d'Aubenton, Y., and Brody, E. N., 1975, A diffusible product of T 4 infected cells which controls delayed early RNA synthesis. *Abs. 1975 Meet. Fed. Eur. Biochem. Soc.*

Dahlberg, J. E., and Blattner, F. R., 1973, *In vitro* transcription products of lambda DNA: Nucleotide sequences and regulatory sites, *in* "Virus Research" (C. F. Fox and W. S. Robinson, eds.), pp. 533–543, Academic Press, New York.

Daniel, V., Sarid, S., and Littauer, U. Z., 1968, Coding by T 4 phage DNA of soluble RNA containing pseudouridylic acid, *Proc. Nat. Acad. Sci. USA* **60**, 1403.

Daniel, V., Sarid, S., and Littauer, U. Z., 1970, Bacteriophage induced transfer RNA in *E. coli, Science* **167**, 1682.

Darlix, J. L., 1973, The function of *rho* in T 7-DNA transcription *in vitro, Eur. J. Biochem.* **35**, 517.

Darlix, J. L., and Fromageot, P., 1972, Discontinuous *in vitro* transcription of DNA, *Biochimie* **54**, 47.

Darlix, J. L., and Horaist, M., 1975, Existence and possible roles of transcriptional barriers in T 7 DNA early region as shown by electron microscopy, *Nature (London)* **256**, 288.

Darlix, J. L., Sentenac, A., Ruet, A., and Fromageot, P., 1969, Role of RNA polymerase stimulating factor on chain initiation, *Eur. J. Biochem.* **11**, 43.

Das, A., Court, D., and Adhya, S., 1976, Isolation and characterization of conditional lethal mutants of *Escherichia coli* defective in transcription termination factor *rho, Proc. Nat. Acad. Sci. USA* **73**, 1959.

Dausse, J.-P., Sentenac, A., and Fromageot, P., 1972*a*, Interaction of RNA polymerase from *Escherichia coli* with DNA. Selection of initiation sites on T 7 DNA, *Eur. J. Biochem.* **26**, 43.

Dausse, J.-P., Sentenac, A., and Fromageot, P., 1972*b*, Interaction of RNA polymerase from *Escherichia coli* with DNA. Influence of DNA scissions on RNA polymerase binding and chain initiation, *Eur. J. Biochem.* **31,** 394.

Davis, B. D., Dulbecco, R., Eisen, H. N., Ginsberg, H. S., Wood, W. B., and McCarty, M., 1973, "Microbiology," Harper and Row, Hagerstown, Maryland.

Davis, R. W., and Hyman, R. W., 1970, Physical location of the *in vitro* RNA initiation site and termination sites of T 7M DNA, *Cold Spring Harbor Symp. Quant. Biol.* **35,** 269.

Davis, R. W., and Hyman, R. W., 1971, A study in evolution: The DNA base sequences homology between coliphages T 7 and T 3, *J. Mol. Biol.* **62,** 287.

Davison, J., Brookman, K., Pilarski, L., and Echols, H., 1970, The stimulation of RNA synthesis by M factor, *Cold Spring Harbor Symp. Quant. Biol.* **35,** 95.

Davison, P. F., Freifelder, D., and Holloway, B. W., 1964, Interruptions in the polynucleotide strands in bacteriophage DNA, *J. Mol. Biol.* **8,** 1.

Dean, W. W., and Lebowitz, J., 1971, Partial alteration of secondary structure in native superhelical DNA, *Nature (London) New Biol.* **231,** 5.

De Crombrugghe, B., Adhya, S., Gottesman, M., and Pastan, I., 1973, Effect of rho on transcription of bacterial operons, *Nature (London) New Biol.* **241,** 260.

Delbrück, M., 1946*a*, Bacterial viruses or bacteriophages, *Biol. Rev. Cambridge Philos. Soc.* **21,** 30.

Delbrück, M., 1946*b*, Experiments with bacterial viruses (bacteriophage), *Harvey Lect. Ser.* **XLI,** 161.

Delius, H., Howe, C., and Kozinski, A. W., 1971, Structure of the replicating DNA from bacteriophage T 4, *Proc. Nat. Acad. Sci. USA* **68,** 3049.

Demerec, M., and Fano, U., 1945, Bacteriophage-resistant mutants in *Escherichia coli*, *Genetics* **30,** 119.

Dennis, P. P., and Bremer, H., 1973, Regulation of ribonucleic acid synthesis in *Escheria coli* B/r: An analysis of a shift-up. 1. Ribosomal RNA chain growth rates, *J. Mol. Biol.* **75,** 145.

Dennis, P. P., and Bremer, H., 1974, Macromolecular composition during steady-state growth of *E. coli* B/r, *J. Bacteriol.* **119,** 270.

Depew, R. E., and Cozzarelli, N. R., 1974, Genetics and physiology of bacteriophage T 4 3′-phosphatase: Evidence for involvement of the enzyme in T 4 DNA metabolism, *J. Virol.* **13,** 888.

Depew, R. E., Snopek, T. J., and Cozzarelli, N. R., 1975, Characterization of a new class of deletions of the D region of the bacteriophage T 4 genome, *Virology* **64,** 144.

Dharmgrongartama, B., Mahadik, S. P., and Srinivasan, P. R., 1973, Modification of RNA polymerase after T 3 phage infection of *Escherichia coli* B, *Proc. Nat. Acad. Sci. USA* **70,** 2845.

Dickson, R. C., Barnes, S. L., and Eiserling, F. A., 1970, Structural proteins of bacteriophage T 4, *J. Mol. Biol.* **53,** 461.

Dickson, R. C., Abelson, J., Barnes, W. M., and Reznikoff, W. S., 1975, Genetic regulation: The *lac* control region, *Science* **187,** 27.

Dicou, L., 1975, Bacteriophage T 4 DNA synthesis *in vitro*, Ph.D. Thesis, Harvard University.

diMauro, F., Snyder, L., Marino, P., Lamberti, A., Coppo, A., and Tocchini-Valentini, G. P., 1969, Rifampicin sensitivity of the components of DNA-dependent RNA polymerase, *Nature (London)* **222,** 533.

di Porzio, U., 1973, unpublished observations.

Doermann, A. H., 1973, T 4 and the rolling circle model of replication, *Annu. Rev. Genet.* **7**, 325.

Doermann, A. H., Eiserling, F. A., and Boehner, L., 1973, Capsid length in bacteriophage T 4 and its genetic control, *in* "Virus Research" (C. F. Fox and W. S. Robinson, eds.), pp. 243–258, Academic Press, New York.

Donini, P., and Edlin, G., 1972, RNA synthesis in T 4-infected *Escherichia coli* during amino acid starvation *Virology* **50**, 273.

Dottin, R. P., and Pearson, M. L., 1973, Regulation by *N* gene protein of phage lambda of anthranilate synthetase synthesis *in vitro*, *Proc. Nat. Acad. Sci. USA* **70**, 1078.

Dougan, A. H., and Glaser, D. A., 1974, Rate of chain elongation of ribosomal RNA molecules in *Escherichia coli*, *J. Mol. Biol.* **87**, 775.

Dube, S. K., and Rudland, P. S., 1970, Control of translation by T 4 phage: Altered binding of disfavoured messengers, *Nature* (*London*) **226**, 820.

Dubin, S. B., Benedek, G. B., Bancroft, F. C., and Freifelder, D., 1970, Molecular weights of coliphages and coliphage DNA. II. Measurement of diffusion coefficients using optical mixing spectroscopy, and measurement of sedimentation coefficients, *J. Mol. Biol.* **54**, 547.

Duffy, J. J., and Geiduschek, E. P., 1973a, Characteristics of a RNA polymerase from phage SPO 1-infected *B. subtilis*, *Fed. Proc.* **32**, 646.

Duffy, J. J., and Geiduschek, E. P., 1973b, Transcription specificity of an RNA polymerase fraction from bacteriophage SPO 1-infected *B. subtilis*, *FEBS* (*Fed. Eur. Biochem. Soc.*) *Lett* **34**, 172.

Duffy, J. J., and Geiduschek, E. P., 1975, RNA polymerase from phage SPO 1 infected and uninfected *B. subtilis*, *J. Biol. Chem.* **250**, 4530.

Duffy, J. J., and Geiduschek, E. P., 1976, The virus-specified subunits of a modified *B. subtilis* RNA polymerase are determinants of DNA binding and RNA chain initiation, *Cell* **8**, 595.

Duffy, J. J., Chaney, S. G., and Boyer, P. D., 1972, Incorporation of water oxygens into intracellular nucleotides and RNA. I. Predominantly nonhydrolytic RNA turnover in *Bacillus subtilis*, *J. Mol. Biol.* **64**, 565.

Duffy, J. J., Petrusek, R. L., and Geiduschek, E. P., 1975, *In vitro* conversion of *B. subtilis* RNA polymerase activity by a phage SPO 1 induced protein, *Proc. Nat. Acad. Sci. USA* **72**, 2366.

Dunn, J. J., ed., 1975, "Processing of RNA," Brookhaven National Laboratory, New York.

Dunn, J. J., and Studier, F. W., 1973a, T 7 early RNAs are generated by site-specific cleavages, *Proc. Nat. Acad. Sci. USA* **70**, 1559.

Dunn, J. J., and Studier, F. W., 1973b, T 7 early RNAs and *Escherichia coli* ribosomal RNAs are cut from large precursor RNAs *in vivo* by ribonuclease III, *Proc. Nat. Acad. Sci. USA* **70**, 3296.

Dunn, J. J., and Studier, F. W., 1975a, Processing, transcription, and translation of bacteriophage T 7 messenger RNAs, *in* "Processing of RNA," pp. 267–276, Brookhaven National Laboratory, New York.

Dunn, J. J., and Studier, F. W., 1975b, Effect of RNase III cleavage on translation of bacteriophage T 7 messenger RNAs, *J. Mol. Biol.*, **99**, 487.

Dunn, J. J., Bautz, F. A., and Bautz, E. K. F., 1971, Different template specificities of phage T 3 and T 7 RNA polymerases, *Nature* (*London*) *New Biol.* **230**, 94.

Dunn, J. J., McAllister, W. T., and Bautz, E. K. F., 1972a, *In vitro* transcription of T 3 DNA by *Escherichia coli* and T 3 polymerases, *Virology* **48**, 112.

Dunn, J. J., McAllister, W. T., and Bautz, E. K. F., 1972*b*, Transcription *in vitro* of T 3 DNA by *Escherichia coli* and T 3 RNA polymerases. Analysis of the products in cell-free protein synthesizing system, *Eur. J. Biochem.* **29,** 500.

Earhart, C. F., Sauri, C. J., Fletcher, G., and Wulff, J. L., 1973, Effect of inhibition of macromolecule synthesis on the association of bacteriophage T 4 DNA with membrane, *J. Virol.* **11,** 527.

Echols, H., 1971, Regulation of lytic development, *in* "The Bacteriophage Lambda" (A. D. Hershey, ed.), pp. 247–270, Cold Spring Harbor Laboratory, New York.

Echols, H., 1974, Some unsolved general problems of phage λ development, *Biochimie* **56,** 1491.

Echols, H., and Green, L., 1971, Establishment and maintenance of repression by bacteriophage lambda: The role of the cI, cII, and cIII proteins, *Proc. Nat. Acad. Sci. USA* **68,** 2190.

Edgar, R. S., and Mosig, G., 1970, Linkage map of bacteriophage T 4, *in* "Handbook of Biochemistry," 2nd ed. (H. A. Sober, ed.), pp. I-32–34, Chemical Rubber Co., Cleveland, Ohio.

Edgar, R. S., and Wood, W. B., 1966, Morphogenesis of bacteriophage T 4 in extracts of mutant-infected cells, *Proc. Nat. Acad. Sci. USA* **55,** 498.

Eiserling, F. A., 1963, *Bacillus subtilis* bacteriophages: Structure, intracellular development and conditions of lysogeny, Ph.D. Thesis, University of California, Los Angeles.

Eiserling, F. A., and Dickson, R. C., 1972, Assembly of viruses, *Annu. Rev. Biochem.* **41,** 467.

Eiserling, F. A., Geiduschek, E. P., Epstein, R. H., and Metter, E. J., 1970, Capsid size and deoxyribonucleic acid length: The petite variant of bacteriophage T 4, *J. Virol.* **6,** 865.

Emanuel, B. S., 1973, Replicative hybrid of T 4 bacteriophage DNA, *J. Virol.* **12,** 408.

Englesberg, E., and Wilcox, G., 1974, Regulation: Positive control, *Annu. Rev. Genet.* **8,** 219.

Eoyang, L., and August, L. T., 1974, Reproduction of RNA bacteriophages, *in* "Comprehensive Virology" (H. Fraenkel-Conrat and R. R. Wagner, eds.), Vol. 2, pp. 1–60, Plenum Press, New York and London.

Epp, C., and Pearson, M. L., 1976, Association of bacteriophage lambda proteins with *E. coli* RNA polymerase, *in* "RNA Polymerases" (M. Chamberlin, and R. Losick, eds.), pp. 667–691, Cold Spring Harbor Laboratory, New York.

Epstein, R. H., Bolle, A., Steinberg, C. M., Kellenberger, E., Boy de la Tour, E., Chevalley, R., Edgar, R. S., Susman, M., Denhardt, G. H., and Lielausis, A., 1963, Physiological studies of conditional lethal mutations of bacteriophage T 4D, *Cold Spring Harbor Symp. Quant. Biol.* **28,** 375.

Falco, S. C., and Rothman-Denes, L. B., 1975*a*, Transcriptional activities induced by bacteriophage N 4. Abstract, Meeting on "RNA polymerases," Cold Spring Harbor, August 1975.

Falco, S. C., and Rothman-Denes, L. B., 1975*b*, personal communication.

Fiser, I., Scheit, K. H., Stöffler, G., and Kuechler, E., 1974, Identification of protein S1 at the messenger RNA binding site of the *E. coli* ribosome, *Biochem. Biophys. Res. Commun.* **60,** 1112.

Fox, T. D., and Pero, J., 1974, New phage-SPO 1-induced polypeptides associated with *Bacillus subtilis* RNA polymerase, *Proc. Nat. Acad. Sci. USA* **71,** 2761.

Fox, T. D., Losick, R., and Pero, J., 1976, Phage-induced subunit of RNA polymerase is coded by SPO 1 regulatory gene 28, *J. Mol. Biol.* **101,** 427.

Frankel, F. R., 1968, Evidence for long DNA strands in the replicating pool after T 4 infection, *Proc. Nat. Acad. Sci. USA* **59**, 131.

Frankel, F. R., Batcheler, M. L., and Clark, C. K., 1971, The role of gene 49 in DNA replication and head morphogenesis in bacteriophage T 4, *J. Mol. Biol.* **62**, 439.

Franklin, N. C., 1971, The *N* operon of lambda: extent and regulation as observed in fusions to the tryptophan operon of *Escherichia coli*, *in* "The Bacteriophage Lambda" (A. D. Hershey, ed.), pp. 621–638, Cold Spring Harbor Laboratory, New York.

Franklin, N. C., 1974, Altered reading of genetic signals fused to the *N* operon of bacteriophage λ: Genetic evidence for modification of polymerase by the protein product of the *N* gene, *J. Mol. Biol.* **89**, 33.

Freifelder, D., 1970, Molecular weights of coliphages and coliphage DNA. IV. Molecular weights of DNA from bacteriophages T 4, T 5, and T 7 and the general problem of determination of M, *J. Mol. Biol.* **54**, 567.

Frenkel, G. D., and Richardson, C. C., 1971a, The deoxyribonuclease induced after infection of *Escherichia coli* by bacteriophage T 5. I. Characterization of the enzyme as a 5′-exonuclease, *J. Biol. Chem.* **246**, 4839.

Frenkel, G. D. and Richardson, C. C., 1971b, The deoxyribonuclease induced after infection of *Escherichia coli* by bacteriophage T 5. II. Role of the enzyme in replication of the phage deoxyribonucleic acid. *J. Biol. Chem.* **246**, 4848.

Friesen, J. D., and Fiil, N., 1973, Accumulation of guanosine tetraphosphate in T 7 bacteriophage-infected *Escherichia coli*, *J. Bacteriol.* **113**, 697.

Fuchs, E., Millette, R. L., Zillig, W., and Walter, G., 1967, Influence of salts on RNA synthesis by DNA-dependent RNA polymerase from *Escherichia coli*, *Eur. J. Biochem.* **3**, 183.

Fuhrman, S. A., 1971, unpublished observations.

Fujita, D. J., 1971, Studies on conditional lethal mutants. Ph.D. Thesis, University of Chicago.

Fujita, D. J., Ohlsson-Wilhelm, B. M., and Geiduschek, E. P., 1971, Transcription during bacteriophage SPO 1 development: Mutations affecting the program of viral transcription, *J. Mol. Biol.* **57**, 301.

Fukuda, R., Iwakura, Y., and Ishihama, A., 1974, Heterogeneity of RNA polymerase in *Escherichia coli*. I. A new holoenzyme containing a new sigma factor, *J. Mol. Biol.* **83**, 353.

Fukuda, R., Keilman, G., McVey, E., and Doi, R. H., 1975, Ribonucleic acid polymerase pattern of sporulating *Bacillus subtilis* cells, *in* "Spores VI" (P. Gerhardt, R. N. Costilow, and H. L. Sadoff, eds.), pp. 213–220, American Society for Microbiology, Washington, D.C.

Fukada, K., Abelson, J., and Velten, J., 1976, personal communication.

Gage, L. P., 1969, Transcription in bacteriophage development, Ph.D. Thesis, The University of Chicago.

Gage, L. P., and Geiduschek, E. P., 1967, Repression of early messenger transcription in the development of a bacteriophage, *J. Mol. Biol.* **30**, 435.

Gage, L. P., and Geiduschek, E. P., 1971a, RNA synthesis during bacteriophage SPO 1 development: six classes of SPO 1 RNA, *J. Mol. Biol.* **57**, 279.

Gage, L. P., and Geiduschek, E. P., 1971b, RNA synthesis during bacteriophage SPO 1 development. II. Some modulations and prerequisites of the transcription program, *Virology* **44**, 200.

Galluppi, G., Grimley, C., Lowery, C., and Richardson, J. P., 1976, Rho termination factor: ATPase activity and altered presence in *suA* mutants, *in* "RNA

Polymerases" (M. Chamberlin and R. Losick, eds.), pp. 657–665, Cold Spring Harbor, New York.

Gausing, K., 1972, Efficiency of protein and messenger RNA synthesis in bacteriophage T 4-infected cells of *Escherichia coli, J. Mol. Biol.* **71**, 529.

Geiduschek, E. P., and Grau, O., 1970, T 4 anti-messenger, in "RNA Polymerase and Transcription" (L. Silvestri, ed.), pp. 190–203, North-Holland Publishing Company, Amsterdam, London.

Geiduschek, E. P., Snyder, L., Colvill, A. J. E., and Sarnat, M., 1966, Selective synthesis of T-even bacteriophage early messengeger *in vitro, J. Mol. Biol.* **19**, 541.

Geiduschek, E. P., Brody, E. N., and Wilson, D. L., 1968, Some aspects of RNA transcription, in "Molecular Associations in Biology" (B. Pullman, ed.) pp. 163–180, Academic Press, New York.

Geiduschek, E. P., Wilson, D. L., and Gage, L. P., 1969, Determinants of gene expression during viral development, *J. Cell Physiol.* **74**, 81.

Georgopoulos, C. P., 1971, Bacterial mutants in which the gene *N* function of bacteriophage Lambda is blocked have an altered RNA polymerase, *Proc. Nat. Acad. Sci. USA* **68**, 2977.

Georgopoulos, C. P., and Revel, H. R., 1971, Studies with glucosyl transferase mutants of the T-even bacteriophages, *Virology* **44**, 271.

Georgopoulos, C. P., Hendrix, R. W., Kaiser, A. D., and Wood, W. B., 1972, Role of the host cell in bacteriophage morphogenesis: Effects of a bacterial mutation on T 4 head assembly, *Nature (London) New Biol.* **239**, 38.

Ghosh, S., and Echols, H., 1972, Purification and properties of D protein: A transcription factor of *E. coli, Proc. Nat. Acad. Sci. USA* **69**, 3660.

Ghysen, A., and Pironio, M., 1972, Relationship between the *N* function of bacteriophage λ and host RNA polymerase, *J. Mol. Biol.* **65**, 259.

Giacomoni, P. U., Talaer, J. Y. le, and Pecq, J. B. le, 1974, *Escherichia coli* RNA-polymerase binding sites on DNA are only 14 base pairs long and are located between sequences that are very rich in A + T, *Proc. Nat. Acad. Sci. USA* **71**, 3091.

Gilbert, W., and Maxam, A., 1973, The nucleotide sequence of the lac operator, *Proc. Nat. Acad. Sci. USA* **70**, 3581.

Gill, D. M., 1975, Involvement of nicotinamide adenine dinucleotide in the action of cholera toxin *in vitro, Proc. Nat. Acad. Sci. USA* **72**, 2064.

Glatron, M. F., and Rapoport, G., 1972, Biosynthesis of the parasporal inclusion of *Bacillus thuringiensis:* half-life of its corresponding messenger RNA, *Biochimie* **54**, 1291.

Goff, C. G., 1974, Chemical structure of a modification of the *Escherichia coli* ribonucleic acid polymerase α polypeptides induced by bacteriophage T 4 infection, *J. Biol. Chem.* **249**, 6181.

Goff, C. G., and Minkley, E. G., Jr., 1970, The RNA polymerase sigma factor: A specificity determinant, in "Lepetit Colloquia on RNA polymerase and transcription" (L. Silvestri, ed.), pp. 124–147, North-Holland Publishing Co., Amsterdam.

Goff, C. G., and Weber, K., 1970, A T 4-induced RNA polymerase α subunit modification, *Cold Spring Harbor Symp. Quant. Biol.* **35**, 101.

Gold, L. M., 1975, personal communication.

Gold, L. M., and Schweiger, M., 1969, Synthesis of phage-specific α- and β-glucosyl transferase directed by T-even DNA *in vitro, Proc. Nat. Acad. Sci. USA* **62**, 892.

Gold, L. M., and Schweiger, M., 1970, Control of β-glucosyltransferase and lysozyme synthesis during T 4 deoxyribonucleic acid-dependent ribonucleic acid and protein synthesis *in vitro, J. Biol. Chem.* **245**, 2255.

Gold, L. M., O'Farrell, P. Z., Singer, B., and Stormo, G., 1973, Bacteriophage T 4 gene expression, *in* "Virus Research" (C. F. Fox and W. S. Robinson, eds.), pp. 205–225, Academic Press, New York and London.

Gold, L., O'Farrell, P. Z., and Russel, M., 1976, Regulation of gene 32 expression during bacteriophage T 4 infection of *Escherichia coli, J. Biol. Chem.,* in press.

Goldberg, A. R., 1970, Termination of *in vitro* RNA synthesis by ρ factor, *Cold Spring Harbor Symp. Quant. Biol.* **35,** 157.

Goldman, E., and Lodish, H. F., 1971, Inhibition of replication of RNA bacteriophage f 2 by superinfection with bacteriophage T 4, *J. Virol.* **8,** 417.

Goldman, E., and Lodish, H. F., 1972, Specificity of protein synthesis by bacterial ribosomes and initiation factors: Absence of change after phage T 4 infection, *J. Mol. Biol.* **67,** 35.

Golomb, M., and Chamberlin, M., 1974a, A preliminary map of the major transcription units read by T 7 RNA polymerase on the T 7 and T 3 bacteriophage chromosomes, *Proc. Nat. Acad. Sci. USA* **71,** 760.

Golomb, M., and Chamberlin, M., 1974b, Characterization of T 7 specific ribonucleic acid polymerase. IV. Resolution of the major *in vitro* transcripts by gel electrophoresis, *J. Biol. Chem.* **249,** 2858.

Gralla, J. D., 1976, Characterization and nucleotide sequence of the tight RNA polymerase binding site in a mutant in the UV5 *lac* promoter, to be published.

Gralla, J., Steitz, J. A., and Crothers, D. M., 1974, Direct physical evidence for secondary structure in an isolated fragment of R 17 bacteriophage mRNA, *Nature (London)* **248,** 204.

Grasso, R. J., and Buchanan, J. M., 1969, Synthesis of early RNA in bacteriophage T 4-infected *Escherichia coli* B, *Nature (London)* **224,** 882.

Grau, O., Ohlsson-Wilhelm, B. M., and Geiduschek, E. P., 1970, Transcription specificity in bacteriophage SPO 1 development, *Cold Spring Harbor Symp. Quant. Biol.* **35,** 221.

Grausz, J. D., 1972, Control of gene action of SPO 1 infection, Ph.D. Thesis, The University of Chicago.

Green, D. M., and Ln, D. M., and Laman, D., 1972, Organization of gene function in *Bacillus subtilis* bacteriophage SP 82G, *J. Virol.* **9,** 1033.

Greene, R., and Korn, D., 1967, The instability of T 4 messenger RNA, *J. Mol. Biol.* **28,** 435.

Griffith, J. D., 1976, Visualization of prokaryotic DNA in a regularly condensed chromatin-like fiber, *Proc. Nat. Acad. Sci. USA* **73,** 563.

Groner, Y., Pollack, Y., Berissi, H., and Revel, M., 1972a, Cistron specific translation control protein in *Escherichia coli, Nature (London) New Biol.* **239,** 16.

Groner, Y., Scheps, R., Kamen, R., Kolakofsky, D., and Revel, M., 1972b, Host subunit of Qβ replicase is translation control factor i, *Nature (London) New Biol.* **239,** 19.

Guha, A., and Szybalski, W., 1968, Fractionation of the complementary strands of coliphage T 4 DNA based on the asymmetric distribution of the polyU and polyUG binding sites, *Virology* **34,** 608.

Guha, A., Szybalski, W., Salser, W., Bolle, A., Geiduschek, E. P., and Pulitzer, J. F., 1971, Controls and polarity of transcription during bacteriophage T 4 development, *J. Mol. Biol.* **59,** 329.

Gussin, G. N., 1966, Three complementation groups in bacteriophage R 17, *J. Mol. Biol.* **21,** 435.

Guthrie, C., and McClain, W. H., 1973, Conditionally lethal mutants of bacteriophage T 4 defective in production of a transfer RNA, *J. Mol. Biol.* **81**, 137.

Guthrie, C., Seidman, J. G., Altman, S., Barrell, B. G., Smith, J. D., and McClain, W. H., 1973, Identification of tRNA precursor molecules made by phage T 4, *Nature (London) New Biol.* **246**, 6.

Guthrie, C., Seidman, J. G., Comer, M. M., Bock, R. M., Schmidt, F. J., Barrell, B. G., and McClain, W. H., 1975, The biology of bacteriophage T 4 transfer RNAs, *in* "Processing of RNA" (J. J. Dunn, ed.), pp. 106–123, Brookhaven National Laboratory, New York.

Guthrie, G. D., 1971, Translational regulation of T 4 messenger RNA metabolism, *Biochim. Biophys. Acta.* **232**, 324.

Guthrie, G. D., and Buchanan, J. M., 1966, Control of phage-induced enzymes in bacteria, *Fed. Proc.* **25**, 864.

Hagen, F., and Young, E. T., 1973, Regulation of synthesis of bacteriophage T 7 lysozyme mRNA, *Virology* **55**, 231.

Hagen, F. S., and Young, E. T., 1974, Preparative polyacrylamide gel electrophoresis of RNA. Identification of multiple molecular species of T 7 lysozyme mRNA, *Biochemistry* **13**, 3394.

Hager, G., Hall, B. D., and Fields, K., 1970, Transcription factors from T 4-infected *E. coli, Cold Spring Harbor Symp. Quant. Biol.* **35**, 233.

Hall, B. D., and Crouch, R., 1968, Transcription of the late genes of bacteriophage T 4, *in* "The Biochemistry of Virus Replication" (S. G. Laland and L. O. Fröholm, eds.), pp. 49–59, Universitetsforlaget, Oslo.

Hall, B. D., and Spiegelman, S., 1961, Sequence complementarity of T 2-DNA and T 2-specific RNA, *Proc. Nat. Acad. Sci. USA* **47**, 137.

Hall, B. D., Green, M. H., Nygaard, A. P., and Boezi, J. A., 1963, Copying of DNA in T 2-infected *E. coli, Cold Spring Harbor Symp. Quant. Biol.* **28**, 201.

Hall, B. D., Nygaard, A. P., and Green, M. H., 1964, Control of T 2-specific RNA synthesis, *J. Mol. Biol.* **9**, 143.

Hall, R. H., 1971, "The Modified Nucleosides in Nucleic Acids," Columbia University Press, New York.

Hamlett, N. V., and Berger, H., 1975, Mutations altering genetic recombination and repair of DNA in bacteriophage T 4, *Virology* **63**, 539.

Hansen, M. T., Bennett, P. M., and Von Meyenburg, K., 1973, Intracistronic polarity during dissociation of translation from transcription in *Escherichia coli, J. Mol. Biol.* **77**, 589.

Harm, W., 1963, Mutants of phage T 4 with increased sensitivity to ultraviolet, *Virology* **19**, 66.

Haselkorn, R., and Rothman-Denes, L. B., 1973, Protein synthesis, *Annu. Rev. Biochem.* **42**, 397.

Haselkorn, R., and Rouvière-Yaniv, J., 1976, Cyanobacterial DNA-binding protein related to *Escherichia coli* HU, *Proc. Nat. Acad. Sci. USA* **73**, 1517.

Haselkorn, R., Baldi, M. I., and Doskocil, J., 1968, RNA synthesis in T 4-infected *Escherichia coli* B, *in* "The Biochemistry of Virus Replication" (S. G. Laland and L. O. Fröholm, eds.), pp. 79–92, Universitetsforlaget, Oslo.

Haselkorn, R., Vogel, M., and Brown, R. D., 1969, Conservation of the rifamycin sensitivity during T 4 development, *Nature (London)* **221**, 836.

Haseltine, W. A., 1972, *In vitro* transcription of *Escherichia coli* ribosomal RNA genes, *Nature (London)* **235**, 329.

Hattman, S., 1970, Influence of T 4 superinfection on the formation of RNA bacteriophage coat protein, *J. Mol. Biol.* **47,** 599.

Hausmann, R., 1968, Sedimentation analysis of phage T 7-directed DNA synthesized in the presence of a dominant conditional lethal phage gene, *Biochem. Biophys. Res. Commun.* **31,** 609.

Hausmann, R., 1973, The genetics of T-odd phages, *Annu. Rev. Microbiol.* **27,** 51.

Hausmann, R., and Gomez, B., 1967, *Amber* mutants of bacteriophages T 3 and T 7 defective in phage-directed DNA synthesis, *J. Virol.* **1,** 779.

Hausmann, R., and Härle, 1971, Expression of the genomes of the related bacteriophages T 3 and T 7, *in* "Proceedings 1st European Biophysics Congress" (E. Broda, A. Locker, and H. Springer-Lederer, eds.), Vol. 1, pp. 467–488. Verlag der Wiener Medizinischen Akademie, Vienna, Austria.

Hausmann, R., Gomez, B., and Moody, B., 1968, Physiological and genetic aspects of abortive infection of a *Shigella sonnei* strain by coliphage T 7, *J. Virol.* **2,** 335.

Hayashi, Y., and Hayashi, M., 1971, Template activities of the ϕX-174 replicative allomorphic deoxyribonucleic acids, *Biochemistry* **10,** 4212.

Hayward, G. S., 1974, Unique double-stranded fragments of bacteriophage T 5 DNA resulting from preferential shear-induced breakage at nicks, *Proc. Nat. Acad. Sci. USA* **71,** 2108.

Hayward, G. S., and Smith, M. G., 1972a, The chromosome of bacteriophage T 5. I. Analysis of the single-stranded DNA fragments by agarose gel electrophoresis, *J. Mol. Biol.* **63,** 383.

Hayward, G. S., and Smith, M. G., 1972b, The chromosome of bacteriophage T 5. II. Arrangement of the single-stranded DNA fragments in the T 5^+ and T 5 st(o) chromosomes, *J. Mol. Biol.* **63,** 397.

Hayward, S. D., and Smith, M. G., 1973, The chromosome of bacteriophage T 5. III. Patterns of transcription from the single-stranded DNA fragments, *J. Mol. Biol.* **80,** 345.

Heinemann, S. F., and Spiegelman, W. G., 1970, Role of the gene N product in phage lambda, *Cold Spring Harbor Symp. Quant. Biol.* **35,** 315.

Helser, T. L., Davies, J. E., and Dahlberg, J. E., 1972, Mechanism of kasugamycin resistance in *Escherichia coli, Nature (London) New Biol.* **233,** 12.

Hemphill, H. E., and Whiteley, H. R., 1975, Bacteriophages of *Bacillus subtilis, Bacteriol. Rev.* **39,** 257.

Hendrickson, H. E., and Bujard, H., 1973, Structure and function of the genome of coliphage T 5. 2. Regions of transcription of the chromosome, *Eur. J. Biochem.* **33,** 529.

Hendrickson, H. E., and McCorquodale, D. J., 1971, Genetic and physiological studies of bacteriophage T 5. I. An expanded genetic map of T 5, *J. Virol.* **7,** 612.

Hendrickson, H. E., and McCorquodale, D. J., 1972, Genetic and physiological studies of bacteriophage T 5. III. Patterns of deoxyribonucleic acid synthesis induced by mutants of T 5 and the identification of genes influencing the appearance of phage-induced dihydrofolate reductase and deoxyribonuclease, *J. Virol.* **9,** 981.

Hendrix, R., 1971, Identification of proteins coded in phage lambda, *in* "The Bacteriophage Lambda" (A. D. Hershey, ed.), pp. 355–370, Cold Spring Harbor Laboratory, New York.

Hercules, K., and Sauerbier, W., 1973, Transcription units in bacteriophage T 4, *J. Virol.* **12,** 872.

Hercules, K., and Sauerbier, W., 1974, Two modes of *in vivo* transcription for genes 43 and 45 of phage T 4, *J. Virol.* **14,** 341.

Herman, R. C., and Moyer, R. W., 1974, *In vivo* repair of single-strand interruptions contained in bacteriophage T 5 DNA, *Proc. Nat. Acad. Sci. USA* **71**, 680.

Herman, R. C., and Moyer, R. W., 1975, *In vivo* repair of bacteriophage T 5 DNA: An assay for viral growth control, *Virology,* **66**, 393.

Herrlich, P., and Schweiger, M., 1971, Host- and phage-RNA polymerase mediated synthesis of T 7 lysozyme *in vivo, Mol. Gen. Genet.* **112**, 152.

Herrlich, P., Rahmsdorf, H. J., Pai, S. H., and Schweiger, M., 1974, Translational control induced by bacteriophage T 7, *Proc. Nat. Acad. Sci. USA* **71**, 1088.

Hershey, A. D., and Chase, M., 1952, Independent functions of viral protein and nucleic acid in growth of bacteriophage, *J. Gen. Physiol.* **36**, 39.

Herskowitz, I., and Signer, E., 1970, A site essential for the expression of all late genes in bacteriophage λ, *J. Mol. Biol.* **47**, 545.

Herzfeld, F., and Zillig, W., 1971, Subunit composition of DNA-dependent RNA polymerase of *Anacystis nidulans, Eur. J. Biochem.* **24**, 242.

Hesselbach, B. A., and Nakada, D., 1975, *E. coli* RNA polymerase and T 7 phage "host shut-off" function. Abstract, Meeting on "RNA polymerases," Cold Spring Harbor, August 1975.

Hesselbach, B. A., Yamada, Y., and Nakada, D., 1974, Isolation of an inhibitor protein of *E. coli* RNA polymerase from T 7 phage infected cells, *Nature (London)* **252**, 71.

Heyden, B., Nüsslein, C., and Schaller, H., 1972, Single RNA polymerase binding site isolated, *Nature (London) New Biol.* **240**, 9.

Hinkle, D. C., and Chamberlin, M., 1970, The role of sigma subunit in template site selection by *E. coli* RNA polymerase, *Cold Spring Harbor Symp. Quant. Biol.* **35**, 65.

Hinkle, D. C., Ring, J., and Chamberlin, M. J., 1972, Studies of the binding of *Escherichia coli* RNA polymerase to DNA. III. Tight binding to single-strand breaks in T 7 DNA, *J. Mol. Biol.* **70**, 197.

Hoffman, D. J., and Niyogi, S. K., 1973, RNA initiation with dinucleoside monophosphates during transcription of bacteriophage T 4 DNA with RNA polymerase of *Escherichia coli, Proc. Nat. Acad. Sci. USA* **70**, 574.

Homyk, T. Jr., and Weil, J., 1974, Deletion analysis of two nonessential regions of the T 4 genome, *Virology* **61**, 505.

Honjo, T., Ueda, K., Tanabe, T., and Hayaishi, O., 1972, Diphtheria toxin-catalyzed adenosine diphosphoribosylation of aminoacyl transferase II from rat liver, *in* "Metabolic Interconversion of Enzymes" (E. Helmreich, E. Holzer, and O. Wieland, eds.), pp. 193–207, Springer-Verlag, Berlin–Heidelberg–New York.

Horvitz, H. R., 1973, Polypeptide bound to the host RNA polymerase is specified by T 4 control gene 33, *Nature (London) New Biol.* **244**, 137.

Horvitz, H. R., 1974a, Control by bacteriophage T 4 of two sequential phosphorylations of the alpha subunit of *Escherichia coli* RNA polymerase, *J. Mol. Biol.* **90**, 727.

Horvitz, H. R., 1974b, Bacteriophage T 4 mutants deficient in alteration and modification of the *Escherichia coli* RNA polymerase, *J. Mol. Biol.* **90**, 739.

Horvitz, H. R., Ratner, D. I., and Poteete, A. R., 1975, personal communication.

Hosoda, J., and Cone, R., 1970, Analysis of T 4 phage proteins. I. Conversion of precursor proteins into lower molecular weight peptides during normal capsid formation, *Proc. Nat. Acad. Sci. USA* **66**, 1275.

Hosoda, J., and Levinthal, C., 1968, Protein synthesis by *Escherichia coli* infected with bacteriophage T 4D, *Virology* **34**, 709.

Howard, G. W., Wolin, M. L., and Champe, S. P., 1972, *Trans. N.Y. Acad. Sci.* **34,** 36.

Howe, C. C., Buckley, P. J., Carlson, K. M., and Kozinski, A. W., 1973, Multiple and specific initiation of T 4 DNA replication, *J. Virol.* **12,** 130.

Hsu, W.-T., and Weiss, S. B., 1969, Selective translation of T 4 template RNA by ribosomes from T 4-infected *Escherichia coli, Proc. Nat. Acad. Sci. USA* **64,** 345.

Hsu, W.-T., Foft, J. W., and Weiss, S. B., 1967, Effect of bacteriophage infection on the sulfur-labeling of sRNA, *Proc. Nat. Acad. Sci. USA* **58,** 2028.

Huang, W. M., 1975, Membrane-associated proteins of T 4-infected *Escherichia coli, Virology* **66,** 508.

Huberman, J., 1968, Visualization of replicating mammalian and T 4 bacteriophage DNA, *Cold Spring Harbor Symp. Quant. Biol.* **28,** 509.

Huberman, J. A., Kornberg, A., and Alberts, B. M., 1971, Stimulation of T 4 bacteriophage DNA polymerase by the protein product of T 4 gene 32, *J. Mol. Biol.* **62,** 39.

Hunter, B. I., Yamagishi, H., and Takahashi, I., 1967, Molecular weight of bacteriophage PBS 1 deoxyribonucleic acid, *J. Virol.* **1,** 841.

Hyman, R. W., 1971, Physical mapping of T 7 messenger RNA, *J. Mol. Biol.* **61,** 369.

Hyman, R. W., and Summers, W. C., 1972, Isolation and physical mapping of T 7 gene 1 messenger RNA, *J. Mol. Biol.* **71,** 573.

Hyman, R. W., Brunovskis, I., and Summers, W. C., 1973, DNA base sequence homology between coliphages T 7 and Φ II and between T 3 and Φ II as determined by heteroduplex mapping in the electron microscope, *J. Mol. Biol.* **77,** 189.

Hyman, R. W., Brunovskis, I., and Summers, W. C., 1974, a biochemical comparison of the related bacteriophages T 7, Φ I, Φ II, W 31, H, and T 3, *Virology* **57,** 189.

Iglewski, B. H., and Kabat, D., 1975, NAD-dependent inhibition of protein synthesis by *Pseudomonas aeruginosa* toxin, *Proc. Nat. Acad. Sci. USA.* **72,** 2284.

Ihler, G., and Nakada, D., 1971, Role of initiation factors in the control of T 4-specific protein synthesis, *J. Mol. Biol.* **62,** 419.

Iida, Y., and Matsukage, A., 1974, Effect of KCl concentration on the transcription by *E. coli* RNA polymerase. III. Starting nucleotide sequences of RNA synthesized on T 7 DNA, *Mol. Gen. Genet.* **129,** 27.

Ikemura, T., and Ozeki, H., 1975, Two-dimensional polyacrylamide-gel electrophoresis for purification of small RNAs specified by virulent coliphages T 4, T 5, T 7 and BF 23, *Eur. J. Biochem.* **51,** 117.

Imamoto, F., 1970, Evidence for premature termination of transcription of the tryptophan operon in polarity mutants of *Escherichia coli, Nature (London)* **228,** 232.

Imamoto, F., 1973a, Translation and transcription of the tryptophan operon, *in* "Progress in Nucleic Acid Research and Molecular Biology" (J. N. Davidson and W. E. Cohn, eds.), Vol 13, pp. 339–407, Academic Press, New York.

Imamoto, F., 1973b, Diversity of regulation of genetic transcription. I. Effect of antibiotics which inhibit the process of translation on RNA metabolism in *Escherichia coli, J. Mol. Biol.* **74,** 113.

Inoko, H., and Imai, M., 1976, Isolation and genetic characterization of the *nit*A mutants of *Escherichia coli* affecting the termination factor rho, *Mol. Gen. Genetics* **143,** 211.

Inouye, H., Pollack, Y., and Petre, J., 1974, Physical and functional homology between ribosomal protein S1 and interference factor i, *Europ. J. Biochem.* **45,** 109.

Issinger, O.-G., and Hausmann, R., 1973, Synthesis of bacteriophage-coded gene products during infection of *Escherichia coli* with amber mutants of T 3 and T 7 defective in gene 1, *J. Virol.* **11**, 465.

Issinger, O.-G., Beier, H., and Hausmann, R., 1973, *In vivo* and *in vitro* "phenotype mixing" with amber mutants of phages T 3 and T 7, *Mol. Gen. Genet.* **122**, 81.

Iwakura, Y., Fukuda, R., and Ishihama, A., 1974, Heterogeneity of RNA polymerase in *Escherichia coli*. II. Polyadenylate · polyuridylate synthesis by holoenzyme II, *J. Mol. Biol.* **83**, 369.

Jacob, F., and Monod, J., 1961, Genetic regulatory mechanisms in the synthesis of proteins, *J. Mol. Biol.* **3**, 318.

Jacquemin-Sablon, A., and Richardson, C. C., 1970, Analysis of the interruptions in bacteriophage T 5 DNA, *J. Mol. Biol.* **47**, 477.

Jay, G., and Kaempfer, R., 1974, Host interference with viral gene expression: Mode of action of bacterial factor i, *J. Mol. Biol.* **82**, 193.

Jayaraman, R., 1972, Transcription of bacteriophage T 4 DNA by *Escherichia coli* RNA polymerase *in vitro*: Identification of some immediate-early and delayed-early genes, *J. Mol. Biol.* **70**, 253.

Jayaraman, R., and Goldberg, E. B., 1969, A genetic assay for mRNA's of phage T 4, *Proc. Nat. Acad. Sci. USA* **64**, 198.

Jayaraman, R., and Goldberg, E. B., 1970, Transcription of bacteriophage T 4 genome *in vivo*, *Cold Spring Harbor Symp. Quant. Biol.* **35**, 197.

Jobe, A., and Bourgeois, S., 1972, *lac* repressor–operator interaction. VI. The natural inducer of the *lac* operon, *J. Mol. Biol.* **69**, 397.

Johnson, G. G., 1972, TF 1 binding to SPO 1 DNA, *Fed. Proc.* **31**, 472.

Johnson, G. G., and Geiduschek, E. P., 1972, Purification of the bacteriophage SPO 1 transcription factor 1, *J. Biol. Chem.* **247**, 3571.

Johnson, G. G., and Geiduschek, E. P., 1977, Specificity of the weak binding between the basic phage SP01 transcription-inhibitory protein, TF 1, and SPO 1 DNA, *Biochemistry*, in press.

Johnson, J. C., DeBacker, M., and Boezi, J. A., 1971, Deoxyribonucleic acid-dependent ribonucleic acid polymerase of *Pseudomonas putida*, *J. Biol. Chem.* **246**, 1222.

Johnson, S. R., and Hall, D. H., 1974, Characterization of new regulatory mutants of bacteriophage T 4, *J. Virol.* **13**, 666.

Johnston, D. E., and McClure, W. R., 1976, Abortive initiation on bacteriophage λ DNA, *in* "RNA Polymerases" (M. Chamberlin and R. Losick, eds.), pp. 413–428, Cold Spring Harbor Laboratory, New York.

Jorgenson, S., Buch, L., and Nierlich, D., 1969, Nucleoside triphosphate termini from RNA synthesized *in vivo* by *E. coli*, *Science* **164**, 1067.

Jörstad, K., Haarr, L., and Nygaard, A. P., 1973, Selected genes of T 2 DNA associated with *E. coli* membrane in the early period of T 2 infection. Abstracts, Ninth International Congress of Biochemistry, Stockholm.

Jurale, C., Kates, J. R., and Colby, C., 1970, Isolation of double-stranded RNA from T 4 phage infected cells, *Nature (London)* **226**, 1027.

Kahan, E., 1966, A genetic study of temperature-sensitive mutants of the *subtilis* phage SP 82, *Virology*, **30**, 650.

Kahan, E., 1971, Early and late gene function in bacteriophage SP 82, *Virology*, **46**, 634.

Kamen, R., 1970, Characterization of the subunits of Qβ replicase, *Nature (London)* **228**, 527.

Kamen, R., Kondo, M., Romer, W., and Weissmann, C., 1972, Reconstitution of Qβ replicase lacking subunit α with protein synthesis intereference factor i, *Eur. J. Biochem.* **31**, 44.

Kan, J., Kano-Sueoka, T., and Sueoka, N., 1968, Characterization of leucine transfer RNA in *E. coli* following infection with bacteriophage T 2, *J. Biol. Chem.* **243**, 5584.

Kaplan, D. A., and Nierlich, D. P., 1975, Cleavage of nonglucosylated bacteriophage T 4 DNA by restriction endonuclease *Eco* RI, *J. Biol. Chem.* **250**, 2395.

Karam, J. D., and Bowles, M. G., 1974, Mutation to overproduction of bacteriophage T 4 gene products, *J. Virol.* **13**, 438.

Kasai, T., 1974, Regulation of the expression of the histidine operon in *Salmonella typhimurium, Nature (London)* **249**, 523.

Kasai, T., and Bautz, E. K. F., 1969, Regulation of gene-specific RNA synthesis in bacteriophage T 4, *J. Mol. Biol.* **41**, 401.

Kates, J. R., and Mc Auslan, B. R., 1967, Messenger RNA synthesis by a "coated" viral genome, *Proc. Nat. Acad. Sci. USA* **57**, 314.

Kawamura, F. and Ito, J., 1975, Phage gene expression in sporulating cells of *B. subtilis* 168, *Virology* **62**, 414.

Kay, D., 1952, The effect of divalent metals on the multiplication of *coli* bacteriophage T5st, *Brit. J. Exp. Pathol.* **33**, 228.

Keil, T. U., and Hofschneider, P. H., 1973, Secondary structure of RNA phage M 12 replicative intermediates *in vivo, Biochim. Biophys. Acta* **312**, 297.

Kemper, B., and Hurwitz, J., 1973, Studies on T 4-induced nucleases. Isolation and characterization of a manganese-activated T 4-induced endonuclease, *J. Biol. Chem.* **248**, 91.

Kennell, D., 1968*a*, Inhibition of host protein synthesis during infection of *Escherichia coli* B by bacteriophage T 4. I. Continued synthesis of host RNA, *J. Virol.* **2**, 1262.

Kennell, D., 1968*b*, Titration of the gene sites on DNA by DNA–RNA hybridization, II. The *Escherichia coli* chromosome, *J. Mol. Biol.* **34**, 85.

Kennell, D., 1970, Inhibition of host protein synthesis during infection of *Escherichia coli* by bacteriophage T 4. II. Induction of host mRNA and its exclusion from polysomes, *J. Virol.* **6**, 208.

Kennell, D., and Bicknell, I., 1973, Decay of messenger ribonucleic acid from the lactose operon of *Escherichia coli* as a function of growth temperature, *J. Mol. Biol.* **74**, 21.

Kenner, R. A., 1973, A protein–nucleic acid crosslink in 30 S ribosomes, *Biochem. Biophys. Res. Commun.* **51**, 932.

Khesin, R. B., 1970, Studies on the RNA synthesis and RNA polymerase in normal and phage infected *E. coli* cells, in Lepetit Symposium, 1969, on "RNA Polymerase and Transcription" (L. Silvestri, ed.), pp. 167–189, North Holland Publishing Company, Amsterdam, London.

Khesin, R. B., Shemyakin, M. F., Gorlenko, J. M., Bogdanova, S. L., and Afanasieva, T. P., 1962, RNA polymerase in *E. coli* cells infected with T 2 phage, *Biokhimiya* **27**, 1092.

Khesin, R. B., Gorlenko, Zh. M., Shemyakin, M. F., Bass, I. A., and Prozorov, A. A., 1963, Connection between protein synthesis and regulation of messenger RNA's formation in *E. coli* B. cells upon development of T 2 phage, *Biokhimiya* **28**, 1070.

Khesin, R. B., Bogdanova, E. S., Goldfarb, A. D., Jr., and Zograff, Yu. N., 1972, Competition for the DNA template between RNA polymerase molecules from normal and phage-infected *E. coli, Mol. Gen. Genet.* **119**, 299.

Kim, J.-S., and Davidson, N., 1974, Electron microscope heteroduplex study of sequence relations of T 2, T 4, and T 6 bacteriophage DNAs, *Virology* **57,** 93.

Kjeldgaard, N. O., and Gausing, K., 1974, Regulation of biosynthesis of ribosomes, *in* "Ribosomes" (M. Nomura, A. Tissieres, and P. Lengyel, eds.), pp. 369–392, Cold Spring Harbor Laboratory, New York.

Klem, E. B., Hsu, W.-I., and Weiss, S. B., 1970, The selective inhibition of protein initiation by T 4 phage-induced factors, *Proc. Nat. Acad. Sci. USA* **67,** 696.

Kleppe, R. K., 1975, Influence of salt on transcription by T 4 core RNA polymerase, *FEBS (Fed. Eur. Biochem. Soc.) Lett.* **51,** 237.

Kleppe, R. K., and Nygaard, A. P., 1969, Messenger RNA concentration and protein synthesis during T 4 infection of *Escherichia coli* B, *Biochim. Biophys. Acta.* **190,** 202.

Klier, A. F., and Lecadet, M. M., 1974, Evidence in favour of the modification *in vivo* of RNA polymerase subunits during sporogenesis in *Bacillus thuringiensis, Eur. J. Biochem.* **47,** 111.

Klier, A. F., Lecadet, M. M., and Dedonder, R., 1973, Sequential modifications of DNA-dependent RNA polymerase during sporogenesis in *Bacillus thuringiensis, Eur. J. Biochem.* **36,** 317.

Knopf, K.-W., and Bujard, H., 1975, Structure and function of the genome of coliphage T 5: Transcription *in vitro* of the "nicked" and "nick-free" T 5⁺ DNA, *Eur. J. Biochem.* **53,** 371.

Koch, G., and Hershey, A. D., 1959, Synthesis of phage-precursor protein in bacteria infected with T 2, *J. Mol. Biol.* **1,** 260.

Kolakofsky, D., and Weissmann, C., 1971, Possible mechanism for transition of viral RNA from polysome to replication complex, *Nature (London) New Biol.* **231,** 42.

Kondo, M., Gallerani, G., and Weissmann, C., 1970, Subunit structure of Qβ replicase, *Nature (London)* **228,** 525.

Konrad, M. W., 1974, Apparent average length of the transcriptional unit in bacteria, *J. Bacteriol.* **119,** 228.

Konrad, M. W., Toivonen, J., and Cook, J., 1976, The 5′ ends of bacterial RNA. II. The triphosphate terminated ends of primary gene transcripts, *Biochim. Biophys. Acta,* **425,** 63.

Kornberg, A., 1974, "DNA synthesis," Freeman, San Francisco.

Kossman, C. R., Stamato, T. D., and Pettijohn, D. E., 1971, Tandem synthesis of the 16 S and 23 S ribosomal RNA sequences of *E. coli, Nature (London) New Biol.* **234,** 102.

Kozinski, A. W., and Doermann, A. H., 1975, Repetitive DNA replication of the incomplete genomes of phage T 4 petite particles, *Proc. Nat. Acad. Sci. USA* **72,** 1734.

Kozloff, K. M., 1968, Biochemistry of the T-even bacteriophages of *Escherichia coli, in* "Molecular Basis of Virology" (H. Fraenkel-Conrat, ed.), pp. 436–496, Reinhold Book Co., New York.

Krakow, J. S., and Von Der Helm, K., 1970, Azotobacter RNA polymerase transitions and the release of sigma, *Cold Spring Harbor Symp. Quant. Biol.* **35,** 73.

Krakow, J. S., Daley, K., and Karstadt, M., 1969, *Azotobacter vinelandii* RNA polymerase. VII. Enzyme transitions during unprimed r[I-C]synthesis, *Proc. Nat. Acad. Sci. USA* **62,** 432.

Kramer, R. A., Rosenberg, M., and Steitz, J. A., 1974, Nucleotide sequences of the 5′ and 3′ termini of bacteriophage T7 early messenger RNAs synthesized *in vivo:* Evidence for sequence specificity in RNA processing, *J. Mol. Biol.* **89,** 767.

Krisch, H. M., Bolle, A., and Epstein, R. H., 1974, The regulation of the synthesis of bacteriophage T 4 gene 32 protein, *J. Mol. Biol.* **88,** 89.

Krylov, V. N., and Yankofsky, N. K., 1975, Mutations in the new gene *st*III of bacteriophage T 4B suppressiong the lysis defect of gene *st*II and gene *e* mutants, *J. Virol.* **15,** 22.

Küpper, H. A., McAllister, W. T., and Bautz, E. K. F., 1973, Comparison of *Escherichia coli* and T 3 RNA polymerases. Differential inhibition of transcription by various drugs, *Eur. J. Biochem.* **38,** 581.

Kurland, C. G., 1974, Functional organization of the 30 S ribosomal subunit, *in* "Ribosomes" (M. Nomura, A. Tissieres, and P. Lengyel, eds.) pp. 309–331, Cold Spring Harbor Laboratory, New York.

Kutter, E. M., and Wiberg, J. S., 1969, Biological effects of substituting cytosine for 5-hydroxy methyl cytosine in the deoxyribonucleic acid of bacteriophage T 4, *J. Virol.* **4,** 439.

Kutter, E. M., Beug, A., Sluss, R., Jensen, L., and Bradley, D., 1975, The production of undergraded cytosine-containing DNA by bacteriophage T 4 in the absence of dCTPase and endonucleases II and IV, and its effects on T 4-directed protein synthesis, *J. Mol. Biol.,* **99,** 591.

Kuwano, M., Schlessinger, D., and Morse, D. E., 1971, Loss of dispensable endonuclease activity in relief of polarity by *su*A, *Nature (London) New Biol.* **231,** 214.

Labedan, B., and Legault-Demare, J., 1973, Penetration into host cells of naked, partially injected (post FST) DNA of bacteriophage T 5, *J. Virol.* **12,** 226.

Labedan, B., Crochet, M., Legault-Demare, J., and Stevens, B. J., 1973, Location of the first-step transfer fragment and single-strand interruptions in T 5 st(o) bacteriophage DNA, *J. Mol. Biol.* **75,** 213.

Laemmli, U. K., 1970, Cleavage of structural proteins during the assembly of the head of bacteriophage T 4, *Nature (London)* **227,** 680.

Lamfrom, H., Sarabhai, A., Nierlich, D. P., and Abelson, J., 1973, Synthesis of tRNA in cell-free extracts, *Nature (London) New Biol.* **246,** 11.

Landy, A., and Spiegelman, S., 1968, Exhaustive hybridization and its application to an analysis of the RNA synthesized in T 4-infected cells, *Biochemistry* **7,** 585.

Lang, D., 1970, Molecular weights of coliphages and coliphage DNA III. Contour length and molecular weight of DNA from bacteriophages T 4, T 5 and T 7, and from Bovine Papilloma Virus, *J. Mol. Biol.* **54,** 557.

Lanni, Y. T., 1960*a*, Invasion by bacteriophage T 5. I. Some basic kinetic features, *Virology* **10,** 501.

Lanni, Y. T., 1960*b*, Invasion by bacteriophage T 5. II. Dissociation of calcium-independent and calcium-dependent processes, *Virology* **10,** 514.

Lanni, Y. T., 1965, DNA transfer from phage T 5 to host cells: Dependence on intercurrent protein synthesis, *Proc. Nat. Acad. Sci. USA* **53,** 969.

Lanni, Y. T., 1968, First-step-transfer deoxyribonucleic acid of bacteriophage T 5, *Bact. Rev.* **32,** 227.

Lanni, Y. T., 1969, Functions of two genes in the first-step-transfer DNA of bacteriophage T 5, *J. Mol. Biol.* **44,** 173.

Lanni, Y. T., and McCorquodale, D. J., 1963, DNA metabolism in T 5-infected *Escheri ia coli*: Biochemical function of a presumptive genetic fragment of the phage. *Virology* **19,** 72.

Lavi, U., and Marcus, M., 1972, Arrest of host DNA synthesis in *Bacillus subtilis* infected with phage φe, *Virology* **49,** 668.

Lavi, U., Nattenberg, A., Ronen, A., and Marcus, M., 1974, *Bacillus subtilis* DNA polymerase III is required for the replication of the virulent bacteriophage φe, *J. Virol.* **14**, 1337.

Lee-Huang, S., and Ochoa, S., 1971, Messenger discriminating species of initiation factor F3, *Nature (London) New Biol.* **234**, 236.

Lee-Huang, S., and Ochoa, S., 1972, Specific inhibitors of MS 2 and late T 4 RNA translation in *E. coli*, *Biochem. Biophys. Res. Commun.* **49**, 371.

Lee-Huang, S., and Ochoa, S., 1973, Purification and properties of two messenger discriminating species of *E. coli* initiation factor 3, *Arch. Biochem. Biophys.* **156**, 84.

Leffler, S., and Szer, W., 1973, Messenger selection by bacterial ribosomes, *Proc. Nat. Acad. Sci. USA* **70**, 2364.

Lehman, I. R., and Pratt, E. A., 1960, On the structure of the glucosylated hydroxymethyl cytosine nucleotides of coliphages T 2, T 4, and T 6, *J. Biol. Chem.* **235**, 3254.

Leive, L., 1965, A nonspecific increase in permeability in *Escherichia coli* produced by EDTA, *Proc. Nat. Acad. Sci. USA* **53**, 745.

Lembach, K. J., Kuninaka, A., and Buchanan, J. M., 1969, The relationship of DNA replication to the control of protein synthesis in protoplasts of T 4-infected *Escherichia coli* B, *Proc. Nat. Acad. Sci. USA* **62**, 446.

Levinthal, C., Hosoda, J., and Shub, D., 1967, The control of protein synthesis after phage infection, *in* "The Molecular Biology of Viruses" (J. S. Colter and W. Paranchych, eds.), pp. 71–87, Academic Press, New York.

Levner, M. H., and Cozzarelli, N. R., 1972, Replication of viral DNA in SPO 1-infected *Bacillus subtilis*. I. Replicative intermediates, *Virology* **48**, 402.

Lewin, B., 1974, "Gene Expression." Volume 1. Bacterial genomes, Chapters 1–5, John Wiley and Sons, New York.

Liljemark, W. F., and Anderson, D. L., 1970, Morphology and physiology of the intracellular development of *Bacillus subtilis* bacteriophage φ 25, *J. Virol.* **6**, 114.

Linial, M., and Malamy, M. H., 1970a, The effects of F factors on RNA synthesis and degradation after infection of *E. coli* with phage Φ II, *Cold Spring Harbor Symp. Quant. Biol.* **35**, 263.

Linial, M., and Malamy, M. H., 1970b, Studies with bacteriophage Φ II. Events following infection of male and female derivatives of *Escherichia coli* K-12′, *J. Virol.* **5**, 72.

Linn, T. G., Greenleaf, A. L., Shorenstein, R. G., and Losick, R., 1973, Loss of the sigma activity of RNA polymerase of *B. subtilis* during sporulation, *Proc. Nat. Acad. Sci. USA* **70**, 1865.

Littauer, U. Z., and Inouye, H., 1973, Regulation of tRNA, *Annu. Rev. Biochem.* **42**, 439.

Little, J. W., 1973, Mutants of bacteriophage T 4 which allow *amber* mutants of gene 32 to grow in *ochre*-suppressing hosts, *Virology* **53**, 47.

Little, J. W., Zimmerman, S. B., Oshinsky, C. K., and Gellert, M., 1967, Enzymatic joining of DNA strands. II. An enzyme-adenylate intermediate in the DPN-dependent DNA ligase reaction, *Proc. Nat. Acad. Sci. USA* **58**, 2004.

Losick, R., 1972, *In vitro* transcription, *Annu. Rev. Biochem.* **41**, 409.

Losick, R., and Sonenshein, A. L., 1969, Change in the template specificity of RNA polymerase during sporulation of *Bacillus subtilis*, *Nature (London)* **224**, 35.

Losick, R., Sonenshein, A. L., Shorenstein, R. G., and Hussey, C., 1970a, Role of RNA polymerase in sporulation, *Cold Spring Harbor Symp. Quant. Biol.* **35**, 443.

Losick, R., Shorenstein, R. G., and Sonenshein, A. L., 1970*b*, Structural alteration of RNA polymerase during sporulation, *Nature (London)* **227**, 910.

Lowery-Goldhammer, C., and Richardson, J. P., 1974, An RNA-dependent nucleoside triphosphate phosphohydrolase (ATPase) associated with rho termination factor, *Proc. Nat. Acad. Sci. USA* **71**, 2003.

Lucas-Lenard, J., and Beres, L., 1974, Protein synthesis—peptide chain elongation, *in* "The Enzymes" (P. D. Boyer, ed.), pp. 53–86, Academic Press, New York.

Lucas-Lenard, J., and Lipmann, F., 1971, Protein biosynthesis, *Annu. Rev. Biochem.* **40**, 409.

Lund, E., Dahlberg, J. E., Lindahl, L., Jaskunas, S. R., Dennis, P. P., and Nomura, M., 1976, Transfer RNA genes between 16 S and 23 S rRNA transcription units of E. coli, *Cell* **7**, 165.

Luria, S. E., and Human, M. L., 1952, A nonhereditary, host-induced variation of bacterial viruses, *J. Bacteriol.* **64**, 557.

Lutkenhaus, J., Ryan, J., and Konrad, M., 1973, Kinetics of phosphate incorporation into adenosine triphosphate and guanosine triphosphate in bacteria, *J. Bacteriol.* **116**, 1113.

McAllister, W. T., Küpper, H. A., and Bautz, E. K. F., 1973, Kinetics of transcription by the bacteriophage-T 3 RNA polymerase *in vitro, Eur. J. Biochem.* **34**, 489.

McClain, W. H., 1970, UAG suppressor coded by bacteriophage T 4, *FEBS (Fed Eur. Biochem. Soc.) Lett.* **6**, 99.

McClain, W. H., Guthrie, C., and Barrell, B. G., 1972, Eight transfer RNAs induced by infection of *Escherichia coli* with bacteriophage T 4, *Proc. Nat. Acad. Sci. USA* **69**, 3703.

McClain, W. H., Guthrie, C., and Barrell, B. G., 1973, The psu† amber suppressor gene of bacteriophage T 4: Identification of its amino acid and transfer RNA, *J. Mol. Biol.* **81**, 157.

McCorquodale, D. J., 1975, The T-odd bacteriophages, *Crit. Rev. Microbiol.*, **4**, 101.

McCorquodale, D. J., and Buchanan, J. M., 1968, Patterns of protein synthesis in T 5-infected *Escherichia coli, J. Biol. Chem.* **243**, 2550.

McCorquodale, D. J., and Lanni, Y. T., 1964, Molecular aspects of DNA transfer from phage T 5 to host cells. I. Characterization of first-step-transfer material, *J. Mol. Biol.* **10**, 10.

McCorquodale, D. J., and Lanni, Y. T., 1970, Patterns of protein synthesis in *Escherichia coli* infected by *amber* mutants in the first-step-transfer DNA of T 5. *J. Mol. Biol.* **48**, 133.

Mahadik, S. P., and Srinivasan, P. R., 1971, Stimulation of DNA-dependent RNA synthesis by a protein associated with ribosomes, *Proc. Nat. Acad. Sci. USA* **68**, 1898.

Mahadik, S. P., Dharmgrongartama, B., and Srinivasan, P. R., 1972, An inhibitory protein of *Escherichia coli* RNA polymerase in bacteriophage T 3-infected cells, *Proc. Nat. Acad. Sci. USA* **69**, 162.

Mahadik, S. P., Dharmgrongartama, B., and Srinivasan, P. R., 1974, Regulation of host RNA synthesis in bacteriophage T 3-infected cells. Properties of an inhibitory protein of *E. coli* ribonucleic acid polymerase, *J. Biol. Chem.* **249**, 1787.

Mailhammer, R., Yang, H., Zubay, G., and Reiness, G., 1975, Effects of bacteriophage T 4 induced modification of *E. coli* RNA polymerase on gene expression *in vitro, Proc. Nat. Acad. Sci. USA,* **72**, 4928.

Maitra, U., 1971, Induction of a new RNA polymerase in *Escherichia coli* infected with bacteriophage T 3, *Biochem. Biophys. Res. Commun.* **43**, 443.

Maitra, U., and Huang, H. H., 1972, Initiation, release, and reinitiation of RNA chains by bacteriophage T 3-induced polymerase from T 3 DNA templates, *Proc. Nat. Acad. Sci. USA* **69**, 55.

Maitra, U., and Hurwitz, J., 1965, The role of DNA in RNA synthesis. IX. Nucleoside triphosphate termini in RNA polymerase products, *Proc. Nat. Acad. Sci. USA* **54**, 815.

Maitra, U., Cohen, S. N., and Hurwitz, J., 1966, Specificity of initiation and synthesis of RNA from DNA templates, *Cold Spring Harbor Symp. Quant. Biol.* **31**, 113.

Maitra, U., Nakata, Y., and Hurwitz, J., 1967, The role of deoxyribonucleic acid in ribonucleic acid synthesis. XIV. A study of the initiation of RNA synthesis, *J. Biol. Chem.* **242**, 4908.

Maitra, U., Lockwood, A. H., Dubnoff, J. S., and Guha, A., 1970, Termination, release, and reinitiation of RNA chains from DNA templates by *Escherichia coli* RNA polymerase, *Cold Spring Harbor Symp. Quant. Biol.* **35**, 143.

Maitra, U., Salvo, R. A., and Chakraborty, P. R., 1974, Specificity of RNA chain initiation by phage T 3-induced RNA polymerase, *J. Biol. Chem.* **249**, 5835.

Maizels, N. M., 1973, The nucleotide sequence of the lactose messenger ribonucleic acid transcribed from the UV5 promoter mutant of *Escherichia coli*, *Proc. Nat. Acad. Sci. USA* **70**, 3585.

Majors, J., 1975a, Specific binding of CAP factor to *lac* promoter DNA, *Nature (London)*, **256**, 672.

Majors, J., 1975b, Initiation of *in vitro* mRNA synthesis from the wild-type lac promoter, *Proc. Nat. Acad. Sci. USA* **72**, 4394.

Mäkela, O. P., Mäkela, P. H., and Soikkeli, S., 1964, Sex-specificity of the bacteriophage T 7, *Ann. Med. Exp. Biol. Fenn.* **42**, 188.

Mandel, M., 1968, Some properties of viral deoxyribonucleic acids: Buoyant densities, melting temperatures and GC content, "Handbook of Biochemistry," 2nd ed. (H. A. Sober, ed.), p. H-9, Chemical Rubber Co., Cleveland.

Mangel, W. F., and Chamberlin, M. J., 1974a, Studies of RNA chain initiation by *E. coli* RNA polymerase bound to T 7 DNA. I. An assay for the role and extent of RNA chain initiation, *J. Biol. Chem.* **249**, 2995.

Mangel, W. F., and Chamberlin, M. J., 1974b, Studies of RNA chain initiation by *E. coli* RNA polymerase bound to T 7 DNA. II. The effect of alterations in ionic strength in chain initiation and on the conformation of binary complexes, *J. Biol. Chem.* **249**, 3002.

Mangel, W. F., and Chamberlin, M. J., 1974c, Studies of RNA chain initiation by *E. coli* RNA polymerase bound to T 7 DNA. III. The effect of temperature on RNA chain initiation and on the conformation of binary complexes, *J. Biol. Chem.* **249**, 3007.

Maniatis, T., and Ptashne, M., 1973, Multiple repressor binding at the operators in bacteriophage λ, *Proc. Nat. Acad. Sci. USA* **70**, 1531.

Maor, G., and Shalitin, C., 1974, Competence of membrane-bound T4rII: DNA for *in vitro* "late" mRNA transcription, *Virology* **62**, 500.

Marchin, G. L., Comer, M. M., and Neidhardt, F. C., 1972, Viral modification of the valyl transfer ribonucleic acid synthetase of *Escherichia coli*, *J. Biol. Chem.* **247**, 5132.

Marchin, G. L., Müller, U. R., and Al-Khateeb, G. H., 1974, The effect of transfer ribonucleic acid on virally modified valyl transfer ribonucleic acid synthetase of *Escherichia coli*, *J. Biol. Chem.* **249**, 4705.

Marcus, M., and Newlon, M. C., 1971, Control of DNA synthesis in *Bacillus subtilis* by phage φe, *Virology* **44,** 82.

Mark, K.-L., 1970, The relationship between the synthesis of DNA and the synthesis of phage lysozyme in *Escherichia coli* infected by bacteriophage T 4, *Virology* **42,** 20.

Marsh, R. C., Breschkin, A. M., and Mosig, G., 1971, Origin and direction of bacteriophage T 4 DNA replication. II. A gradient of marker frequencies in partially replicated T 4 DNA as assayed by transformation, *J. Mol. Biol.* **60,** 213.

Mason, W. S., and Haselkorn, R., 1972, Product of T 4 gene 12, *J. Mol. Biol.* **66,** 445.

Mathews, C. K., 1970, T-even bacteriophage-tolerant mutants of *Escherichia coli* B. I. Isolation and preliminary characterization, *J. Virol.* **6,** 163.

Mathews, C. K., 1971, "Bacteriophage Biochemistry," Van Nostrand Reinhold, New York.

Mathews, C. K., 1972, Biochemistry of deoxyribonucleic acid-defective *amber* mutants of bacteriophage T 4. III. Nucleotide pools, *J. Biol. Chem.* **247,** 7430.

Mathews, C. K., 1977, Reproduction of large virulent bacteriophages, in "Comprhehensive Virology" (H. Fraenkel-Conrat and R. Wagner, eds.), Volume 7, Plenum Press, New York.

Mathews, C. K., and Kessin, R. H., 1967, Control of bacteriophage-induced enzyme synthesis in cells infected with a temperature-sensitive mutant, *J. Virol.* **1,** 92.

Mathews, C. K., Crosby, L. K., and Kozloff, L. M., 1973, Inactivation of T 4D bacteriophage by antiserum against bacteriophage dihydrofolate reductase, *J. Virol.* **12,** 74.

Matson, T., Richardson, J., and Goodin, D., 1974, Mutant of bacteriophage T 4D affecting expression of many early genes, *Nature (London)* **250,** 48.

Maynard-Smith, S., and Symonds, N., 1973, The unexpected location of a gene conferring abnormal radiation sensitivity on phage T 4, *Nature (London)* **241,** 395.

Mével-Ninio, M. T., and Valentine, R. C., 1975, Energy requirement for biosynthesis of DNA in *Escherichia coli.* Role of membrane-bound energy-transducing ATPase (coupling factor), *Biochim. Biophys. Acta* **376,** 485.

Milanesi, G., Brody, E. N., and Geiduschek, E. P., 1969, Sequence of the *in vitro* transcription of T 4 DNA, *Nature (London)* **221,** 1014.

Milanesi, G., Brody, E. N., Grau, O., and Geiduschek, E. P., 1970, Transcription of the bacteriophage T 4 template *in vitro*: Separation of "delayed early" from "immediate early" transcription, *Proc. Nat. Acad. Sci. USA* **66,** 181.

Miller, M. J., and Wahba, A. J., 1974, Inhibition of synthetic and natural messenger translation. II. Specificity and mechanism of action of a protein isolated from *E. coli* MRE 600 ribosomes, *J. Biol. Chem.* **249,** 3808.

Miller, M. J., Niveleau, A., and Wahba, A. J., 1974, Inhibition of synthetic and natural messenger translation. I. Purification and properties of a protein isolated from *Escherichia coli* MRE 600 ribosomes, *J. Biol. Chem.* **249,** 3803.

Miller, O. L. Jr., Beatty, B. R., Hamkalo, B. A., and Thomas, C. A., Jr., 1970, Electron microscopic visualization of transcription, *Cold Spring Harbor Symp. Quant. Biol.* **35,** 505.

Miller, R. C., Jr., 1972, Association of replicative T 4 deoxyribonucleic acid and bacterial membranes, *J. Virol.* **10,** 920.

Miller, R. C., Jr., and Kozinski, A. W., 1970, Early intracellular events in the replication of bacteriophage T 4 deoxyribonucleic acid. V. Further studies on the T 4 protein–deoxyribonucleic acid complex, *J. Virol.* **5,** 490.

Miller, R. C., Jr., Kozinski, A. W., and Litwin, S., 1970, Molecular recombination in bacteriophage deoxyribonucleic acid. III. Formation of long single strands during recombination, *J. Virol.* **5**, 368.

Millet, J., Kerjan, P., Aubert, J. P., and Szulmajster, J., 1972, Proteolytic conversion *in vitro* of *B. subtilis* vegetative RNA polymerase into the homologous spore enzyme, *FEBS (Fed. Eur. Biochem. Soc.) Lett.* **23**, 47.

Millette, R. L., Trotter, C. D., Herrlich, P., and Schweiger, M., 1970, *In vitro* synthesis, termination, and release of active messenger RNA, *Cold Spring Harbor Symp. Quant. Biol.* **35**, 135.

Minkley, E. G., Jr., 1973, Functional form of RNA synthesis termination factor rho, *J. Mol. Biol.* **78**, 577.

Minkley, E. G., Jr., 1974*a*, Transcription of the early region of bacteriophage T 7: Characterization of the *in vivo* transcripts, *J. Mol. Biol.* **83**, 289.

Minkley, E. G., Jr., 1974*b*, Transcription of the early region of bacteriophage T 7: Specificity and selectivity in the *in vitro* initiation of RNA synthesis, *J. Mol. Biol.* **83**, 305.

Minkley, E. G., Jr., and Pribnow, D., 1973, Transcription of the early region of bacteriophage T 7: Selective initiation with dinucleotides, *J. Mol. Biol.* **77**, 255.

Mitchison, J. M., 1971, "The Biology of the Cell Cycle," Cambridge University Press, London and New York.

Mizobuchi, K., and McCorquodale, D. J., 1974, Abortive infection by bacteriophage BF 23 due to the colicin Ib factor. II. Involvement of pre-early proteins. *J. Mol. Biol.* **85**, 67.

Mizobuchi, K., Anderson, G. C., and McCorquodale, D. J., 1971, Abortive infection of bacteriophage BF 23 due to the colicin Ib factor. I. Genetic studies of nonrestricted and *amber* mutants of bacteriophage BF 23, *Genetics* **68**, 323.

Mizuno, S., and Nitta, K., 1969, Effect of streptovaricin on RNA synthesis in phage T 4-infected *Escherichia coli*, *Biochem. Biophys. Res. Commun.* **35**, 127.

Montgomery, D. L., and Snyder, L. R., 1973, A negative effect of β-glucosylation on T 4 growth in certain RNA polymerase mutants of *Escherichia coli*: Genetic evidence implicating pyrimidine-rich sequences of DNA in transcription, *Virology* **53**, 349.

Morrison, T. G., and Malamy, M. H., 1971, T 7 translational control mechanisms and their inhibition by F factors, *Nature (London) New Biol.* **231**, 37.

Morrison, T. G., Blumberg, D. D., and Malamy, M. H., 1974, T 7 protein synthesis in F′ episome-containing cells: Assignment of specific proteins to three translational groups, *J. Virol.* **13**, 386.

Morse, D. E., 1970, "Delayed early" mRNA for the tryptophan operon? An effect of chloramphenicol, *Cold Spring Harbor Symp. Quant. Biol.* **35**, 495.

Morse, D. E., 1971, Polarity induced by chloramphenicol and relief by *su*A, *J. Mol. Biol.* **55**, 113.

Morse, D. E., and Morse, A. N. C., 1976, Dual-control of the tryptophan operon by the repressor and tryptophanyl-tRNA synthetase, *J. Mol. Biol.* **103**, 209.

Morse, D. E., and Yanofsky, C., 1969, Polarity and the degradation of mRNA, *Nature (London)* **224**, 329.

Mosig, G., 1968, A map of distances along the DNA molecule of bacteriophage T 4, *Genetics* **59**, 137.

Mosig, G., 1970, A preferred origin and direction of bacteriophage T 4 DNA replication, *J. Mol. Biol.* **53**, 503.

Mosig, G., and Werner, R., 1969, On the replication of incomplete chromosomes of phage T 4, *Proc. Nat. Acad. Sci. USA* **64**, 747.

Mosig, G., Carnighan, J. R., Bibring, J. B. Cole, R., Bock, H.-G. O., and Bock, S., 1972, Coordinate variation in lengths of deoxyribonucleic acid molecules and head lengths in morphological variants of bacteriophage T 4, *J. Virol.* **9**, 857.

Moyer, R. W., and Buchanan, J. M., 1969, Patterns of RNA synthesis in T 5-infected cells. I. As studied by the technique of DNA–RNA hybridization-competition, *Proc. Nat. Acad. Sci. USA* **64**, 1249.

Moyer, R. W., and Buchanan, J. M., 1970a, Effect of calcium ions on synthesis of T 5-specific proteins. *J. Biol. Chem.* **245**, 5897.

Moyer, R. W., and Buchanan, J. M., 1970b, Effect of calcium ions on synthesis of T 5-specific ribonucleic acid, *J. Biol. Chem.* **245**, 5904.

Moyer, R. W., Fu, A. S., and Szabo, C., 1972, Regulation of bacteriophage T 5 development by ColI factors, *J. Virol.* **9**, 804.

Müller, U. R., and Marchin, G. L., 1975, Temporal appearance of bacteriophage T 4-modified valyl tRNA synthetase in *Escherichia coli, J. Virol.* **15**, 238.

Murialdo, H., and Siminovitch, L., 1972, The morphogenesis of bacteriophage lambda. IV. Identification of gene products and control of the expression of the morphogenetic information, *Virology* **48**, 785.

Murooka, Y., and Lazzarini, R. A., 1973, *In vitro* synthesis of ribosomal RNA by a DNA–protein complex isolated from *Escherichia coli, J. Biol. Chem.* **248**, 6248.

Nakazato, H., Venkatesan, S., and Edmonds, M., 1975, Polyadenylic acid sequences in *E. coli* messenger RNA, *Nature (London)* **256**, 144.

Naot, Y., and Shalitin, C., 1973, Role of gene 52 in bacteriophage T 4 DNA synthesis, *J. Virol.* **11**, 862.

Natale, P. J., and Buchanan, J. M., 1972, DNA-directed synthesis *in vitro* of T 4 phage-specific enzymes, *Proc. Nat. Acad. Sci. USA* **69**, 2513.

Natale, P. J., and Buchanan, J. M., 1974, Initiation characteristics for the synthesis of five T 4 phage-specific messenger RNAs *in vitro, Proc. Nat. Acad. Sci. USA* **71**, 422.

Neidhardt, F. C., and Earhart, C. F., 1966, Phage-induced appearance of a valyl tRNA synthetase activity in *E. coli, Cold Spring Harbor Symp. Quant. Biol.* **31**, 557.

Neidhardt, F. C., Marchin, G. L., McClain, W. H., Boyd, R. F., and Earhart, C. F., 1969, Phage-induced modification of valyl-tRNA synthetase, *J. Cell. Physiol. Suppl.* **1**, 87.

Nelson, E. T., and Buller, C. S., 1974, Phospholipase activity in bacteriophage-infected *Escherichia coli.* I. Demonstration of a T 4 bacteriophage-associated phospholipase, *J. Virol.* **14**, 479.

Nelson, M., and Gold, L. M., 1975, personal communication.

Nierlich, D. P., 1974, Regulation of bacterial growth, *Science* **184**, 1043.

Nierlich, D. P., Lamfrom, H., Sarabhai, A., and Abelson, J., 1973, Transfer RNA synthesis *in vitro, Proc. Nat. Acad. Sci. USA* **70**, 179.

Nikolaev, N., Silengo, L., and Schlessinger, D., 1973a, Synthesis of a large precursor to ribosomal RNA in a mutant of *Escherichia coli, Proc. Nat. Acad. Sci. USA* **70**, 3361.

Nikolaev, N., Silengo, L., and Schlessinger, D., 1973b, A role for RNase III in processing of rRNA and mRNA precursors in *Escherichia coli. J. Biol. Chem.* **248**, 7967.

Nikolaev, N., Birge, C. H., Gotoh, S., Glazier, K., and Schlessinger, D., 1975, Primary processing of high molecular weight preribosomal RNA in *Escherichia coli* and HeLa cells, *in* "Processing of RNA" (J. J. Dunn, ed.), pp. 175–193, Brookhaven National Laboratory, New York.

Niles, E. G., and Condit, R. C., 1975, Translational mapping of T 7 RNAs synthesized *in vitro* by purified T 7 RNA polymerase, *J. Mol. Biol.*, **98**, 57.

Niles, E. G., Conlon, S. W., and Summers, W. C., 1974, Purification and physical characterization of T 7 RNA polymerase from T 7-infected *E. coli* B, *Biochemistry* **13**, 3904.

Nisioka, T., and Ozeki, H., 1968, Early abortive lysis by phage BF 23 in *Escherichia coli* K-12 carrying the colicin Ib factor, *J. Virol.* **2**, 1249.

Niyogi, S., 1972, Effect of sigma factor and oligoribonucleotides on the transcription of well-defined templates with the ribonucleic acid polymerase of *Escherichia coli, J. Mol. Biol.* **64**, 609.

Niyogi, S. K., and Underwood, B. H., 1975, The isolation and properties of the specific binding sites for *Escherichia coli* RNA polymerase on T 4 and T 7 bacteriophage DNAs, *J. Mol. Biol.* **94**, 527.

Noller, H. F., 1975, personal communication.

Noller, H. F., Chang, C., Thomas, G., and Aldridge, J., 1971, Chemical modification of the transfer RNA and polyuridylic acid binding sites of *Escherichia coli* 30 S ribosomal subunits, *J. Mol. Biol.* **61**, 669.

Nomura, M., Hall, B. D., and Spiegelman, S., 1960, Characterization of RNA synthesized in *E. coli* after bacteriophage T 2 infection, *J. Mol. Biol.* **2**, 306.

Nomura, M., Witten, C., Mantei, N., and Echols, H., 1966, Inhibition of host nucleic acid synthesis by bacteriophage T 4: Effect of chloramphenicol at various multiplicities of infection, *J. Mol. Biol.* **17**, 273.

Nomura, M., Tissières, A., and Lengyel, P. (eds.), 1974, "Ribosomes," Cold Spring Harbor Laboratory, New York.

Nossal, N. G., 1974, DNA synthesis on a *double-stranded* DNA template by the T 4 bacteriophage DNA polymerase and the T 4 gene 32 DNA unwinding protein, *J. Biol. Chem.* **249**, 5668.

Notani, G. W., 1973, Regulation of bacteriophage T 4 gene expression, *J. Mol. Biol.* **73**, 231.

Ochoa, S., and Mazumder, R., 1974, Polypeptide chain initiation, *in* "The Enzymes" (P. D. Boyer, ed.), pp. 1–51, Academic Press, New York.

Oda, T., and Takanami, M., 1972, Observations on the structure of the termination factor rho and its attachment to DNA, *J. Mol. Biol.* **71**, 799.

O'Farrell, P., 1975a, in preparation.

O'Farrell, P. Z., 1975b, High resolution two-dimensional electrophoresis of proteins, *J. Biol. Chem.* **250**, 4007.

O'Farrell, P. Z., and Gold, L. M., 1973a, Bacteriophage T 4 gene expression; evidence for two classes of prereplicative cistrons, *J. Biol. Chem.* **248**, 5502.

O'Farrell, P. Z., and Gold, L. M., 1973b, Transcription and translation of prereplicative bacteriophage T 4 genes *in vitro, J. Biol. Chem.* **248**, 5512.

O'Farrell, P. Z., Gold, L. M., and Huang, W. M., 1973, The identification of prereplicative bacteriophage T 4 proteins, *J. Biol. Chem.* **248**, 5499.

Ohta, N., Sanders, M., and Newton, A., 1975, Poly(adenylic acid) sequences in the RNA of *Caulobacter crescentus, Proc. Nat. Acad. Sci. USA* **72**, 2343.

Okamoto, K., and Yutsudo, M., 1974, Participation of the *s* gene product of phage T 4 in the establishment of resistance to T 4 ghosts, *Virology* **58**, 369.

Okamoto, T., Sugiura, M., and Takanami, M., 1972, RNA polymerase binding sites of phage fd replicative form DNA, *Nature (London) New Biol.* **237**, 108.

Okazaki, R., Okazaki, T., Hirose, S., Sugino, A., Ogawa, T., Kurosawa, Y., Shinozaki, K., Tamanoi, F., Seki, T., Machida, Y., Fujiyama, A., and Kohara, Y.,

1975, Discontinuous replication in prokaryotic systems, *in* "DNA Synthesis and Its Regulation" (M. Goulian, P. Hanawalt, and C. F. Fox, eds.), pp. 832–862, W. A. Benjamin, Inc., California.

Okubo, S., Yanagida, T., Fujita, D. J., and Ohlsson-Wilhelm, B. M., 1972, The genetics of bacteriophage SPO 1, *Biken J.* **15**, 81.

Okuyama, A. N., Machiyama, N., Kinoshita, T., and Tanaka, N., 1971, Inhibition by kasugamycin of initiation complex formation on 30 S ribosomes, *Biochem. Biophys. Res. Commun.* **43**, 196.

Ozeki, H., Sakano, H., Yamada, S., Ikemura, T., and Shimura, Y., 1975, Temperature-sensitive mutants of *Escherichia coli* defective in tRNA biosynthesis, *in* "Processing of RNA" (J. J. Dunn, ed.), pp. 89–105, Brookhaven National Laboratory, New York.

Paddock, G., and Abelson, J., 1973, Sequence of T 4, T 2, and T 6 bacteriophage species I RNA and specific cleavage by an *E. coli* endonuclease, *Nature (London) New Biol.* **246**, 2.

Pao, C.-C., and Speyer, J. F., 1973, Order of injection of T 7 bacteriophage DNA, *J. Virol.* **11**, 1024.

Paul, A. V., and Lehman, I. R., 1966, The deoxyribonucleases of *Escherichia coli*. VII. A deoxyribonuclease induced by infection with phage T 5, *J. Biol. Chem.* **241**, 3441.

Pène, J., 1968, Host macromolecular synthesis in phage-infected *Bacillus subtilis*, *Bact. Rev.* **32**, 379.

Pène, J. J., and Barrow-Carraway, J., 1972, Initiation of *Bacillus subtilis* RNA polymerase on DNA from bacteriophages 2C, ϕ 29, T 4 and lambda, *J. Bacteriol.* **111**, 15.

Pène, J. J., and Marmur, J., 1967, Deoxyribonucleic acid replication and expression of early and late bacteriophage functions in *Bacillus subtilis*, *J. Virol.* **1**, 86.

Pène, J. J., Murr, P. C., and Barrow-Carraway, J., 1973, Synthesis of bacteriophage ϕ 29 proteins in *Bacillus subtilis*, *J. Virol.* **12**, 61.

Pero, J., Nelson, J., and Fox, T. D., 1975*a*, Highly asymmetric transcription by RNA polymerase containing phage SPO 1-induced polypeptides and a new host protein, *Proc. Nat. Acad. Sci. USA* **72**, 1589.

Pero, J., Tjian, R., Nelson, J., and Losick, R., 1975*b*, *In vitro* transcription of a late class of phage SPO 1 genes, *Nature (London)* **257**, 248.

Pero, J., Nelson, J., and Losick, R., 1975*c*, *In vitro* and *in vivo* transcription by vegetative and sporulating *Bacillus subtilis*, *in* "Spores VI" (P. Gerhardt, R. N. Costilow, and H. L. Sadoff, eds.), pp. 202–212, American Society for Microbiology, Washington, D.C.

Pesce, A., Satta, G., and Schito, G. C., 1969, Factors in lysis-inhibition by N 4 coliphage, *G. Microbiol.* **17**, 119.

Peters, G. G., and Hayward, R. S., 1974, Dinucleotide sequences in the region of T 7 DNA coding for termination of early transcription, *Eur. J. Biochem.* **48**, 19.

Peterson, R. F., Cohen, P. S., and Ennis, H. L., 1972, Properties of phage T 4 messenger RNA synthesized in the absence of protein synthesis, *Virology* **48**, 201.

Petit-Glatron, M.-F., and Rapoport, G., 1975, *In vivo* and *in vitro* evidence for existence of stable messenger ribonucleic acids in sporulating cells of *Bacillus thuringiensis*, *in* "Spores VI" (P. Gerhardt, R. N. Costilow, and H. L. Sadoff, eds.), pp. 255–264, American Society for Microbiology, Washington, D.C.

Petrusek, R., Duffy, J. J., and Geiduschek, E. P., 1976*a*, The action of phage-specific modifiers of transcription in "RNA Polymerases" (M. J. Chamberlin and R. Losick, eds.), pp. 567–585, Cold Spring Harbor Laboratory.

Petrusek, R., Armelin, M. C. S., Geiduschek, E. P., Johnson, G. G., and Semler, B., 1976*b*, in preparation.

Pettijohn, D. E., 1974, personal communication.

Pettijohn, D. E., and Hecht, R., 1973, RNA molecules bound to the folded bacterial genome stabilize DNA folds and segregate domains of supercoiling, *Cold Spring Harbor Symp. Quant. Biol.* **38**, 31.

Pinkerton, T. C., Paddock, G., and Abelson, J., 1972, Bacteriophage T 4 tRNA[Leu], *Nature (London) New Biol.* **240**, 88.

Pispa, J. P., and Buchanan, J. M., 1971, Synthesis of bacteriophage T 5 specific RNA *in vitro, Biochim. Biophys. Acta* **247**, 181.

Pispa, J. P., Sirbasku, D. A., and Buchanan, J. M., 1971, Patterns of ribonucleic acid synthesis in T 5′-infected *Escherichia coli*. IV. Examination of the role of deoxyribonucleic acid replication, *J. Biol. Chem.* **246**, 1658.

Pollack, Y., Groner, Y., Aviv (Greenshpan), H., and Revel, M., 1970, Role of initiation factor B (F3) in the preferential translation of T 4 late messenger RNA in T 4 infected *E. coli*, FEBS *(Fed. Eur. Biochem. Soc.) Lett.* **9**, 218.

Ponta, H., Rahmsdorf, H. J., Pai, S. H., Herrlich, P., and Schweiger, M., 1974*a*, Control of gene expression in bacteriophage T 7. Isolation of a new control protein and mechanism of action, *Mol. Gen. Genet.* **134**, 29.

Ponta, H., Rahmsdorf, H. J., Pai, S. H., Hirsch-Kauffmann, M., Herrlich, P., and Schweiger, M., 1974*b*, Control of gene expression in bacteriophage T 7: Transcriptional controls, *Mol. Gen. Genet.* **134**, 281.

Pribnow, D., 1975*a*, Nucleotide sequence of an RNA polymerase binding site at an early T 7 promoter, *Proc. Nat. Acad. Sci. USA* **72**, 784.

Pribnow, D., 1975*b*, Bacteriophage T 7 early promoters: Nucleotide sequences of two RNA polymerase binding sites, *J. Mol. Biol.* **99**, 419.

Pribnow, D., 1976, Promoter structure. Abstracts, Meeting on "RNA polymerases," Cold Spring Harbor, August 1975.

Price, A., and Frabotta, M., 1972, Resistance of bacteriophage PBS 2 infection to rifampicin, an inhibitor of *Bacillus subtilis* RNA synthesis, *Biochem. Biophys. Res. Commun.* **48**, 1578.

Price, A. R., Hitzeman, R., Frato, J., and Lombardi, K., 1974, Rifampicin-resistant bacteriophage BPS 2 infection and RNA polymerase in *Bacillus subtilis, Nucleic Acids Res.* **1**, 1497.

Puga, A., and Tessman, I., 1973*a*, Mechanism of transcription of bacteriophage S 13. I. Dependence of messenger RNA synthesis on amount and configuration of DNA, *J. Mol. Biol.* **75**, 83.

Puga, A., and Tessman, I., 1973*b*, Mechanisms of transcription of bacteriophage S 13. II. Inhibition of phage-specific transcription by nalidixic acid, *J. Mol. Biol.* **75**, 99.

Pulitzer, J. F., 1970, Function of T 4 gene 55. I. Characterization of temperature-sensitive mutations in the "maturation" gene 55, *J. Mol. Biol.* **49**, 473.

Pulitzer, J. F., and Geiduschek, E. P., 1970, Function of T 4 gene 55. II. RNA synthesis by temperature-sensitive gene 55 mutants, *J. Mol. Biol.* **49**, 489.

Pulitzer, J. F., and Yanagida, M., 1971, Inactive T 4 progeny virus formation in a temperature-sensitive mutant of *E. coli* K12, *Virology* **45**, 539.

Qasba, P. K., and Zillig, W., 1969, Nature of RNA synthesized at low and high ionic strength by DNA-dependent RNA-polymerase from *Escherichia coli, Eur. J. Biochem.* **7**, 315.

Rabussay, D., 1970, "Zur Expression der genetischen Information," Thesis, University of Munich, Munich.

Rabussay, D., and Geiduschek, E. P., 1973, Transcription of T 4 late RNA *in vitro,* *Hoppe-Seyler's Z. Physiol. Chem.* **354,** 1231.

Rabussay, D., Mailliammer, R., and Zillig, W., 1972, Regulation of transcription by T 4 phage induced chemical alteration and modification of transcriptase (EC 2.7.7.6), *in* "Metabolic Interconversion of Enzymes" (O. Wieland, E. Helmreich, and H. Holzer, eds.), pp. 213–227, Springer-Verlag, Berlin, Heidelberg, New York.

Radding, C. M., 1969, The genetic control of phage-induced enzymes, *Annu. Rev. Genet.* **3,** 363.

Rahmsdorf, H. J., Deusser, E., Herrlich, P., Schweiger, M., Stoeffler, G., and Wittmann, H. G., 1972, Ribosomes after T 4-infection, *Hoppe-Seyler's Z. Physiol. Chem.* **353,** 746.

Rahmsdorf, H. J., Herrlich, P., Pai, S. H., Schweiger, M., and Wittmann, H. G., 1973, Ribosomes after infection with bacteriophage T 4 and T 7, *Mol. Gen. Genet.* **127,** 259.

Rahmsdorf, H. J., Pai, S. H., Ponta, H., Herrlich, P., Roskoski, R. Jr., Schweiger, M., and Studier, F. W., 1974, Protein kinase induction in *Escherichia coli* by bacteriophage T 7, *Proc. Nat. Acad. Sci. USA* **71,** 586.

Ramakrishnan, T., and Echols, H., 1973, Purification and properties of M protein: An accessory factor for RNA polymerase, *J. Mol. Biol.* **78,** 67.

Ratner, D., 1974*a,* The interaction of bacterial and phage proteins with immobilized *Escherichia coli* RNA polymerase, *J. Mol. Biol.* **88,** 373.

Ratner, D., 1974*b,* Bacteriophage T 4 transcriptional control gene 55 codes for a protein bound to *Escherichia coli* RNA polymerase, *J. Mol. Biol.* **89,** 803.

Ratner, D., 1976, The interaction of proteins with immobilized *E. coli* RNA polymerase (and the genetic mapping of termination factor rho) *in* "RNA Polymerases" (M. Chamberlin and R. Losick, eds.), pp. 645–655, Cold Spring Harbor Laboratory, New York.

Ray, P. N., and Pearson, M. L., 1974, Evidence for post-transcriptional control of the morphogenetic genes of bacteriophage lambda, *J. Mol. Biol.* **85,** 163.

Reichardt, L. F., 1975*a,* Control of bacteriophage lambda repressor synthesis after phage infection: The role of the *N, c*II, *c*III and cro products, *J. Mol. Biol.* **93,** 267.

Reichardt, L. F., 1975*b,* Control of bacteriophage lambda repressor synthesis: Regulation of the maintenance pathway by the cro and cI products, *J. Mol. Biol.* **93,** 289.

Reichardt, L., and Kaiser, A. D., 1971, Control of λ repressor synthesis, *Proc. Nat. Acad. Sci. USA* **68,** 2185.

Remold-O'Donnell, E., and Zillig, W., 1969, Purification and properties of DNA-dependent RNA-polymerase from *Bacillus stearothermophilus, Eur. J. Biochem.* **7,** 318.

Reuben, R. C., and Gefter, M. L., 1974, A DNA-binding protein induced by phage T 7. Purification and properties, *J. Biol. Chem.* **249,** 3843.

Revel, H. R., 1967, Restriction of nonglucosylated T-even bacteriophage: Properties of permissive mutants of *Escherichia coli* B and K 12, *Virology* **31,** 688.

Revel, M., Pollak, Y., Groner, Y., Scheps, R., Inouye, H., Berissi, H., and Zeller, H., 1973*a,* IF3-interference factors: Protein factors in *E. coli* controlling initiation of mRNA translation, *Biochimie* **55,** 41.

Revel, M., Groner, Y., Pollack, Y., Zeller, H., Canaani, D., and Nudel, U., 1973*b,* The control of protein synthesis by initiation factors. Abstracts, Ninth International Congress of Biochemistry in Stockholm, p. 133.

Rhoades, M., and Rhoades, E. A., 1972, Terminal repetition in the DNA of bacteriophage T 5, *J. Mol. Biol.* **69,** 187.

Ricard, B., and Salser, W., 1974, Size and folding of the messenger for phage T 4 lysozyme, *Nature (London)* **248**, 314.

Richardson, J. P., 1970*a*, Rates of bacteriophage T 4 RNA chain growth *in vitro, J. Mol. Biol.* **49**, 235.

Richardson, J. P., 1970*b*, Rho factor function in T 4 RNA transcription, *Cold Spring Harbor Symp. Quant. Biol.* **35**, 127.

Richardson, J. P., 1975, Initiation of transcription by *Escherichia coli* RNA polymerase from supercoiled and nonsupercoiled bacteriophage PM 2 DNA, *J. Mol. Biol.* **91**, 477.

Richardson, J. P., Grimley, C., and Lowery, C., 1975, Transcription termination factor rho activity is altered in *Escherichia coli* with *suA* gene mutations, *Proc. Nat. Acad. Sci. USA* **72**, 1725.

Riggs, A. D., Suzuki, H., and Bourgeois, S., 1970, *lac* repressor–operator interaction. I. Equilibrium studies, *J. Mol. Biol.* **48**, 67.

Riggs, A. D., Reiness, G., and Zubay, G., 1971, Purification and DNA-binding properties of the catabolite gene activator protein, *Proc. Nat. Acad. Sci. USA* **68**, 1222.

Rima, B. K., and Takahashi, I., 1973, The synthesis of nucleic acids in *Bacillus subtilis* infected with phage PBS 1, *Can. J. Biochem.* **51**, 1219.

Ritchie, D. A., and White, F. E., 1972, The inability of T-even phages to produce heat-stable density mutants and its bearing on chromosome maturation, *J. Gen. Virol.* **16**, 91.

Ritchie, D. A., Thomas, C. A. Jr., MacHattie, L. A., and Wensink, P. C., 1967, Terminal repetition in nonpermuted T 3 and T 7 bacteriophage DNA molecules, *J. Mol. Biol.* **23**, 365.

Riva, S., Cascino, A., and Geiduschek, E. P., 1970*a*, Coupling of late transcription to viral replication in bacteriophage T 4 development, *J. Mol. Biol.* **54**, 85.

Riva, S., Cascino, A., and Geiduschek, E. P., 1970*b*, Uncoupling of late transcription from DNA replication in bacteriophage T 4 development, *J. Mol. Biol.* **54**, 103.

Roberts, J. W., 1969, Termination factor for RNA synthesis, *Nature (London)* **224**, 1168.

Roberts, J. W., 1970, The ρ factor: Termination and antitermination in lambda, *Cold Spring Harbor Symp. Quant. Biol.* **35**, 121.

Roberts, J. W., and Roberts, C. W., 1975, Proteolytic cleavage of bacteriophage lambda repressor in induction, *Proc. Nat. Acad. Sci. USA* **72**, 147.

Robertson, H. D., Webster, R. E., and Zinder, N. D., 1968*a*, Bacteriophage coat protein as repressor, *Nature (London)* **218**, 533.

Robertson, H. D., Webster, R. E., and Zinder, N. D., 1968*b*, Purification and properties of ribonuclease III from *Escherichia coli, J. Biol. Chem.* **243**, 82.

Rohrer, H., Zillig, W., and Mailhammer, R., 1975, ADP-ribosylation of DNA-dependent RNA polymerase of *E. coli* by an NAD^+: protein ADP-ribosyl-transferase from bacteriophage T 4, *Eur. J. Biochem.*, **60**, 227.

Roscoe, D. H., 1969, Synthesis of DNA in phage-infected *Bacillus subtilis, Virology* **38**, 527.

Roscoe, D. H., and Tucker, R. G., 1966, The biosynthesis of 5-hydroxymethyl-deoxyuridylic acid in bacteriophage-infected *Bacillus subtilis, Virology* **29**, 157.

Rose, J. K., Mosteller, R. D., and Yanofsky, C., 1970, Tryptophan messenger ribonucleic acid elongation rates and steady-state levels of tryptophan operon enzymes under various growth conditions, *J. Mol. Biol.* **51**, 541.

Rosenberg, M., and Weissman, S., 1975, Termination of transcription in bacteriophage λ. Heterogeneous, 3′-terminal oligo-adenylate additions and the effects of ρ factor, *J. Biol. Chem.* **250**, 4755.

Rosenberg, M., Kramer, R. A., and Steitz, J. A., 1974, T 7 early messenger RNAs are the direct products of ribonuclease III cleavage, *J. Mol. Biol.* **89**, 777.

Rosenberg, M., de Crombrugghe, B., and Musso, R., 1976, Determination of nucleotide sequences beyond the sites of transcriptional termination, *Proc. Nat. Acad. Sci. USA* **73**, 717.

Rosenkranz, H. S., 1973, RNA in coliphage T 5, *Nature (London)* **242**, 327.

Rothman-Denes, L. B., 1975, personal communication.

Rothman-Denes, L. B., and Schito, G. C., 1974, Novel transcribing activities in N 4-infected *E. coli, Virology* **60**, 65.

Rothman-Denes, L. B., Haselkorn, R. and Schito, G. C., 1972, Selective shutoff of catabolite-sensitive host syntheses by bacteriophage N 4, *Virology* **50**, 95.

Rothman-Denes, L., Muthukrishnan, S., Haselkorn, R., and Studier, F. W., 1973, A T 7 gene function required for shut-off of host and early T 7 transcription, *in* "Virus Research" (C. F. Fox and W. S. Robinson, eds.), pp. 227–242, Academic Press, New York.

Rottman, F., Shatkin, A. J., and Perry, R. P., 1974, Sequences containing methylated nucleotides at the 5′ termini of messenger RNAs: Possible implications for processing, *Cell* **3**, 197.

Rouvière-Yaniv, J., and Gros, F., 1975, Characterization of a novel, low molecular weight DNA-binding protein from *Escherichia coli, Proc. Nat. Acad. Sci. USA,* **72**, 3428.

Russel, M., 1973, Control of bacteriophage T 4 DNA polymerase synthesis, *J. Mol. Biol.* **79**, 83.

Russel, M., Gold, L., Morrissett, H., and O'Farrell, P. Z., 1976, Translational, autogenous regulation of gene 32 expression during bacteriophage T4 infection, *J. Biol. Chem.,* in press.

Sabol, S., and Ochoa, S., 1971, Ribosomal binding of labelled initiation factor F$_3$, *Nature (London) New Biol.* **234**, 233.

Sabol, S., Sillero, M. A. G., Iwasaki, K., and Ochoa, S., 1970, Purification and properties of initiation factor F$_3$, *Nature (London)* **228**, 1269.

Sadler, J. R., and Smith, T. F., 1971, Mapping of the lactose operator, *J. Mol. Biol.* **62**, 139.

Sadowski, P. D., and I. Bakyta, 1972, T 4 endonuclease IV. Improved purification procedure and resolution from T 4 endonuclease III, *J. Biol Chem.* **247**, 405.

Sakano, H., Shimura, Y., and Ozeki, H., 1974, Studies on T 4-tRNA biosynthesis: Accumulation of precursor tRNA molecules in a temperature-sensitive mutant of *Escherichia coli, FEBS (Fed. Eur. Biochem. Soc.) Lett.* **40**, 312.

Sakiyama, S., and Buchanan, J. M., 1971, *In vitro* synthesis of deoxynucleotide kinase programmed by bacteriophage T 4-RNA, *Proc. Nat. Acad. Sci. USA* **68**, 1376.

Sakiyama, S., and Buchanan, J. M., 1972, Control of the synthesis of T 4 phage deoxynucleotide kinase messenger ribonucleic acid *in vivo, J. Biol. Chem.* **247**, 7806.

Sakiyama, S., and Buchanan, J. M., 1973, Relationship between molecular weight of T 4 phage-induced deoxynucleotide kinase and the size of its messenger ribonucleic acid, *J. Biol. Chem.* **248**, 3150.

Salser, W., Gesteland, R. F., and Bolle, A., 1967, *In vitro* synthesis of bacteriophage lysozyme, *Nature (London)* **215**, 588.

Salser, W., Bolle, A., and Epstein, R., 1970, Transcription during bacteriophage T 4 development: A demonstration that distinct subclasses of the "early" RNA appear at different times and that some are "turned off" at late times, *J. Mol. Biol.* **49**, 271.

Salvo, R. A., Chakraborty, R., and Maitra, U., 1973, Studies on T 3-induced ribonucleic acid polymerase. IV. Transcription of denatured deoxyribonucleic acid preparations by T 3 ribonucleic acid polymerase, *J. Biol. Chem.* **248**, 6647.

Sarocchi, M.-T., and Darlix, J.-L., 1974, A spectroscopic approach to DNA transcription and protein binding, *Eur. J. Biochem.* **46**, 481.

Saucier, J. M., and Wang, J. C., 1972, Angular alteration of the DNA helix by *E. coli* RNA polymerase, *Nature (London) New Biol.* **239**, 167.

Sauerbier, W. and Hercules, K., 1973, Control of gene function in bacteriophage T 4. IV. Post-transcriptional shutoff of expression of early genes, *J. Virol.* **12**, 538.

Sauerbier, W., Puck, S. M., Bräutigam, A. R., and Hirsch-Kaufmann, M., 1969, Control of gene function in bacteriophage T 4. I. Ribonucleic acid and deoxyribonucleic acid metabolism in T 4 rII-infected lambda-lysogenic hosts, *J. Virol.* **4**, 742.

Sauerbier, W., Millette, R. L., and Hackett, P. B., Jr., 1970, The effects of ultraviolet irradiation on the transcription of T 4 DNA, *Biochim. Biophys. Acta* **209**, 368.

Schachner, M., and Zillig, W., 1971, Fingerprint maps of tryptic peptides from subunits of *Escherichia coli* and T 4-modified DNA-dependent RNA polymerases, *Eur. J. Biochem.* **22**, 513.

Schachner, M., Seifert, W., and Zillig, W., 1971, A correlation of changes in host and T 4 bacteriophage specific RNA synthesis with changes of DNA-dependent RNA polymerase in *Escherichia coli* infected with bacteriophage T 4, *Eur. J. Biochem.* **22**, 520.

Schachtele, C. F., DeSain, C. V., and Anderson, D. L., 1973, Transcription durigg the development of bacteriophage ϕ 29: definition of "early" and "late" ϕ 29 ribonucleic acid, *J. Virol.* **11**, 9.

Schäfer, R., 1972, Über Mechanismen der Initiation, Elongation und Termination in der Transcription, Thesis, University of Munich, Munich.

Schäfer, R., and Zillig, W., 1973a, Kappa, a novel factor for the arrest of transcription *in vitro* by DNA-dependent RNA polymerase from *Escherichia coli* at specific sites of natural templates. *Eur. J. Biochem.* **33**, 201.

Schäfer, R., and Zillig, W., 1973b, The effects of ionic strength on termination of transcription of DNAs from bacteriophages T 4, T 5 and T 7 by DNA-dependent RNA polymerase from *Escherichia coli* and the nature of termination by factor ρ, *Eur. J. Biochem.* **33**, 215.

Schäfer, R., Zillig, W., and Zechel, K., 1973a, A model for the initiation of transcription by DNA-dependent RNA polymerase from *E. coli, Eur. J. Biochem.* **33**, 207.

Schäfer, R., Krämer, R., Zillig, W., and Cudny, H., 1973b, On the initiation of transcription by DNA-dependent RNA polymerase from *Escherichia coli, Eur. J. Biochem.* **40**, 367.

Schaller, H., Otto, B., Nüsslein, V., Huf, J., Herrmann, R., and Bonhoeffer, F., 1972, Deoxyribonucleic acid replication *in vitro, J. Mol. Biol.* **63**, 183.

Schaller, H., Gray, C., and Herrmann, K., 1975, Nucleotide sequence of an RNA polymerase binding site from the DNA of bacteriophage fd, *Proc. Nat. Acad. Sci. USA* **72**, 737.

Schedl, P. D., Singer, R. E., and Conway, T. W., 1970, A factor required for the translation of bacteriophage f2 RNA in extracts of T 4-infected cells, *Biochem. Biophys. Res. Commun.* **38**, 631.

Schedl, P., Primakoff, P., and Roberts, J., 1975, Processing of *E. coli* tRNA precursors, *in* "Processing of RNA" (J. J. Dunn, ed.), pp. 53–76, Brookhaven National Laboratory, New York.

Schekman, R., Weiner, A., and Kornberg, A., 1974, Multienzyme systems of DNA replication, *Science* **186**, 987.

Scherberg, N. H., and Weiss, S. B., 1970, Detection of bacteriophage T 4- and T 5-coded transfer RNAs. *Proc. Nat. Acad. Sci. USA* **67**, 1164.

Scherberg, N. H., and Weiss, S. B., 1972, T 4 transfer RNAs: Codon recognition and translational properties, *Proc. Nat. Acad. Sci. USA* **69**, 1114.

Scherberg, N. H., Guha, A., Hsu, W.-T., and Weiss, S. B., 1970, Evidence for the early synthesis of T 4 bacteriophage-coded transfer RNA, *Biochem. Biophys. Res. Commun.* **40**, 919.

Scherzinger, E., Herrlich, P., Schweiger, M., and Schuster, H., 1972, The early region of the DNA of bacteriophage T 7, *Eur. J. Biochem.* **25**, 341.

Schiff, N., Miller, M. J., and Wahba, A., 1974, Purification and properties of chain initiation factor 3 from T 4-infected and uninfected *E. coli* MRE 600, *J. Biol. Chem.* **249**, 3797.

Schito, G. C., 1973, The genetics and physiology of coliphage N 4, *Virology* **55**, 254.

Schito, G. C., 1974, Development of coliphage N 4: Ultrastructural studies, *J. Virol.* **13**, 186.

Schito, G. C., 1975, Lecture delivered at the 3rd International Congress of Virology, Madrid, September 1975.

Schito, G. C., Rialdi, G., and Pesce, A., 1966*a*, Biophysical properties of N 4 coliphage, *Biochim. Biophysical properties of N_4 coliphage, Biochim. Biophys. Acta* **129**, 482.

Schito, G. C., Rialdi, G., and Pesce, A., 1966*b*, The physical properties of the deoxyribonucleic acid from N 4 coliphage, *Biochim. Biophys. Acta* **129**, 491.

Schleicher, M., and Bautz, E. K. F., 1972, Stability of phage T 7 gene 1 mRNA in *E. coli* cells infected with wild-type phage and with gene 1 amber mutants, *FEBS* (*Fed. Eur. Biochem. Soc.*) *Lett.* **28**, 139.

Schmidt, D. A., Mazaitis, A. J., Kasai, T., and Bautz, E. K. F., 1970, Involvement of a phage T 4 σ factor and an anti-terminator protein in the transcription of early T 4 genes *in vivo*, *Nature* (*London*) **225**, 1012.

Schweiger, M., and Gold, L. M., 1970, *Escherichia coli* and *Bacillus subtilis* phage deoxyribonucleic acid-directed deoxycytidylate deaminase synthesis in *Escherichia coli* extracts, *J. Biol. Chem.* **245**, 5022.

Schweiger, M., Herrlich, P., and Millette, R. L., 1971, Gene expression *in vitro* from deoxyribonucleic acid of bacteriophage T 7, *J. Biol. Chem.* **246**, 6707.

Searcy, D. G., 1975, Histone-like protein in the prokaryote *Thermoplasma acidophilum*, *Biochim. Biophys. Acta* **95**, 535.

Sederoff, R., Bolle, A., Goodman, H. M., and Epstein, R. H., 1971, Regulation of rII and region D transcription in T 4 Bacteriophage: A sucrose gradient analysis, *Virology*, **46**, 817.

Segawa, T., and Imamoto, F., 1974, Diversity of regulation of genetic transcription. II. Specific relaxation of polarity in read-through transcription of the translocated *trp* operon in bacteriophage lambda *trp*, *J. Mol. Biol.* **87**, 741.

Seifert, W., Qasba, P., Walter, G., Palm, P., Schachner, M., and Zillig, W., 1969, Kinetics of the alteration and modification of DNA-dependent RNA polymerase in T 4-infected *E. coli* cells, *Eur. J. Biochem.* **9**, 319.

Seifert, W., Rabussay, D., and Zillig, W., 1971, On the chemical nature of alteration and modification of DNA-dependent RNA polymerase of *E. coli* after T 4 infection, *FEBS (Fed Eur. Biochem. Soc.) Lett.* **16**, 175.

Sekiguchi, M., 1966, Studies on the physiological defect in r11 mutants of bacteriophage T 4, *J. Mol. Biol.* **16**, 503.

Sekiguchi, M., and Cohen, S. S., 1964, The synthesis of messenger RNA without protein synthesis. II. Synthesis of phage-induced RNA and sequential enzyme production, *J. Mol. Biol.* **8**, 638.

Shah, D. B., and Berger, H., 1971, Replication of gene 46-47 *amber* mutants of bacteriophage T 4D, *J. Mol. Biol.* **57**, 17.

Shalitin, C., and Naot, Y., 1971, Role of gene 46 in bacteriophage T 4 deoxyribonucleic acid synthesis, *J. Virol.* **8**, 142.

Sheldon, M. V., 1971, "Bacteriophage T 4 Gene Expression," Thesis, University of Chicago, Chicago.

Shimizu, K., and Sekiguchi, M., 1974, Biochemical studies on the *x* mutation of bacteriophage T 4: Differential inhibition of x^+ and x DNA synthesis by mitomycin C, *J. Virol.* **13**, 1.

Shine, J., and Dalgarno, L., 1974, The 3'-terminal sequence of *Escherichia coli* 16 S ribosomal RNA: Complementarity to nonsense triplets and ribosome binding sites, *Proc. Nat. Acad. Sci. USA* **71**, 1342.

Shine, J., and Dalgarno, L., 1975, Determinant of cistron specificity in bacterial ribosomes, *Nature (London)* **254**, 34.

Shub, D. A., 1966, Stability of messenger RNA during bacteriophage development, Ph.D. Thesis, Massachusetts Institute of Technology.

Shub, D. A., 1975, Bacteriophage SPO 1 DNA- and RNA-directed protein synthesis *in vitro*: Comparison with *in vivo* control, *Mol. Gen. Genet.* **137**, 171.

Shub, D. A., and Johnson, G. G., 1975, Bacteriophage SPO 1 DNA- and RNA-directed protein synthesis *in vitro*: The effect of TF1, a template-selective transcription inhibitor, *Mol. Gen. Genet.* **137**, 161.

Shub, D. A., Swanton, M., and Smith, D. H., 1976, DNA-binding specificity of rifampicin-inhibited RNA polymerase of *Escherichia coli*, to be published.

Siegel, P. S., and Schaechter, M., 1973, The role of the host cell membrane in the replication and morphogenesis of bacteriophages, *Annu. Rev. Microbiol.* **27**, 261.

Siegel, R. B., and Summers, W. C., 1970, The process of infection with coliphage T 7. III. Control of phage-specific RNA synthesis *in vivo* by an early phage gene, *J. Mol. Biol.* **49**, 115.

Silber, R., Malathi, V. G., and Hurwitz, J., 1972, Purification and properties of bacteriophage T 4-induced RNA ligase, *Proc. Nat. Acad. Sci. USA* **69**, 3009.

Simon, L. D., and Snover, Dale, 1974, Bacterial mutation affecting T 4 phage DNA synthesis and tail production, *Nature (London)* **252**, 451.

Simon, M. N., and Studier, F. W., 1973, Physical mapping of the early region of bacteriophage T 7 DNA, *J. Mol. Biol.* **79**, 249.

Singer, B. S., Gold, L. M., and Pribnow, D., 1976, personal communication.

Sippel, A. E., and Hartmann, G. R., 1970, Rifampicin resistance of RNA polymerase in the binary complex with DNA, *Eur. J. Biochem.* **16**, 152.

Sirbasku, D. A., and Buchanan, J. M., 1970a, Patterns of ribonucleic acid synthesis in T 5-infected *Escherichia coli*. II. Separation of high molecular weight ribonucleic acid species by disc electrophoresis on acrylamide gel columns, *J. Biol. Chem.* **245**, 2679.

Sirbasku, D. A., and Buchanan, J. M., 1970b, Patterns of ribonucleic acid synthesis in

T 5-infected *Escherichia coli*. III. Separation of low molecular weight ribonucleic acid species by disc electrophoresis on acrylamide gel columns. *J. Biol. Chem.* **245**, 2693.

Sirbasku, D. A., and J. M. Buchanan, 1971, Patterns of ribonucleic acid synthesis in T 5-infected *Escherichia coli*. V. Formation of stable, discrete, degradation products during turnover of phage-specific ribonucleic acid, *J. Biol. Chem.* **246**, 1665.

Skare, J., Niles, E. G., and Summers, W. C., 1974, Localization of the leftmost site for T 7 late transcription *in vivo* and *in vitro*, *Biochemistry* **13**, 3912.

Smith, D. W. E., and Russell, N. L., 1970, Transfer RNA synthesis, methylation and thiolation in *Escherichia coli* cells infected with phage T 2, *Biochim. Biophys. Acta* **209**, 171.

Smith, F. L., and Haselkorn, R., 1969, Proteins associated with ribosomes in T 4-infected *E. coli*, *Cold Spring Harbor Symp. Quant. Biol.* **34**, 91.

Snustad, P., 1968, Dominance interactions in *Escherichia coli* cells mixedly infected with bacteriophage T 4D wild-type and *amber* mutants and their possible implications as to type of gene-product function: Catalytic vs. stochiometric, *Virology* **35**, 550.

Snyder, L., 1972, An RNA polymerase mutant of *Escherichia coli* defective in the T 4 viral transcription program, *Virology* **50**, 396.

Snyder, L., 1973, Change in RNA polymerase associated with the shutoff of host transcription by T 4, *Nature (London) New Biol.* **243**, 131.

Snyder, L., 1975 personal communication.

Snyder, L., 1976, personal communication.

Snyder, L., and Geiduschek, E. P., 1968, *In vitro* synthesis of T 4 late messenger RNA, *Proc. Nat. Acad. Sci. USA* **59**, 459.

Snyder, L. R., and Montgomery, D. L., 1974, Inhibition of T 4 growth by an RNA polymerase mutation of *E. coli*: Physiological and genetic analysis of the effects during phage development, *Virology* **62**, 184.

Snyder, L., Gold, L., and Kutter, E., 1976, Production of viable T 4 coliphage whose DNA has almost all cytosine instead of 5-hydroxymethylcytosine, *Proc. Nat. Acad. Sci. USA* **73**, 3098.

Sonenshein, A. L., 1970, Trapping of unreplicated phage DNA into spores of *Bacillus subtilis* and its stabilization against damage by ^{32}P decay, *Virology*, **42**, 488.

Sonenshein, A. L., and Roscoe, D. H., 1969, The course of phage ϕe infection in sporulating cells of *Bacillus subtilis* strain 3610, *Virology* **39**, 265.

Speyer, J. F., Chao, J., and Chao, L., 1972, Ribonucleotides covalently linked to deoxyribonucleic acid in T 4 bacteriophage, *J. Virol.* **10**, 902.

Spiegelman, G. B., and Whiteley, H. R., 1974*a*, Purification of ribonucleic acid polymerase from SP 82-infected *Bacillus subtilis*, *J. Biol. Chem.* **249**, 1476.

Spiegelman, G. B., and Whiteley, H. R., 1974*b*, *In vivo* and *in vitro* transcription by ribonucleic acid polymerase from SP 82-infected *Bacillus subtilis*, *J. Biol. Chem.* **249**, 1483.

Spiegelman, G. B., and Whiteley, H. R., 1975, personal communication.

Srinivasan, P. R., 1973, personal communication.

Srinivasan, P. R., and Dharmgrongartama, B., 1975, Transcriptional control in T 3 phage infected *E. coli* B, Abstract, Meeting on "RNA polymerases," Cold Spring Harbor, August 1975.

Stahl, F. W., Edgar, R. S., and Steinberg, C. M., 1964, The linkage map of bacteriophage T 4, *Genetics* **50**, 539.

Stahl, F. W., Crasemann, J. M., Yegian, C. D., Stahl, M. M., and Nakata, A., 1970, Cotranscribed cistrons in bacteriophage T 4, *Genetics* **64**, 157.

Stahl, S., Paddock, G., and Abelson, J., 1973, T 4 bacteriophage tRNA^{gly}, *Biochem. Biophys. Res. Commun.* **54**, 567.

Steitz, J. A., 1975, personal communication.

Steitz, J. A., and Jakes, K., 1975, How ribosomes select initiator regions in messenger RNA: Direct evidence for the formation of base pairs between the 3′ terminus of 16 S rRNA and the mRNA during initiation of protein synthesis in *E. coli*, *Proc. Nat. Acad. Sci. USA*, **72**, 4734.

Steitz, J. A., Dube, S. K., and Rudland, P. S., 1970, Control of translation by T 4 phage: Altered ribosome binding at R17 initiation sites, *Nature (London)* **226**, 824.

Steitz, T. A., Richmond, T. J., Wise, D., and Engelman, D., 1974, The *lac* repressor protein: Molecular shape, subunit structure, and proposed model for operator interaction based on structural studies of microcrystals, *Proc. Nat. Acad. Sci. USA* **71**, 593.

Stent, G. S., 1963, "Molecular Biology of Bacterial Viruses," W. S. Freeman and Co., San Francisco.

Sternberg, N., 1973a, Properties of a mutant of *E. coli* defective in bacteriophage λ head formation (gro E): I. Initial characterization, *J. Mol. Biol.* **76**, 1.

Sternberg, N., 1973b, Properties of a mutant of *E. coli* defective in bacteriophage λ head formation (gro E): II. The propagation of phage λ, *J. Mol. Biol.* **76**, 25.

Stevens, A., 1970, An isotopic study of DNA-dependent RNA polymerase of *E. coli* following T 4 infection, *Biochem. Biophys. Res. Commun.* **41**, 367.

Stevens, A., 1972, New small polypeptides associated with DNA-dependent RNA polymerase of *Escherichia coli* after infection with bacteriophage T 4, *Proc. Nat. Acad. Sci. USA* **69**, 603.

Stevens, A., 1974, Deoxyribonucleic acid dependent ribonucleic acid polymerases from two T 4 phage-infected systems, *Biochemistry* **13**, 493.

Stewart, C. R., and Tole, M. F., 1972, A host mutation affecting the synthesis of late proteins during infection of *Bacillus subtilis* by bacteriophage SP 82, *Virology* **50**, 733.

Stewart, C. R., Cater, M., and Click, B., 1971, Lysis of *Bacillus subtilis* by bacteriophage SP 82 in the absence of DNA synthesis, *Virology* **46**, 327.

Stewart, C. R., Click, B. and Tole, M. F., 1972, DNA replication and late protein synthesis during SP 82 infection of *Bacillus subtilis*, *Virology* **50**, 653.

Streisinger, G., Edgar, R. S., and Denhardt, G. H., 1964, Chromosome structure in phage T 4. I. Circularity of the linkage map, *Proc. Nat. Acad. Sci. USA* **51**, 775.

Streisinger, G., Emrich, J., and Stahl, M. M., 1967, Chromosome structure in phage T 4. III. Terminal redundancy and length determination, *Proc. Nat. Acad. Sci. USA* **57**, 292.

Strobel, M., and Nomura, M., 1966, Restriction of bacteriophage BF 23 by a colicin I factor, *Virology* **28**, 763.

Studier, F. W., 1969, The genetics and physiology of bacteriophage T 7, *Virology* **39**, 562.

Studier, F. W., 1972, Bacteriophage T 7, *Science* **176**, 367.

Studier, F. W., 1973a, Genetic analysis of non-essential bacteriophage T 7 genes, *J. Mol. Biol.* **79**, 227.

Studier, F. W., 1973b, Analysis of bacteriophage T 7 early RNAs and proteins on slab gels, *J. Mol. Biol.* **79**, 237.

Studier, F. W., 1975a, Gene 0.3 of bacteriophage T 7 acts to overcome the DNA restriction system of the host, *J. Mol. Biol.* **94**, 283.

Studier, F. W., 1975b, Gene expression after bacteriophage T 7 infection, *in* "Organization and Expression of the Viral Genome" (Proc. 10th F.E.B.S. meeting, Vol. 39; F. Chapeville, and M. Grunberg-Manago, eds.), pp. 45–53, North Holland, The Netherlands.

Studier, F. W., and Maizel, J. W. Jr., 1969, T 7-directed protein synthesis, *Virology* **39**, 575.

Stryer, L., 1975, Chapter 26, Protein Synthesis, *in* "Biochemistry," W. H. Freeman and Company, California.

Subramanian, A. R.,and Davis, B. D., 1970, Activity of initiation factor F_3 in dissociating *Escherichia coli* ribosomes, *Nature (London)* **228**, 1273.

Sueoka, N., and Kano-Sueoka, T., 1964, A specific modification of leucyl-sRNA of *Escherichia coli* after phage T 2 infection, *Proc. Nat. Acad. Sci. USA* **52**, 1535.

Sugiyama, T., and Nakada, D., 1967, Control of translation of MS 2 RNA cistrons by MS 2 coat protein, *Proc. Nat. Acad. Sci. USA* **57**, 1744.

Summers, W. C., 1969, The process of infection with coliphage T 7. I. Characterization of T 7 RNA by polyacrylamide gel electrophoretic analysis, *Virology,* **39**, 175.

Summers, W. C., 1970, The process of infection with coliphage T 7. IV. Stability of RNA in bacteriophage-infected cells, *J. Mol. Biol.* **51**, 671.

Summers, W. C., 1972, Regulation of RNA metabolism of T 7 and related phages, *Annu. Rev. Genet.* **6**, 191.

Summers, W. C., and Jakes, K., 1971, Phage T 7 lysozyme mRNA transcription and translation *in vivo* and *in vitro, Biochem. Biophys. Res. Commun.* **45**, 315.

Summers, W. C., and Siegel, R. B., 1969, Control of template specificity of *E. coli* RNA polymerase by a phage-coded protein, *Nature (London)* **223**, 1111.

Summers, W. C., and Siegel, R. B., 1970, Transcription of late phage RNA by T 7 RNA polymerase, *Nature (London)* **228**, 1160.

Summers, W. C., and Szybalski, W., 1968, Totally asymmetric transcription of coliphage T 7 *in vivo*: Correlation with polyG binding sites, *Virology* **34**, 9.

Summers, W. C., Brunovskis, I., and Hyman, R. W., 1973, The process of infection with coliphage T 7. VII. Characterization and mapping of the major *in vivo* transcription products of the early region, *J. Mol. Biol.* **74**, 291.

Sussman, R., and Zeev, H. B., 1975, Proposed mechanism of bacteriophage lambda induction: Acquisition of binding sites for lambda repressor by DNA of the host, *Proc. Nat. Acad. Sci. USA* **72**, 1973.

Swanton, M., Smith, D. H., and Shub, D. A., 1975, Synthesis of specific functional messenger RNA in vitro by phage-SPO 1-modified RNA polymerase of *Bacillus subtilis, Proc. Nat. Acad. Sci. USA* **72**, 4886.

Szabo, C., and Moyer, R. W., 1975, Purification and properties of a bacteriophage T 5 modified form of *Escherichia coli* RNA polymerase, *J. Virol.* **15**, 1042.

Szabo, C., Dharmgrongartama, B., and Moyer, R. W., 1975, The regulation of transcription in bacteriophage T 5-infected *Escherichia coli, Biochem.* **14**, 989.

Szer, W., and Leffler, S., 1974, Interaction of *Escherichia coli* 30 S ribosomal subunits with MS 2 phage RNA in the absence of initiation factors, *Proc. Nat. Acad. Sci. USA* **71**, 3611.

Tai, P.-C., and Davis, B. D., 1974, Activity of colicin E3-treated ribosomes in initiation and chain elongation, *Proc. Nat. Acad. Sci. USA* **71**, 1021.

Takacs, B. J., and Rosenbusch, J. P., 1975, Modification of *Escherichia coli* membranes in the prereplicative phase of phage T 4 infection, *J. Biol. Chem.* **250,** 2339.

Takahashi, H., Coppo, A., Manzi, A., Martire, G., and Pulitzer, J. F., 1975, Design of a system of conditional lethal mutations (tab/k/com) affecting protein–protein interactions in bacteriophage T 4-infected *Escherichia coli, J. Mol. Biol.* **96,** 563.

Takano, T., and Kakefuda, T., 1972, Involvement of a bacterial factor in morphogenesis of bacteriophage capsid, *Nature (London) New Biol.* **239,** 34.

Tate, W. P., and Caskey, C. T., 1974, Polypeptide chain termination, *in* "The Enzymes" (P. D. Boyer, ed.), pp. 87–118, Academic Press, New York.

Terzi, M., 1967, Studies on the mechanism of bacteriophage T 4 interference with host metabolism, *J. Mol. Biol.* **28,** 37.

Tessman, I., and Greenberg, D. B., 1972, Ribonucleotide reductase genes of phage T 4: Map location of the thioredoxin gene *nrdC, Virology* **49,** 337.

Thermes, C., Daegelen, P., de Franciscis, V., and Brody, E., 1976, An *in vitro* system for induction of T 4 delayed early RNA, *Proc. Nat. Acad. Sci. USA* **73,** 2569.

Thomas, C. A., and Rubenstein, I., 1964, The arrangements of nucleotide sequences in T 2 and T 5 bacteriophage molecules, *Biophys. J.* **4,** 93.

Thomas, R., 1971, Control circuits, *in* "The Bacteriophage Lambda" (A. D. Hershey, ed.), pp. 211–220, Cold Spring Harbor Laboratory, New York.

Tillack, T. W., and Smith, D. E., 1968, The effect of bacteriophage T 2 infection on the synthesis of transfer RNA in *E. coli, Virology* **36,** 212.

Tjian, R., Pero, J., Losick, R., and Fox, T. D., 1976, Positive control of the temporal program of bacteriophage SPO 1 gene expression by phage and host specified subunits of RNA polymerase, *in* "Molecular Mechanisms in the Control of Gene Expression" (D. P. Nierlich, W. J. Rutter, and C. F. Fox, eds.), pp. 89–104, Academic Press, New York.

Tomizawa, J., Anraku, N., and Iwama, Y., 1966, Molecular mechanisms of genetic recombination in bacteriophage. VI. A mutant defective in the joining of DNA molecules, *J. Mol. Biol.* **21,** 247.

Traut, R. R., Heimark, R. L., Sun, T-T., Hershey, J. W. B., and Bollen A., 1974, Protein topography of ribosomal subunit from *Escherichia coli, in,* "Ribosomes" (M. Nomura, A. Tissières, and P. Lengyel, eds.), pp. 271–308, Cold Spring Harbor Laboratory, New York.

Travers, A. A., 1969, Bacteriophage sigma factor for RNA polymerase, *Nature (London)* **223,** 1107.

Travers, A. A., 1970*a,* Positive control of transcription by a bacteriophage sigma factor, *Nature (London)* **225,** 1009.

Travers, A., 1970*b,* RNA polymerase and T 4 development, *Cold Spring Harbor Symp. Quant. Biol.* **35,** 241.

Travers, A., 1971, Control of transcription in bacteria, *Nature (London) New Biol.* **229,** 69.

Travers, A., 1973, Control of ribosomal RNA synthesis *in vitro, Nature (London)* **244,** 15.

Travers, A., 1974, Ribosomal RNA synthesis *in vitro, in* "Ribosomes" (M. Normura, A. Tissières, and P. Lengyel, eds.) pp. 763–770, Cold Spring Harbor Laboratory, New York.

Travers, A., and Burgess, R., 1969, cyclic re-use of the RNA polymerase sigma factor, *Nature (London)* **222,** 537.

Travers, A., Baillie, D., and Pedersen, S., 1973, Effect of DNA conformation on ribosomal RNA synthesis *in vitro*, *Nature* (*London*) *New Biol.* **243**, 161.

Trimble, R. B., Galivan, J., and Maley, F., 1972, The temporal expression of T2r⁺ bacteriophage genes *in vivo* and *in vitro*, *Proc. Nat. Acad. Sci. USA* **69**, 1659.

Truffaut, N., Revet, B., and Soulie, M., 1970, Etude comparative des DNA des phage 2C, SP 8*, SP 82, Φe, SPO 1 et SP 50, *Eur. J. Biochem.* **15**, 391.

Tutas, D. J., Parson, K. A., Wehner, J. M., Harlander, S. K., Koerner, J. F., Warner, H. R., and Snustad, D. P., 1974*a*, Mechanism of nuclear disruption by phage T 4, *Fed. Proc.* **33**, 1487.

Tutas, D. J., Wehner, J. M., and Koerner, J. F., 1974*b*, Unfolding of the host genome after infection of *Escherichia coli* with bacteriophage T 4, *J. Virol.* **13**, 548.

van der Laan, and Rothman-Denes, L. B., 1975, personal communication.

Vanderslice, R. W., and Yegian, C. D., 1974, The identification of late bacteriophage T 4 proteins on sodium dodecyl sulfate polyacrylamide gels, *Virology* **60**, 265.

Van Dieijen, G., Van der Laken, C. J., Van Knippenberg, P. H., and Van Duin, J., 1975, Function of *Escherichia coli* ribosomal protein S1 in translation of natural and synthetic messenger RNA, *J. Mol. Biol.* **93**, 351.

Van Duin J., and Kurland, C. G., 1970, Functional heterogeneity of the 30 S ribosomal subunit of *Escherichia coli.*, *Mol. Gen. Genet.* **109**, 169.

Van Etten, J. L., Vidaver, A. K., Koski, R. K., and Semancik, J. S., 1973, RNA polymerase activity associated with bacteriophage φ6, *J. Virol.* **12**, 464.

Vetter, D., and Sadowski, P. D., 1974, Point mutants in the D2a region of bacteriophage T 4 fail to induce T 4 endonuclease IV, *J. Virol.* **14**, 207.

Vollenweider, H. J., Sogo, J. M., and Koller, T., 1975, A routine method for protein-free spreading of double- and single-stranded nucleic acid molecules, *Proc. Nat. Acad. Sci. USA* **72**, 83.

Von Hippel, P. H., Revzin, A., Gross, C. A., and Wang, A. C., 1975, Interactions of *lac* repressor with non-specific DNA binding sites, *in* "Symposium on Protein–Ligand Interactions" (H. Sund and G. Blauer, eds.), pp. 270–285, de Gruyter, Berlin.

Wahba, A. J., 1975, personal communication.

Wahba, A. J., Miller, M. J., Niveleau, A., Landers, T. A., Carmichael, G. G., Weber, K., Hawley, D. A., and Slobin, L. I., 1974, Subunit I of Qβ replicase and 30 S ribosomal protein S1 of *Escherichia coli*. Evidence for the identity of the two proteins, *J. Biol. Chem.* **249**, 3314.

Wainfan, E., Srinivasan, P. R., and Borek, E., 1965, Alterations in the transfer ribonucleic acid methylases after bacteriophage infection or induction, *Biochemistry* **4**, 2845.

Walter, G., Zillig, W., Palm, P., and Fuchs, E., 1967, Initiation of DNA-dependent RNA synthesis and the effect of heparin on RNA polymerase, *Eur. J. Biochem.* **3**, 194.

Walter, G., Seifert, W., and Zillig, W., 1968, Modified DNA-dependent RNA polymerase from *E. coli* infected with bacteriophage T 4, *Biochem. Biophys. Res. Commun.* **30**, 240.

Walz, A., and Pirotta, V., 1975, Sequence of the P_R promoter of phage λ, *Nature* (*London*) **254**, 118.

Wang, J. C., 1974, Interactions between twisted DNAs and enzymes: The effects of superhelical turns, *J. Mol. Biol.* **87**, 797.

Wang, J. C., Barkley, M. D., and Bourgeois, S., 1974, Measurements of unwinding of lac operator by repressor, *Nature* (*London*) **251**, 247.

Warner, H. R., and Hobbs, M. D., 1967, Incorporation of uracil-^{14}C into nucleic acids in *Escherichia coli* infected with bacteriophage T 4 and T 4 amber mutants, *Virology* **33**, 376.

Warner, H. R., Drong, R. F. and Berget, S. M., 1975, Early events after infection of *Escherichia coli* by bacteriophage T 5. I. Induction of a 5′-nucleotidase activity and excretion of free bases, *J. Virol.* **15**, 273.

Waters, L. C., and Novelli, G. D., 1967, A new change in leucine transfer RNA observed in *Escherichia coli* infected with bacteriophage T 2*, *Proc. Nat. Acad. Sci. USA* **57**, 979.

Waters, L. C., and Novelli, G. D., 1968, The early change in *E. coli* leucine RNA after infection with bacteriophage T 2, *Biochem. Biophys. Res. Commun.* **32**, 971.

Weber, H., Billeter, M. A., Kahane, S., Weissmann, C., Hindley, J., and Porter, A., 1972, Molecular basis for repressor activity of Qβ replicase, *Nature (London) New Biol.* **237**, 166.

Weiss, S. B., Hsu, W., Foft, J. W., and Scherberg, N. H., 1968, Transfer RNA coded by the T 4 bacteriophage genome, *Proc. Nat. Acad. Sci. USA* **61**, 114.

Weissmann, C., Billeter, M. A., Goodman, H. M., Hindley, J., and Weber, H. 1973, Structure and function of phage RNA, *Annu. Rev. Biochem.* **42**, 303.

Werner, R., 1968, Initiation and propagation of growing points in the DNA of phage T 4, *Cold Spring Harbor Symp. Quant. Biol.* **28**, 501.

Whiteley, H. R., 1975, personal communication.

Wiberg, J. S., 1966, Mutants of bacteriophage T 4 unable to cause breakdown of host DNA, *Proc. Nat. Acad. Sci. USA* **55**, 614.

Wiberg, J. S., Dirksen, M.-L., Epstein, R. H., Luria, S. E., and Buchanan, J. M., 1962, Early enzyme synthesis and its control in *E. coli* infected with some amber mutants of bacteriophage T 4, *Proc. Nat. Acad. Sci. USA* **48**, 293.

Wiberg, J. S., Mendelsohn, S., Warner, V., Hercules, K., Aldrich, C., and Munro, J. L., 1973, SP 62, a viable mutant of bacteriophage T 4D defective in regulation of phage enzyme synthesis, *J. Virol.* **12**, 775.

Wickner, W., and Kornberg, A., 1974, A novel form of RNA polymerase from *Escherichia coli*, *Proc. Nat. Acad. Sci. USA* **71**, 4425.

Wilhelm, J. M., and Haselkorn, R., 1971a, *In vitro* synthesis of T 4 proteins: The products of genes 9, 18, 19, 23, 24, and 38, *Virology* **43**, 198.

Wilhelm, J. M., and Haselkorn, R., 1971b, *In vitro* synthesis of T 4 proteins: Control of transcription of gene 57, *Virology* **43**, 209.

Wilhelm, J. M., Johnson, G., Haselkorn, R., and Geiduschek, E. P., 1972, Specific inhibition of bacteriophage SPO 1 DNA-directed protein synthesis by the SPO 1 transcription factor, TF 1, *Biochem. Biophys. Res. Commun.* **46**, 1970.

Wilson, D., and Gage, L. P., 1971, Certain aspects of SPO 1 development, *J. Mol. Biol.* **57**, 297.

Wilson, D. L., and Geiduschek, E. P., 1969, A template-selective inhibitor of *in vitro* transcription, *Proc. Nat. Acad. Sci. USA* **62**, 514.

Wilson, J. H., 1973, Function of the bacteriophage T 4 Transfer RNAs, *J. Mol. Biol.* **74**, 753.

Wilson, J. H., and Abelson, J. N., 1972, Bacteriophage T 4 transfer RNA. II. Mutants of T 4 defective in the formation of functional suppressor transfer RNA, *J. Mol. Biol.* **69**, 57.

Wilson, J. H., and Kells, S., 1972, Bacteriophage T 4 transfer RNA. I. Isolation and characterization of two phage-coded nonsense suppressors, *J. Mol. Biol.* **69**, 39.

Wilson, J. H., Kim, J. S., and Abelson, J. B., 1972, Bacteriophage T 4 transfer RNA. III. Clustering of the genes for the T 4 transfer RNA's, *J. Mol. Biol.* **71**, 547.

Winsten, J. A., and Huang, P. C., 1972, Ribosomal RNA synthesis *in vitro*: A protein-DNA complex from *Bacillus subtilis* active in inhibition of transcription, *Proc. Nat. Acad. Sci. USA* **69**, 1387.

Wood, W., 1972, Caltech Biology Division Annual Report, p. 14, California Institute of Technology, Pasadena.

Wood, W. B., 1974, Bacteriophage T 4, *in* "Handbook of Genetics" (R. C. King, ed.), Vol. 1, pp. 327–331, Plenum Press, New York and London.

Wood, W. B., and Bishop, R. J., 1973, Bacteriophage T 4 tail fibers: Structure and assembly of a viral organelle, *in* "Virus Research" (C. F. Fox and W. S. Robinson, eds.), pp. 303–326, Academic Press, New York and London.

Wood, W. B., Dickson, R. C., Bishop, R. J., and Revel, H. R., 1973, Self-assembly and non-self-assembly in bacteriophage T 4 morphogenesis, *in* "The Generation of Subcellular Structures" (R. Markham, J. B. Bancroft, D. R. Davies, D. A. Hopwood, and R. W. Horne, eds.), pp. 25–58, North-Holland Publishing Co., Amsterdam-London, and American Elsevier Publishing Co., New York.

Worcel, A., and Burgi, E., 1972, On the structure of the folded chromosome of *Escherichia coli, J. Mol. Biol.* **71**, 127.

Wovcha, M. G., Tomich, P. K., Chiu, C. S., and Greenberg, G. R., 1973, Direct participation of dCMP hydroxymethylase in synthesis of bacteriophage T 4 DNA, *Proc. Nat. Acad. Sci. USA* **70**, 2196.

Wu, R., 1975, personal communication.

Wu, C.-W., Yarbrough, L. R., Hillel, Z., and Wu, F. Y.-H., 1975, Sigma cycle during *in vitro* transcription: Demonstration by nanosecond fluorescence depolarization spectroscopy, *Proc. Nat. Acad. Sci. USA* **72**, 3019.

Wu, R., and Geiduschek, E. P., 1974, DNA ligase and the regulation of bacteriophage T 4 late gene expression, *Fed. Proc.* **33**, 1487.

Wu, R., and Yeh, Y.-C., 1974, DNA arrested mutants of gene 59 of bacteriophage T 4. II. Replicative intermediates, *Virology* **59**, 108.

Wu, R., and Geiduschek, E. P., 1975, The role of replication proteins in the regulation of bacteriophage T 4 transcription. I. Gene 45 and hydroxymethyl-C-containing DNA, *J. Mol. Biol.* **96**, 513.

Wu, R., Geiduschek, E. P., Rabussay, D., and Cascino, A., 1973, Regulation of transcription in bacteriophage T 4-infected *E. coli*—a brief review and some recent results, *in* "Virus Research" (C. F. Fox and W. S. Robinson, eds.), pp. 181–204, Academic Press, New York and London.

Wu, R., Geiduschek, E. P., and Cascino, A., 1975, The role of replication proteins in the regulation of bacteriophage T 4 transcription. II. Gene 45 and late transcription uncoupled from replication, *J. Mol. Biol.* **96**, 539.

Yamada, Y., and Nakada, D., 1976, Early to late switch in bacteriophage T 7 development: No translational discrimination between T 7 early messenger RNA and late messenger RNA, *J. Mol. Biol.* **100**, 35.

Yamada, Y., Whitaker, P. A., and Nakada, D., 1974a, Functional instability of T 7 early mRNA, *Nature (London)* **248**, 335.

Yamada, Y., Whitaker, P. A., and Nakada, D., 1974b, Early to late switch in bacteriophage T 7 development: Functional decay of T 7 early messenger RNA, *J. Mol. Biol.* **89**, 293.

Yang, H., and Zubay, G., 1974, A possible termination factor for transcription in *E. coli, Biochem. Biophys. Res. Commun.* **56**, 725.

Yegian, C. D., Mueller, M. Selzer, G., Russo, V., and Stahl, F. W., 1971, Properties of the DNA-delay mutants of bacteriophage T 4, *Virology* **46**, 900.

Yeh, Y.-C., and Tessman, I., 1972, Control by bacteriophage T 4 of the reduction of adenosine nucleotide to deoxyadenosine nucleotide, *J. Biol. Chem.* **247**, 3252.

Yeh, Y.-C., and Wu, J.-R., 1973, Requirement of a functional gene 32 product of bacteriophage T 4 in UV repair, *J. Virol.* **12**, 758.

Yehle, C. O. and Doi, R. H., 1967, Differential expression of bacteriophage genomes in vegetative and sporulating cells in *Bacillus subtilis*, *J. Virol.* **1**, 935.

Yehle, C. O., and Ganesan, A. T., 1972, Deoxyribonucleic acid synthesis in bacteriophage SPO 1-infected *Bacillus subtilis*: I. Bacteriophage deoxyribonucleic acid synthesis and fate of host deoxyribonucleic acid in normal and polymerase-deficient strains, *J. Virol.* **9**, 263.

Yoshida, M., 1972, Lack of discrimination between early and late T 4 mRNA by host initiation factors, *Nature (London) New Biol.* **239**, 178.

Young, E. T., 1970, Control of functional T 4 messenger synthesis, *Cold Spring Harbor Symp. Quant. Biol.* **35**, 189.

Young, E. T., 1975, Analysis of T 4 chloramphenicol RNA by DNA:RNA hybridization and by cell-free protein synthesis, and the effect of *E. coli* polarity-suppressing alleles on its synthesis, *J. Mol. Biol.* **96**, 393.

Young, E. T., and Menard, R. C., 1975, Analysis of the template activity of bacteriophage T 7 messenger RNAs during infection of male and female strains of *Escherichia coli*, *J. Mol. Biol.* **99**, 167.

Young, E. T., and Van Houwe, G., 1970, Control of synthesis of glucosyl transferase and lysozyme messenger after T 4 infection, *J. Mol. Biol.* **51**, 605.

Yudelevich, A., 1971, Specific cleavage of an *Escherichia coli* leucine transfer RNA following bacteriophage T 4 infection, *J. Mol. Biol.* **60**, 21.

Yura, T., and Igarashi, K., 1968, RNA polymerase mutants of *E. coli* I: Mutants resistant to streptovaricin, *Proc. Nat. Acad. Sci. USA* **61**, 1313.

Zabin, I., and Fowler, A. V., 1970, β-galactosidase and thiogalactoside transacetylase, *in* "The Lactose Operon" (J. R. Beckwith and D. Zipser eds.), pp. 27–47, Cold Spring Harbor Laboratory, New York.

Zetter, B. R. and Cohen, P. S., 1974, personal communication.

Zetter, B. R., and Cohen, P. S., 1975, personal communication.

Zillig, W., Zechel, K., Rabussay, D., Schachner, M., Sethi, V. S., Palm, P., Heil, A., and Seifert, W., 1970, On the role of different subunits of DNA-dependent RNA polymerase from *E. coli* in the transcription process, *Cold Spring Harbor Symp. Quant. Biol.* **35**, 47.

Zillig, W., Fujiki, H., Blum, W., Janeković, D., Schweiger, M., Rahmsdorf, H.-J., Ponta, H., and Hirsch-Kauffmann, M., 1975, *In vivo* and *in vitro* phosphorylation of DNA-dependent RNA polymerase of *Escherichia coli* by bacteriophage-T 7-induced protein kinase, *Proc. Nat. Acad. Sci. USA* **72**, 2506.

Zimmer, S. G., and Millette, R. L., 1975, DNA-dependent RNA polymerase from *Pseudomonas* BAL-31. II. Transcription of the allomorphic forms of bacteriophage PM 2 DNA, *Biochemistry* **14**, 300.

Zimmer, S. G., Millette, R. L., and Morse, M. L., 1974, Specific transcription by *Pseudomonas BAL-31* core polymerase directed by *Escherichia coli* sigma factor, Abstracts of the Annual Meeting of the American Society for Microbiology, p. 228, American Society for Microbiology, Washington, D.C.

Zweig, M., and Cummings, D. J., 1973*a*, Cleavage of head and tail proteins during bacteriophage T 5 assembly: Selective host involvement in the cleavage of a tail protein, *J. Mol. Biol.* **80**, 505.

Zweig, M., and Cummings, D. J., 1973*b*, Structural proteins of bacteriophage T 5, *Virology* **51**, 443.

Zweig, M., Rosenkranz, H. S., and Morgan, C., 1972, Development of coliphage T 5: Ultrastructural and biochemical studies, *J. Virol.* **9**, 526.

NOTES ADDED IN PROOF

The following recent developments bear significantly on Chapter 1:

1.   The postulated slowing down of transcription in the region of the ρ-dependent termination signal has been observed by M. Rosenberg, D. Court, and D. Wulff (1976, personal communication).

2.   Direct proof that the SPO 1-coded RNA polymerase binding proteins, peptides V and VI-ν^{13}, are the products of SPO 1 genes 34 and 33, respectively, has now been presented (Fox, T. D., 1976 *Nature* **262**, 748). These two subunits have been dissociated from RNA polymerase and shown to exhibit positively regulating properties when recombined with RNA polymerase core from uninfected *B. subtilis* (Tjian, R., and Pero, J., 1976, *Nature* **262**, 753).

3.   In discussing the possibility that the T4 early regulatory event might be controlled by preformed viral and cellular components, including membrane proteins, we cited as evidence the case of an rII mutation whose phenotype was thought to resemble that of a *mot-far*I mutation. It now appears (Gold, further personal communication) that this phenotype was not due simply to the rII mutation but involved a mutation at a second site. The argument which we made about the ways in which the early regulatory event is controlled remains plausible, even in the absence of what we originally considered to be an item of supporting evidence.

4. α_A and α_M are identical. Both subunits result from ADP ribosylation, evidently at the same site in α (R. Mailhammer, 1976, Ph.D. Thesis, University of Munich).

5.   Evidence that the 15,000 dalton RNA polymerase binding protein is associated with T 4 gene *alc* has now been presented (Sirotkin, K., Wei, J., and Snyder, L. S., 1977, *Nature*, in press).

6.   Further evidence has been presented that the 10,000 dalton RNA polymerase binding protein is a transcription inhibitor. The precise mode of action has yet to be determined (Stevens, A., 1976, in "RNA Polymerases" (R. Losick and M. Chamberlin, eds.), pp. 617–627, Cold Spring Harbor Laboratory, New York).

7.   Further evidence on the location of the genes for the isoaccepting leu tRNA's has been presented (Hunt, C., Hwang, L.-T., and Weiss, S. B., 1976, *J. Virol.*, **20**, 63). Genes for isoaccepting ser tRNA's have also been identified (Henckes, G., Panayotis, O., and Heyman, T., 1976, *J. Virol.* **17**, 316).

8.   The discovery of RNA polymerase activities in N 4 virions has now been claimed (Pesce, A., Casoli, C., and Schito, G. C., 1976, *Nature* **262**, 412; Falco, S. C., Van der Laan, K., and Rothman-Denes, L. B., 1977, *Proc. Nat. Acad. Sci.,* in press). The detailed findings are contradictory, but such an activity certainly exists and has been shown to involve the product of a viral gene which is essential for transcription *in vivo*.

Bacteriophage λ: The Lysogenic Pathway

R. A. Weisberg*

Laboratory of Molecular Genetics
National Institute of Child Health and Human Development
National Institutes of Health
Bethesda, Maryland 20014

S. Gottesman

Department of Biology
Massachusetts Institute of Technology
Cambridge, Massachusetts 02139

and

M. E. Gottesman*

Laboratory of Molecular Biology
National Cancer Institute
National Institutes of Health
Bethesda, Maryland 20014

1. INTRODUCTION

1.1 The Life Cycle

Coliphage λ is a temperate phage, so-called because it can grow in two distinct ways. During lytic growth, common to both temperate and

* This article was written by R. A. Weisberg and M. E. Gottesman in their private capacities. No official support or endorsement by the National Institutes of Health is intended or should be inferred.

intemperate phage, the virus chromosome is usually replicated several hundredfold within the span of a single cellular generation, the replicas are packaged into mature virus particles by newly synthesized virus proteins, and the particles are released as a result of cell lysis. Temperate viruses alone, however, are capable of growing in harmony with their hosts. Lambda, the best studied of the temperate coliphages, accomplishes this by synthesizing a protein that promotes the insertion of the virus chromosome into the host chromosome and by synthesizing a repressor that inhibits the further expression of virus genes. The resulting cell is called a lysogen, the process, lysogenization, and the inserted virus chromosome, prophage. [See Herskowitz (1973) for a recent review.]

Prophage λ is replicated as part of the host chromosome and is thus inherited by both daughter cells at cell division. An inserted prophage initiates lytic growth when repression fails, a normally rare event that can be induced with high efficiency by agents that inhibit the replication of the host chromosome (such as ultraviolet irradiation). After "induction" of a lysogen, virus-encoded proteins are synthesized and excise the prophage from the host chromosome. In this article, we shall review our current understanding of how repression is established and maintained and how the virus DNA promotes its own insertion and excision.

1.2. Relevant Physical Characteristics

The virion consists of approximately equal amounts of protein and DNA. The DNA is a linear, mostly double-stranded molecule with a molecular weight of 31×10^6 (equivalent to 47 kilobase pairs). Single-stranded segments 12 nucleotides long are located at each end; these are complementary to each other in sequence and enable the molecule to circularize by end-joining. Mutants in about 40 λ cistrons have been isolated and located on a genetic map by measurement of recombination frequencies and deletion mapping. The genetic map is colinear with a physical map that has been established by electron-microscopic examination of heteroduplex DNAs formed by annealing single-stranded λ DNA with complementary strands from various λ deletion and substitution mutants. These facts are summarized in Fig. 1. [See also Davidson and Szybalski (1971).]

1.3 Repression, Insertion, and Abortive Lysogeny

Lambda virions, when mixed with susceptible bacteria, adsorb to a specific receptor on the cell surface (Thirion and Hofnung, 1972). The

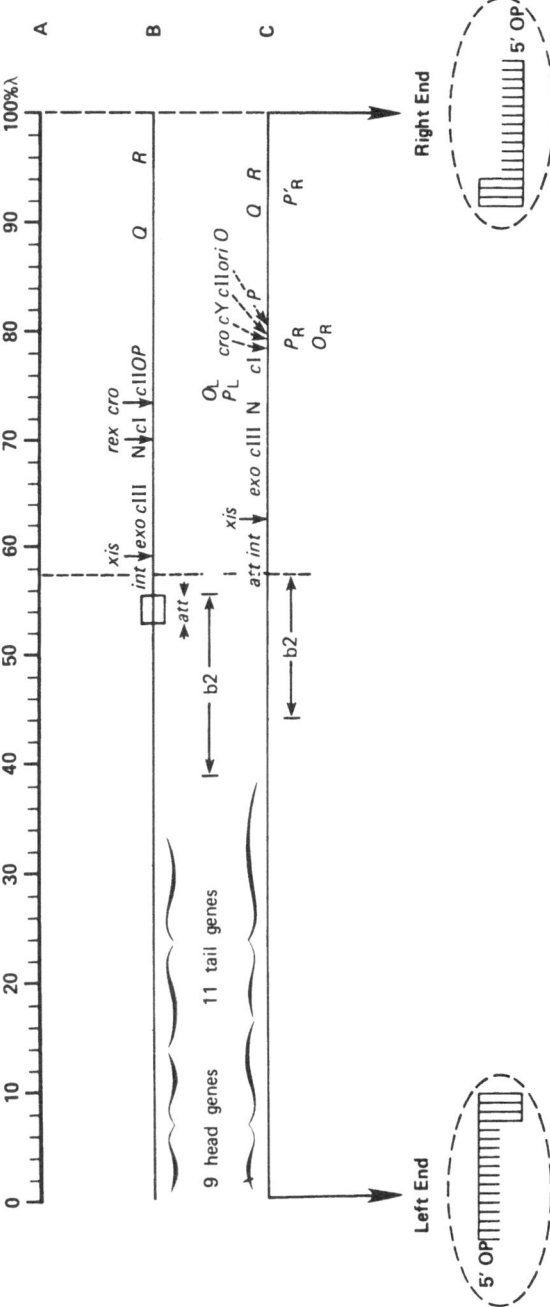

Fig. 1. Genetic and physical maps of λ. (A) Percentage scale of chromosomal distance. (B) Map based on recombination frequency in phage crosses (1% = 0.7 centimorgans; see Campbell, 1971). (C) Map based on electron microscopy of deletions and substitutions (1% = 470 base pairs; see Davidson and Szybalski, 1971). All genes and sites given are described in the text. The left and right ends of the map are shown in expanded scale below part C. Each horizontal line represents a single DNA strand, and each vertical bar a base pair between complementary nucleotides. From Davidson and Szybalski (1971) with the permission of the authors and of Cold Spring Harbor Laboratory.

DNA of the phage is injected through the tail and, once inside the cell, rapidly circularizes by end-joining and by the action of DNA ligase (Bode and Kaiser, 1965a; Gellert, 1967; Tomizawa and Ogawa, 1968). Transcription of the viral genome by the host RNA polymerase then begins. The development of the intracellular virus proceeds in one of two mutually exclusive pathways: the lytic or the lysogenic. In the former, full expression of the viral genome takes place, with cell death and lysis. The lysogenic pathway, however, is characterized by expression of only a limited set of viral genes, the "early" functions, which are in turn repressed. The balance between these two pathways is a delicate one, controlled by both host and viral factors, and will be examined in detail below.

The two components of the lysogenic pathway—the repression of the viral genome and the insertion of the viral DNA into the host chromosome—can be considered separately (Zichichi and Kellenberger, 1963). Each component is defined by a set of nonoverlapping λ mutants, and the appearance of stable lysogens after infection of a bacterial culture requires that both components be operative. A phage mutant unable to repress the expression of its genes will kill every cell it infects. Conversely, a noninserting phage mutant can repress its genes, and this permits survival of a fraction of the cells. However, since repressed λ DNA will only replicate when it is inserted into the *E. coli* chromosome, the mutant genome will soon be diluted as the surviving cells divide. Cells carrying repressed but unintegrated λ DNA are called abortive lysogens. They differ from stable lysogens in that the viral genome is inherited by only one instead of both daughter cells at each cell division.

2. REPRESSION

2.1. The Maintenance of Repression

The prophage expresses very few functions. For example, λ lysogens show no trace of two phage-encoded enzymes abundant after phage infection, λ exonuclease and endolysin (Jacob and Fuerst, 1958; Korn and Weissbach, 1964; Radding, 1964). The absence of viral proteins is a consequence of the low level of transcription of the viral genome (Attardi *et al.*, 1963; Isaacs *et al.*, 1965). The level of λ mRNA in a lysogen is less than 5% that found 2 min after infection (Taylor *et al.*, 1967). It derives principally from a small region of the λ chromosome, the *cI-rex* segment (see Fig. 1). The genetic basis of repression was revealed by the remarkable experiments of F. Jacob, E. Wollman, A. Kaiser, and their collaborators in the period 1953–1961. Biochemical and genetic analysis have since confirmed and extended their proposals with happy consistency.

2.1.1 Lytic Phage Growth Is Inhibited in Lysogens by a Specific Cytoplasmic Repressor

Two models have been considered to explain the absence of phage gene products in a lysogen. In the first model, the prophage synthesizes a cytoplasmic repressor of lytic growth. In the second, the prophage occupies and thereby blocks a specific site on the host chromosome that is required for lytic development. An analysis of the phenomenon of zygotic induction clearly showed that the first model was correct for λ. Prophage can be transferred from lysogenic males to female bacteria by mating. When the female is nonlysogenic, lytic growth of the transferred prophage frequently ensues; this is called zygotic induction (Jacob and Wollman, 1956a). When the female is lysogenic, there is no zygotic induction whether or not the male is lysogenic. Only DNA is transferred from male to female in bacterial mating (Silver, 1963; Rosner *et al.*, 1967). Thus, as predicted by the first model, the cytoplasm of the lysogenic female prevents lytic growth of the transferred prophage (Jacob and Monod, 1961). Lytic growth is also blocked when phage genomes are introduced into the cytoplasm of a lysogen by infection. Clearly, this immunity of lysogens to superinfection can also be explained by the action of a cytoplasmic repressor (Bertani, 1956). Superinfection immunity is seen in all temperate phage, although it is phage specific (Jacob and Monod, 1961). Thus λ lysogens are immune to superinfection by λ, but nonimmune (or sensitive) to superinfection by the closely related temperate phages 21, φ80, and 434.

2.1.2. *c*I Is the Structural Gene for λ Repressor

The restriction of λ repressor to λ lysogens suggests that repressor is encoded by a phage gene. Mutants unable to synthesize active repressor, and thus defective in lysogeny, were isolated quite early. These mutants were distinguished by their clear plaque phenotype. Wild-type λ makes a turbid plaque on a lawn of bacteria growing on agar. The turbidity is due to the survival and growth of repressor-containing cells immune to the phage in the plaque. A phage that cannot make repressor kills all of the cells it infects, and no survivors appear in the center of the plaque.

The λ clear (*c*) mutants were classified into four groups *c*I, *c*II, *c*III, and *c*Y—by complementation tests, map location, and the nature and severity of the lysogenization defect (Kaiser, 1957; Brachet and Thomas, 1969). Complementation between *c* mutants is demonstrated by coinfecting cells with, for example, λ *c*II and λ *c*III, neither of which efficiently lysogenizes by itself. The doubly infected cells are lysogenized well by either or by both of the *c* mutants. The map locations of the four groups are given in Figs. 1 and 2.

Which of the *c* mutations is in the structural gene for repressor? The behavior of the *c*I mutants indicated that the *c*I gene encoded the λ repressor: (1) Lambda *c*II, *c*III, and *c*Y mutants lysogenize with a reduced but measurable frequency (10^{-1} to 10^{-5} per infected cell) (Kaiser, 1957); λ *c*I never lysogenize. (2) Although λ *c*II, *c*III, or *c*Y single lysogens are easily obtained by complementation, complementation of λ *c*I mutants never results in the formation of single λ *c*I lysogens. Cells bearing a *c*I prophage always carry a second cI$^+$ prophage as well (Jacob and Monod, 1961). (3) Lambda *c*II, *c*III, or *c*Y lysogens are indistinguishable from wild-type lysogens: They do not synthesize lytic λ functions, and they are immune to superinfection by λ. Thus, these mutants, unlike *c*I, are defective in the establishment of lysogeny but not in its maintenance (Kaiser, 1957).

The isolation of two unusual types of *c*I mutants strengthened this conclusion. The first, λ *c*Iind$^-$, lysogenizes at normal frequency, but the resulting lysogens are not induced by ultraviolet irradiation. The alteration is dominant: Lysogens carrying both λ *c*Iind$^-$ and λ cI$^+$ prophage are noninducible (Jacob and Campbell, 1959). It was concluded that ultraviolet irradiation induces phage growth by inactivating repressor, and that λ *c*Iind$^-$ synthesizes a radiation-resistant repressor.

The second type of mutant, represented by λ *c*Its857, is a temperature-sensitive mutation of the *c*I gene: It makes clear plaques at elevated (39°C) temperatures only. Lambda *c*Its857 lysogens, formed at low temperature, are induced when shifted to high temperatures (Sussman and Jacob, 1962). This demonstrates that the product of the *c*I gene is required to maintain repression and must be, therefore, continuously produced in a lysogen. The isolation of *c*I amber mutants (Jacob *et al.*, 1962; Thomas and Lambert, 1962) demonstrates that the functional *c*I gene product must be a protein.

That *c*I is the structural gene for λ repressor was confirmed by Ptashne (1967*a*), who isolated the *c*I protein and demonstrated that it interacted specifically with λ DNA to block transcription. We shall

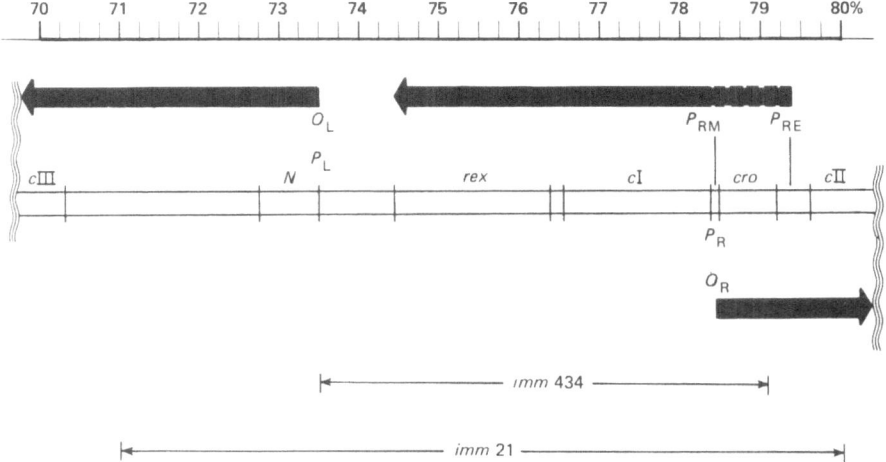

Fig. 2. Physical map of the immunity region. The top scale gives distance in percent from the left end of the map (1% = 470 base pairs). The two solid lines represent complementary DNA strands. The 5′-end of the top strand is assumed to be at the left. Gene sizes and the end points of the *imm*434 and *imm*21 substitutions are taken from Fiandt *et al.* (1971), Hayes and Szybalski (1973), and Szybalski and Szybalski (1974). Site P_{re} is defined by *c*Y mutants (see text). The solid arrows give the origin and direction of various immunity-region transcripts (see text).

return to this point after discussing how repressor blocks lytic phage growth.

Another phage function is expressed by prophage λ. Lambda lysogens do not support the growth of *r*II mutants of phage T4 (Benzer, 1955). The responsible λ gene has been determined by the isolation of mutants (Howard, 1967; Gussin and Peterson, 1972). These mutants, λ *rex⁻*, do not restrict T4*r*II growth but otherwise yield normal lysogens. The *rex* gene is adjacent to *c*I on the λ map (see Fig. 2).

2.1.3. Repressor Directly Blocks the Expression of λ Early Functions

Lambda genes can be placed into three classes according to their time of expression after infection or induction: immediate early, early, and late (see Echols, 1972). The products of these groups of genes not only do appear sequentially, but must appear sequentially; i.e., expression of the early genes requires the expression of an immediate early gene (gene *N*), and the late genes are activated by the product of an early gene (gene *Q*). This pattern of control was suggested by the ex-

periments of Jacob and Wollman (1956b) and Jacob, Fuerst, and
Wollman (1957). They mutagenized lysogens to obtain mutant, defective prophages. These lysogens, which released no active phage, were
induced and the production of various phage proteins (endolysin, phage
antigen, and virion-related structures) as well as phage DNA replication measured. They found that a single mutation could reduce the
activities of several phage functions. These mutations, they proposed,
lay in positive control genes whose expression activated other phage
genes.

   In fact, λ has two positive control genes. The first, the immediate
early gene *N*, is required for the expression of the early genes. It is
thought to act by allowing the transcripts of the immediate early genes
N and *cro* to be extended leftward and rightward into regions containing early genes (Roberts, 1969; see Figs. 1 and 2). Among the early
genes is *Q*, whose product activates late gene expression (Dove, 1966),
perhaps by extending transcripts that have originated at promoter P'_R
(Herskowitz and Signer, 1970; Roberts, 1975; see Fig. 1). Amber mutations have been obtained in both these genes, indicating that their
products are proteins (Campbell, 1961).

   The realization that positive control functions exist in λ greatly
simplified the task of explaining how repression is maintained. Jacob
and Monod (1961) proposed that λ repressor, by analogy to *lac*
repressor, acts by directly preventing the synthesis of a positive control
protein; other λ functions would be blocked indirectly. That λ late
genes are only indirectly under repressor control was demonstrated by
R. Thomas and his collaborators (see Thomas, 1970). They infected λ
lysogens with the closely related phage λ *imm*434 (see the next section).
This phage carries most of the λ genome, including genes *N* and *Q*, but
is not subject to inhibition by λ repressor. It grows in a λ lysogen
without inducing the growth of the prophage (Thomas and Bertani,
1964). Nevertheless, λ *imm*434 activates the expression of most of the λ
late genes. In the case of gene *R* (the gene for endolysin; Campbell and
del Campillo-Campbell, 1963), activation has been shown to require an
active *Q* product (Dambly *et al.*, 1968).

   A similar analysis of the superinfected lysogens for the products of
early genes shows that the expression of the prophage genes *xis, exo,*
*c*III, *c*II, *O*, and *Q* is induced poorly or not at all by λ *imm*434 (Bode and
Kaiser, 1965b; Dambly and Couturier, 1971; Luzzati, 1970; Gottesman
and Weisberg, 1971). An increased amount of the *int* and *P* gene
products can be detected after heteroimmune superinfection (Thomas,
1966; Kaiser and Masuda, 1970; Weisberg, 1970), although the inability
to obtain a quantitative estimate makes interpretation of these results

difficult. It therefore appears that the expression of most, if not all, of the early genes is under dual control: It is blocked by repressor even in the presence of N product and is stimulated by N product in the absence of repressor. Since gene N itself is not expressed in a lysogen (Thomas, 1966), dual control probably contributes to the efficacy of repression.

No extensive replication of the prophage is found after superinfection with λ imm434, even though the functions specifically required for autonomous λ DNA synthesis, the O and P gene products, are the same for the two phage (Thomas and Bertani, 1964; Brooks, 1965). Thus replication, like immediate early and early gene expression, is under the direct control of repressor. This direct control is exerted through transcription: a λ chromosome replicates only when its ori region (see Fig. 1) is transcribed (Dove $et\ al.$, 1969). As the level of many phage gene products depends on the number of copies of the λ genome (Dove, 1966), the tight control of λ replication in a lysogen greatly increases the efficiency of repression.

2.1.4. Repressor Acts at Operators O_L and O_R

The sites of repressor action were located by the analysis of λ variants and mutants insensitive to λ repressor. Such phages are detected by their ability to form plaques on λ lysogens.

The temperate phage 434 is closely related to λ, but grows on λ lysogens. By repeated crosses with λ, a hybrid phage called λ imm434 was constructed (Kaiser and Jacob, 1957). This phage is nearly isogenic with λ, but retains the immunity specificity of 434; i.e., it is sensitive to 434 but not to λ repressor, and λ imm434 lysogens are immune to 434 but not to λ. Crosses between λ imm434 c^+ and cI mutants of λ yielded neither λ imm434 c nor λcI$^+$. This failure to separate cI from imm by recombination suggests that the genetic determinants of 434 immunity include a cI-434 gene functionally analogous to but structurally different from cI-λ (Kaiser and Jacob, 1957). In addition, since the two phages are insensitive to each other's repressors, the sites of action ("operators") of the two repressors must also be different and located within the immunity region (Jacob and Monod, 1961). Subsequent observations of heteroduplexes between the DNAs of λ and λimm434 showed a region of nonhomology between 73.6 and 79.1 on the physical map of the λ chromosome (Davidson and Szybalski, 1971; see Fig. 2). As expected, this region contains gene cI (Fiandt $et\ al.$, 1971) and the operators (Hopkins and Ptashne, 1971).

The operators were identified by the isolation of a virulent mutant of λ (Jacob and Wollman, 1954). Like λ cI, λ vir forms clear plaques. This phenotype results not from the inability to form repressor, but from an inability to respond to repressor. Lambda vir therefore, and not λ cI, forms plaques on a λ lysogen. Lambda vir differs from wild-type λ by three mutations: v1, v2, and v3 (Jacob and Wollman, 1954, Hopkins and Ptashne, 1971). As expected for mutations in the λ operators, v1, v2, and v3 do not recombine with the immunity region of λ imm434 (Hopkins and Ptashne, 1971). The v1 and v3 mutations are closely linked and lie between cI and cro (Pereira da Silva and Jacob, 1968; Ordal, 1971; Hopkins and Ptashne, 1971; Ordal and Kaiser, 1973); v2 lies between cI and gene N (Hopkins and Ptashne, 1971; Fig. 2). The existence of two distinct operator regions and the observation that early transcription of λ diverges from the immunity region (Taylor et al., 1967; Kourilsky et al., 1968; see Fig. 2) suggested that binding of repressor at one operator, O_L, controls leftward transcription while the other, O_R, controls rightward transcription. In fact, the v2 mutation, which should reduce the sensitivity of O_L to repressor, results in the constitutive expression of genes transcribed to the left, e.g., gene N, while the O_R mutant λ v3 is constitutive for rightwardly transcribed genes, e.g., gene O. This was demonstrated by a complementation test (Ptashne and Hopkins, 1968). A culture of a λ lysogen was infected with λ v2. One portion of the infected culture was coinfected with λ imm434 N^-, and the remainder with λ imm434 O^-. Complementation between λ v2 and λ imm434 N^-, but not λ imm434 O^-, was observed. Conversely, λ v3 preferentially enhances the growth of λ imm434 O^- in a lysogen.

2.1.5. The Isolation of Repressor

The rationale for the isolation of the λ repressor was thus established. It was shown (1) to be the product of the cI gene, (2) to be a protein, and (3) to be one of the few phage-encoded proteins synthesized in a lysogen. To detect repressor, a radiochemical technique was employed (Ptashne, 1967a). Cells lysogenic for λ cIind⁻ N^- were heavily irradiated with ultraviolet light to depress gene expression. They were then infected either with λ cI^+N^- or λ cIamber N^- which, having undamaged DNA, could synthesize proteins. The presence of preexisting repressor and the absence of N product ensured that the cI^+ phage would synthesize repressor and very little else. By co-chromatographing ¹⁴C-leucine-labeled proteins made in the cI^+ infection with ³H-leucine-labeled proteins made in the cI amber infection, the product of the cI gene was detected. Infection with cI missense mutants gave a protein of altered structure; this proves that the protein is indeed the product of the cI gene.

To demonstrate that the cI protein was in fact the λ repressor, Ptashne (1967b) showed that it binds strongly to λ DNA, less strongly to λ vir DNA (Ptashne and Hopkins, 1968), and does not bind appreciably to λ imm434 DNA. The v2 mutation and the v1v3 double mutation were separated and shown each to contribute to the loss of repressor binding activity. The product of the cI-434 gene, similarly isolated, binds to λ imm434 DNA but not to λ DNA (Pirotta and Ptashne, 1969). Thus, the cI gene specifies a protein capable of interacting with DNA at the operators.

Using the specific binding of repressor to DNA (see Riggs and Bourgeois, 1968), substantial quantities of pure λ and 434 repressors have been isolated (Pirotta et al., 1970, 1971). The active form of both repressors is an oligomer that binds extremely tightly, but reversibly, to the appropriate DNA (Pirotta et al., 1970; Chadwick et al., 1970): The equilibrium binding constant of λ repressor oligomer is 3×10^{-14} moles-liter^{-1} and the half-life at 20°C of the repressor–DNA complex is 7 min (Chadwick et al., 1970; Ptashne, 1971). Recent electron micrographs of repressor and of repressor–DNA complexes suggest that the active oligomer may be a tetramer, (Brock and Pirotta, 1975), although this conclusion must be accepted with caution in view of the difficulty of determining the structure of small proteins by electron microscopy. The results of earlier binding studies indicated that the dimer was the active form (Pirotta et al., 1970; Chadwick et al., 1970), and the binding constant given above was calculated on that assumption.

2.1.6.   Repressor Specifically Prevents the Transcription of λ DNA

In a λ lysogen, the bulk of the phage-specific mRNA derives from transcription of the immunity region, as shown by its failure to hybridize to λ imm434 DNA (Taylor et al., 1967). The RNA is probably encoded by the cI and rex genes (Hayes and Szybalski, 1973). Denaturation of the repressor results in the rapid appearance of two new λ RNA species corresponding to the immediate early λ genes N and cro (Taylor et al., 1967; Kourilsky et al. 1968). The transcription of the remainder of the λ genome depends on the appearance of the immediate early transcripts; λ double mutants defective both in the promoter for N transcription (P_L) and for cro transcription (P_R) do not transcribe their DNA after induction (Kourilsky et al., 1970). It is thus sufficient to assume that the role of λ repressor is to block transcription only from P_L and P_R.

It is possible to show transcription from P_L and P_R in a purified *in vitro* transcription system (Roberts, 1969); the RNA corresponding to *N* and *cro* can be separated from other RNAs made from a λ DNA template by sedimentation and hybridization to appropriate mutant DNAs. These *in vitro* RNAs correspond to the *in vivo N* and *cro* messages: (1) the putative P_L and P_R transcripts disappear when λ P_L^- and λ P_R^- DNA templates are used; (2) a portion of the 5′-end sequence as well as the general fingerprint pattern of the *in vitro* P_L transcript corresponds to transcripts isolated after λ induction *in vivo* (Lozeron, Funderburgh, Dahlberg, Stark, and Szybalski; cited in Blattner and Dahlberg, 1972).

The action of λ repressor on the transcription of λ DNA is precisely what is expected. It specifically inhibits the synthesis of the *N* and *cro* transcripts (Steinberg and Ptashne, 1971; Wu *et al.*, 1972*b*). The level of repressor required for complete repression *in vitro* is comparable to its biological concentration. Lambda repressor does not block transcription of λ *imm*434 DNA, and it is only partially active on a λ *vir* template (Steinberg and Ptashne, 1971). In particular, the synthesis of *N* RNA is less repressible from a λ *v2* than from a λ *v2*⁺ template, while the synthesis of *cro* RNA in the presence of repressor is increased by the *v3* and *v1v3* mutations (Steinberg and Ptashne, 1971). The establishment of a faithful coupled transcription–translation system dependent on λ DNA has been reported, and λ repressor is active in this system as well (Gesteland and Kahn, 1972).

2.1.7. How Does Repressor Block Transcription?

Since repressor does not degrade λ mRNA (Steinberg and Ptashne, 1971), it must block its synthesis. It might do this by preventing the binding of RNA polymerase to its promoter site on λ DNA. Alternatively, it might block a step subsequent to binding—either the formation of the first phosphodiester bond or the elongation of the RNA chain beyond the first few 5′ nucleotides. Evidence for both models have been presented; nevertheless, experiments supporting the first model are more direct and, therefore, more convincing.

1. If the operators are downstream of the promoters as predicted by the second model, they would be expected to be transcribed. This, however, is not the case. Lambda and λ *imm*434 DNAs, which have different operators, produce *N* transcripts, derived from P_L, which are

identical at their 5′ termini (Blattner and Dahlberg, 1972). Transcription begins about three nucleotides to the left of O_L (Maniatis et al., 1974, 1975b), as determined from sequence analysis of operator DNA (see the next section). A proposal that RNA polymerase can transverse large regions of DNA uncoupled to transcription has been withdrawn (Blattner et al., 1972; and personal communication).

2.   The DNA restriction enzyme Hin II cuts λ DNA at specific sites in O_L and O_R (Maniatis and Ptashne, 1973a; see next section). The P_L mutations sex1 and sex3 and the P_R mutations x3 and x7 (but not x13) eliminate these sites of Hin II action, and a revertant of sex1 to sex^+ has regained the Hin II site in O_L (Maurer et al., 1974; Allet et al., 1974). Furthermore, the cutting at this site can also be blocked by prior binding of RNA polymerase to the DNA. Taken together, these data lead to the conclusion that the operators and the promoters overlap (see also Ordal and Kaiser, 1973), consistent with the first model.

3.   Wu et al. (1972a) have shown that DNA bound with repressor does not efficiently bind RNA polymerase at the early promotors. To do this, λ DNA, bound with repressor, was incubated with RNA polymerase and the purine nucleoside triphosphates. Then RNA polymerase that had not interacted with the DNA at the promoters was inactivated by rifampicin, repressor was removed from the DNA by increasing the salt concentration, pyrimidine nucleoside triphosphates were added, and transcription measured. The amount of transcription was about one-third that found with DNA not bound with repressor.

4.   Certain experiments of Wu et al. (1972b) do not fit comfortably with the first model. These workers conclude that an asymmetry exists in the interaction between repressor and RNA polymerase, and that while repressor blocks RNA polymerase binding, bound RNA polymerase remains partly sensitive to the action of repressor. This conclusion follows from the observation that repressor can still partially inhibit transcription in preformed λ DNA–RNA polymerase complexes. In contrast, Chadwick et al. (1970) have concluded that RNA polymerase interferes with the binding of λ repressor to DNA at O_L. In this case, the amount of bound repressor was estimated from the extent of cosedimentation of labeled repressor with λ DNA in a glycerol gradient. No easy resolution of these conflicting conclusions exists at the moment. Perhaps, since the operators have multiple binding sites for repressor (see next section), RNA polymerase binding blocks only one of the repressor binding sites. If so, it is difficult to see how repressor bound at the remaining sites, which are upstream of the promoter, could still inhibit transcription.

2.1.8. The Structure of the λ Operators

We have seen that the λ operators lie to the left and to the right of the cI gene. They are defined genetically; O_L by the $v2$ mutation, and O_R by the mutations $v1$ and $v3$. These mutations reduce repressor binding. The operators overlap the promoters P_L and P_R defined, respectively, by the mutations $sex1$ and $x3$, $x7$, or $x13$. The sequence of most of this region has now been obtained (Pirotta, 1975; Maniatis *et al.,* 1974, 1975a and b) (Fig. 3) and the base alterations corresponding to some mutations located. The effort required for this work has been clearly rewarded; the sequence has remarkable and unexpected aspects, which we shall consider in detail.

Unlike the *lac* operator, which has a single DNA sequence to which repressor binds (Gilbert and Maxam, 1973), each λ operator consists of at least three tandem sequences with repressor binding capacity. This was first demonstrated by experiments in which λ DNA was protected by bound repressor from digestion by pancreatic DNase (Maniatis and Ptashne, 1973b). It was found that the higher the ratio of repressor to operator, the longer the protected fragment. At excess repressor, the largest protected DNA fragment is over 100 base pairs long (Maniatis and Ptashne, 1973a; Brock and Pirotta, 1975). It has been shown that repressor must bind first to one site in each operator (called O_L1 and O_R1) and then fill in the other nonidentical sites (O_L2, O_R2, etc.) sequentially and in tandem. The existence of these sites and their orientation relative to the surrounding genes was established by the use of endonuclease Hin II, a restriction endonuclease that cleaves within and on either side of each operator (Maniatis and Ptashne, 1973a; Maniatis *et al.,* 1973). The resulting operator-containing fragments can still be protected from DNase I digestion by binding repressor. The length of the protected regions shows that the Hin II cleavage separates O_L1 and O_R1, which have the highest affinity for repressor, from the remaining sites, which have a lower but still measurable affinity. The four operator-containing Hin II fragments can be separated by electrophoresis and hybridized to the DNAs of appropriate λ substitution mutants to give the following order: $-N-O_L1-O_L2-O_L3-rex-c$I$-O_R3-O_R2-O_R1-cro-$. The length of the regions protected from DNase I digestion by repressor binding suggests that there are three or four such sites in each operator, a number consistent with electron microscopic observations of repressor–DNA complexes (Maniatis and Ptashne, 1973b; Brock and Pirotta, 1975).

A number of these binding sites have been sequenced (Maniatis *et al.,* 1974; 1975a and b; Pirotta, 1975; Walz and Pirotta, 1975). The results suggest that the nucleotides recognized by repressor are within the 17 base-pair units shown in brackets in Fig. 3A (Maniatis *et al.*, 1975*b*). The units are separated by AT-rich spacers 3 to 7 base pairs long. Thus, the repeat distance is 20 to 24 base pairs, roughly consistent with the size of the fragments protected from DNase digestion by repressor binding. In addition, six of six operator mutations have been located within these units. The putative repressor binding sites have certain features in common:

1. Each complete site consists of 17 base pairs with an axis of two-fold rotational symmetry drawn through the 9th base. This symmetry may reflect a corresponding symmetry in the bound repressor oligomer. In any event, the potential of these sequences to form hairpin loops is not promoted by repressor: Repressor does not unwind DNA and hence would not stabilize these hypothetical hairpins (Maniatis and Ptashne, 1973*b*).

2. The half-sites are sequence related (see Fig. 3B). Comparison of nine half-sites shows invariance at positions 2, 4, and 6. The operator mutations so far sequenced alter positions 2, 6, or 8. Note that mutations conferring virulence, vN and v1, appear in O_R2, indicating that this sequence is active biologically, as well as in *in vitro* binding studies. Operator O_L1, which has greater affinity for repressor than O_R1, differs from it only at position 5, where there has been an exchange of a T for an A.

3. The binding sites are approximately 50% AT, whereas the spacer sequences are extremely AT rich. The AT portions of the sites are found at the edges; 8 of 11 of the internal base pairs are GC.

The P_L mutations *sex*1 and *sex*3 and the P_R mutations *x*3 and *x*7 are assumed to change the operator sequences recognized by Hin II since mutant DNA was not cleaved at the corresponding site (Allet *et al.*, 1974; Maurer *et al.*, 1974). Unexpectedly, the binding of RNA polymerase did not protect this sequence from DNase I digestion; instead, it protected a region beginning nine base pairs downstream and including the transcription start point (Walz and Pirotta, 1975). Promoter mutations affecting this latter region have not, as yet, been demonstrated. These and similar observations on the *lac* promoter (Dickson *et al.*, 1975) suggest that at least some of the DNA sequences specifically recognized by RNA polymerase are upstream of the stable RNA polymerase–DNA complex that forms following recognition. This suggestion is supported by the existence of homologous sequences, perhaps enzyme recognition sites, upstream of the regions in P_L, P_R, and SV40 DNA that are protected from DNase I digestion by RNA polymerase binding (see Maniatis *et al.*, 1975*b*).

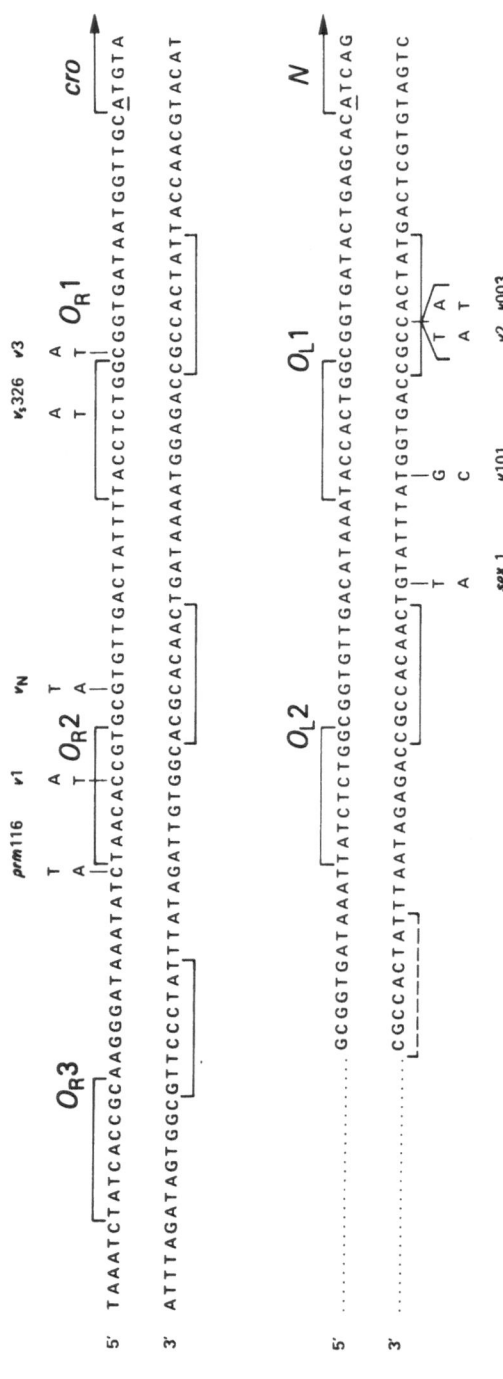

Fig. 3A. Nucleotide sequence of O_L and O_R. O_L is reversed from its usual orientation on the λ map. The segments set off by brackets are thought to contain nucleotides recognized by repressor. The nucleotide changes associated with various operator constitutive and promoter mutations are indicated above and below the main sequences. The two arrows represent the start points of the cro and N transcripts.

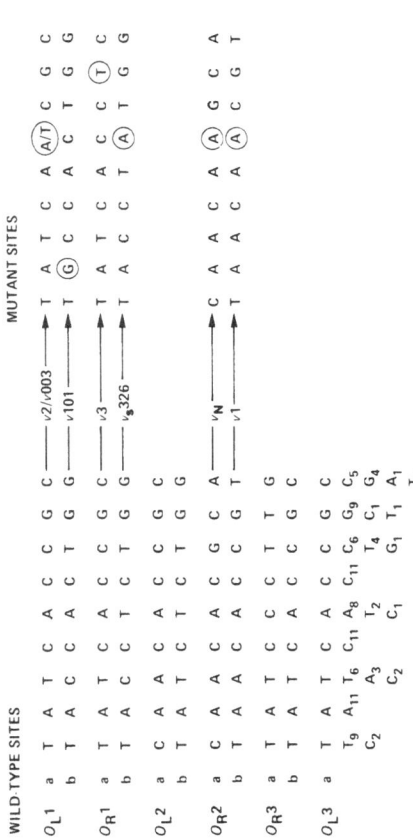

Fig. 3B. The repressor recognition sites indicated by brackets in Fig. 3A are aligned with their 5′-ends at the left for easy comparison. The nucleotide frequencies at each position are given below the sequences, and the changes associated with various operator constitutive mutations are given to the right. From Maniatis *et al.* (1975*b*) by permission of the authors and *Cell*. and Backman, Humayan, Jeffrey, Kleid, Flashman, Maniatis, Maurer, and Ptashne, personal communication.

2.1.9. Derepression

2.1.9a. *rec*A-Dependent Induction

Ultraviolet irradiation of recA$^+$ cells lysogenic for λ cIind^+ results, after a short lag, in repressor inactivation (Jacob and Campbell, 1959; Tomizawa and Ogawa, 1967; Roberts and Roberts, 1975). The effect of irradiation is mimicked by treatment with a variety of agents that interfere with bacterial DNA synthesis. Ultraviolet irradiation does not directly inactivate λ repressor: Irradiated nonlysogenic F$^+$ or F$'$ male strains indirectly induce female lysogens by transfer of photochemically damaged F factor DNA (Borek and Ryan, 1958; George, 1966; Devoret and George, 1967; Rosner *et al.*, 1968). Transfer of irradiated DNA, however, is not a sufficient condition for induction: Irradiated Hfr strains, which usually do not transfer a complete F$^+$ factor, were inactive in indirect induction (Devoret and George, 1967; Rosner, 1967). Rosner *et al.* (1968) and Devoret and George (1967) have proposed that autonomous replicons damaged by ultraviolet light are the responsible agents. Consistent with this hypothesis is the observation that irradiated phage P 1, which is an autonomous replicon in *E coli*, induced prophage λ on infection (Rosner *et al.*, 1968).

What are the steps that lead from a damaged replicon to inactive repressor? The product of the host recA$^+$ gene is involved, as prophage λ is not derepressed by ultraviolet irradiation of recA cells (Brooks and Clark, 1967; Hertman and Luria, 1967; Fuerst and Siminovitch, 1965; Tomizawa and Ogawa, 1967). Since the original characterization of recA, seven other classes of host mutant that block ultraviolet induction have been found (see Bailone *et al.*, 1975). As all of these mutations allow lytic λ growth after infection or after heat induction of a cIts prophage (see below), they must affect processes occurring between the primary photochemical lesion and the inactivation of repressor. Moreover, the multiplicity of mutations suggests that a number of biochemical steps are involved. Witkin (1967), Witkin and George (1973), and Radman (1975) have suggested that photochemically damaged DNA promotes the synthesis of a product required for repressor inactivation. There is, however, disagreement about the dependence of ultraviolet induction on protein synthesis (compare Tomizawa and Ogawa, 1967, and Taylor *et al.*, 1967).

Ultraviolet induction is mimicked in most respects by heating λ cIind^+ lysogens of a host mutant called *tif* (Goldthwait and Jacob, 1964; Castellazzi *et al.*, 1972a). *Tif* induction is blocked by all of the mutations that block ultraviolet induction (Castellazzi *et al.*, 1972b;

Bailone *et al.*, 1975). Thus, if irradiation initiates a sequence of biochemical steps that leads to repressor inactivation, *tif* induction probably triggers an early step. *tif* is located within or very close to the *rec*A gene (Castellazzi *et al.*, 1972a). Therefore, a change in the activity of the *rec*A gene product could be the early step in question. The final step in the process is probably a proteolytic cleavage of repressor: such cleavage has been observed *in vivo* after treatment of *rec*A$^+$ lysogens of λ *c*I*ind*$^+$ with an inducing agent (Roberts and Roberts, 1975).

2.1.9b. Induction by Antirepressor

The product of the *ant* gene of Salmonella phage P 22 derepresses the prophage, probably by directly inactivating the P 22 repressor (Levine *et al.*, 1975; Botstein *et al.*, 1975). The *ant* gene product, called antirepressor, also induces prophage λ and reversibly inactivates the λ repressor *in vitro* (M. Susskind and D. Botstein, personal communication).

2.1.9c. Thermal Induction

Certain mutations in the *c*I gene produce a thermolabile repressor. These prophage are induced by heat treatment and are of two types: type A, in which thermal inactivation is irreversible, and type B (represented by λ*c*I*ts*857), where heat-inactivated repressor is reactivated after a temperature downshift (Green, 1966; Naono and Gros, 1966; Lieb, 1966; Weisberg and Gallant, 1967; Horiuchi and Inokuchi, 1967). The A and B type mutations are clustered rather than interspersed in the map of the *c*I gene (Lieb, 1966); the physiological meaning of this grouping is unknown. L. Reichardt (personal communication) has found that the repressor proteins of some type A mutants are degraded at elevated temperature. Thermal induction is independent of the *rec*A$^+$ pathway (Brooks and Clark, 1967).

2.1.10. Repressor Synthesis Is Controlled in a Lysogen

That the synthesis of repressor in a lysogen might not be constitutive was first demonstrated by Eisen *et al.* (1968 *a* and *b*) in an experiment remarkable for its deceptive simplicity. They constructed a lysogen bearing a λ *N*$^-$*c*I*ts*857 *P*$_R$$^-$ prophage. Because of the *N* and the *P*$_R$ mutations, this prophage is capable of expressing only the *c*I and

rex genes. The prophage defects prevent cell death after thermal induction, and the strain can be studied, therefore, both when the repressor is
active, at 34°C, and at 42°C, when it is not. An immune culture of this
lysogen lost immunity to superinfection by λ when the growth temperature was shifted from 34° to 42°C. If the culture was cooled to 34°C
within a few generations of the upshift, immunity immediately returned
in most cells, presumably by renaturation of heat-inactivated *c*I*ts*857
repressor. If inactive repressor were synthesized at 42°C, this rapid
return of immunity upon cooling should not be altered by increasing
the time of culture growth at 42°C. Contrary to this expectation, the
rate of return of immunity after cooling was slow when the culture was
cultivated for many generations at 42°C. The function of gene *rex* (exclusion of T 4 *r*II mutants) was also lost after prolonged cultivation at
42°C and, similar to *c*I function, was only restored slowly after cooling
of the culture. These results suggest that repressor synthesis is not fully
constitutive, and that it is under the positive control of some function
which is inactivated at high temperature. That function must be a
phage function, since λ*c*$^+$ remains immune at 42°C. It must be *rex*
product, or repressor, since those are the only known functions
expressed in the highly defective prophage tested. Eisen *et al.* (1968 *a*
and *b*) concluded that repressor is its own inducer. Since the *rex* gene
product is active in the absence of repressor (Astrachan and Miller,
1972), the expression of *rex* is controlled in the same way as that of *c*I.

Biochemical confirmation of this model comes from the work of
Reichardt and Kaiser (1971) and Reichardt (1975*a*). They measured the
level of *c*I*ts*857 repressor in crude extracts by its ability to bind to antibody raised against purified repressor. They found that while the
*c*I*ts*857 repressor was antigenic and stable at 42°C, new repressor
antigen synthesis was greatly reduced at this temperature. After
prolonged growth of a λ *N*$^-$ *c*I*ts*857 *P*$_R$$^-$ lysogen at 42°C, the amount
of repressor antigen fell to 3 to 10% of its previous level. The new level
apparently represents autonomous *c*I expression since it was found in
the absence of any known phage gene products. Recently Dottin *et al.*
(1975) have shown that repressor synthesis is self-stimulating in a crude
in vitro protein-synthesizing system dependent on the addition of λ
DNA.

How does repressor promote its own synthesis?

1.   Repressor must bind to *O*$_R$ to activate the adjacent *c*I gene.
Infection of a λ *c*I*ts*857 lysogen at 32°C by λ *c*I$^+$ results in the rapid
appearance of the *c*I$^+$ repressor. If the superinfecting phage carries the
O$_R$ mutations *v*1 and *v*3, however, *c*I$^+$ repressor synthesis is not observed. Most convincingly, Reichardt (1975*a*) demonstrated that a λ

*c*I*ts*857 prophage with a partial O_R defect. λ *c*I*ts*857 *v*3, does not synthesize repressor at 37°C whereas repressor synthesis by λ *c*I*ts*857 O_R^+ is the same at 37°C and at 32°C (Fig. 4). Since both prophage synthesize normal amounts of repressor at 32°C, *v*3 does not affect the efficiency of *c*I gene expression when repressor is bound to O_R. Instead, it acts by reducing the maximum temperature at which the *c*I*ts*857 repressor can bind this operator.

2. Recently, Yen and Gussin (1973) have isolated a λ mutant (λ *prm*116) that cannot maintain the synthesis of repressor and *rex* product. The mutation lies outside the structural gene for repressor and just to the left of *v*3. It does not inactivate O_R since synthesis of gene *O* product by this mutant is shut off normally by repressor. The mutation, then, may define the *promoter* for *repressor maintenance*, P_{rm}, the promoter responsible for *c*I-*rex* synthesis in a lysogen. The *prm*116 muta-

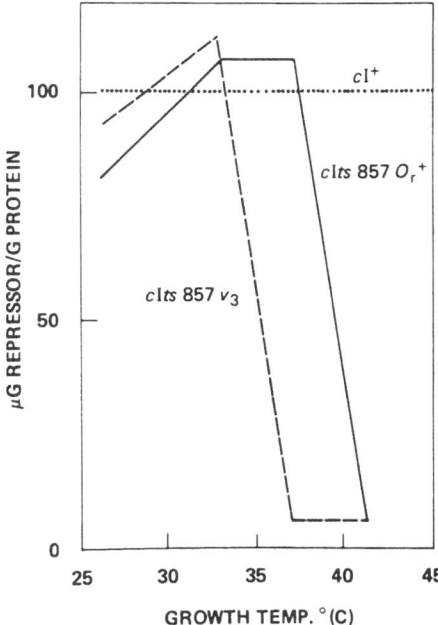

Fig. 4. Effect of repressor level and O_R on repressor synthesis. Cultures lysogenic for λ *c*I$^+$*c*Y$^-$*O*$^-$ (dotted line) or for *cro*$^-$ derivatives of λ *c*I*ts*857 *v*3 (dashed line) or λ *c*I*ts*857 O_R^+ (solid line) were grown at 27°, 33°, 37°, or 42°C for at least 20 generations, harvested, and the level of repressor antigen measured as described in the legend to Table 1. The *c*I*ts*857 prophages were mutant in genes responsible for cell killing. Three different *v*3 lysogens and two O_R^+ lysogens were used. *v*3 is a weakly constitutive mutation of O_R (see text). From Reichardt (1975a) with the permission of the author and of the *Journal of Molecular Biology*.

tion is located between O_R2 and O_R3 (see Fig. 3) and has been shown to depress cI transcription from a DNA fragment containing O_R (Meyer et al., 1975).

Does repressor stimulate transcription from P_{rm}? It is possible to measure directly the transcription of the cI-rex region. This transcription initiates at or near P_{rm} and proceeds leftward through the cI and rex genes (Taylor et al., 1967; Hayes and Szybalski, 1973). cI-rex RNA represents the major λ-specific transcription in λ cIts857 lysogens at 32°C (Taylor et al., 1967; Hayes and Szybalski, 1973). What happens to this transcription when a λ cIts857 lysogen is shifted to high temperature is the subject of controversy. Two groups (Kourilsky et al., 1970; Heineman and Spiegelman, 1970) find a substantial decrease after induction. Another group, however (Kumar et al., 1970; Hayes and Szybalski, 1973), detects no permanent change in immunity-region transcription. Both groups used the same, highly defective, prophage which was N^- and blocked in rightward transcription by promoter mutations or extensive deletions. Until this conflict is resolved, it cannot be excluded that the self-stimulation of repressor synthesis in a lysogen is exerted at a posttranscription stage.

The location of the $prm116$ mutation—in the spacer region between O_R2 and O_R3 (see Fig. 3)—suggests that at high ratios of repressor to operator, repressor might block P_{rm}-derived transcription by binding to O_R3. Meyer et al. (1975) have confirmed this hypothesis by measuring cI transcription in the presence of varying amounts of repressor from a 790 base-pair DNA fragment containing O_R. Tomizawa and Ogawa (1967), whose evidence was less direct, had suggested earlier that repressor represses its own synthesis. Meyer et al. (1975) also found that low ratios of repressor to operator stimulated cI transcription threefold over its level in the absence of repressor. It thus appears that repressor synthesis in a lysogen is autoregulated in two senses: at low repressor concentration cI expression is stimulated, and at high repressor concentration it is depressed. The mechanism of stimulation remains unclear.

2.2. The Regulation of Repressor Synthesis after Infection

The decision between the lysogenic and lytic pathways made by infecting λ is finely balanced and modulated by a number of phage and host gene products. Nonhereditary factors such as temperature, age of the cells, growth medium, and multiplicity of infection affect this modulation. The critical event in the establishment of lysogeny appears to be the early and abundant synthesis of repressor. Lambda mutants, such as λ cII, λ cIII, and λ cY, make very little repressor after infection; they rarely establish lysogeny. They make a clear plaque because they leave no immune lysogenic survivors. Lambda cro mutants, on the other hand, produce excessive repressor after infection; cells infected with these mutants nearly all enter the lysogenic pathway. Lambda cIts857 cro⁻ fails to make a plaque at 32°C since lytic phage growth is repressed. According to our current notions, there are thought to be two distinct promotors involved in the regulation of repressor synthesis (Reichardt and Kaiser, 1971; Castellazzi et al., 1972c). The first of these P_{rm}, defined by the mutation prm116, has already been discussed. The second, P_{re} (promoter for repressor establishment), is involved only in the establishment of repression. The activity of P_{re} is directly modulated by the products of the cII and cIII genes and indirectly by the product of cro gene, which affects the synthesis of cII and cIII products. The two promoters map some distance apart: P_{rm} lies within the immunity region, and the factors that affect its activity, repressor and cro, are immunity specific (see Fig. 2). Promoter P_{re} lies outside the immunity region, at least several hundred base pairs to the right of P_{rm}, and its modulators are common to λ and λ imm434.

2.2.1. Activation of Repressor Synthesis by the cII and cIII Products

The properties of cII and cIII mutants have already been discussed. In summary, these mutants fail to establish immunity efficiently, can be complemented to form lysogens, and produce normal levels of repressor when established as prophage (Reichardt and Kaiser, 1971). This and other evidence (Bode and Kaiser, 1965b) show that cII and cIII gene products are neither required nor synthesized in established lysogens.

The action of cII and cIII products is seen after infection of a sensitive cell as a transient stimulation of cI gene expression (Eisen et al., 1968, 1970). Infection with λ cIII⁺ cII⁺ results in a rate of repressor synthesis, assayed biochemically, 5 to 10 times higher per cI gene copy than in established lysogens (Reichardt and Kaiser, 1971; Echols and Green, 1971). In addition to cII and cIII gene products, rapid repressor synthesis also requires the integrity of a DNA site. This site, located between genes cro and cII (see Fig. 2), is defined by a class of clear plaque mutants, λ cY. Like cII and cIII mutants, cY mutants produce little repressor after infection. But their complementation pattern demonstrates that cY is a cis dominant (or site) mutation: mixed infection with $\lambda\ cI^+cY^-$ and $\lambda\ cI^-cY^+$ does not result in repressor synthesis (see Table 1) or lysogeny. Lambda cY mutants can, however, be complemented by $\lambda\ cI^+cY^+$ and form stable single lysogens with normal repressor levels (Reichardt and Kaiser, 1971). Thus, these mutants have a normal cI gene and a normal P_{rm}. They lack, instead, the site at which cII and $cIII$ products act.

cY mutations could exert their effect in either of two ways. On the one hand, they might inactivate the P_{re} promoter. On the other hand, they might cause premature termination of leftward transcripts originating upstream from the mutations themselves (Honigman *et al.,* 1975). Similarly, cII and $cIII$ products could either activate P_{re} or prevent premature termination of transcripts originating at P_{re}. These models will hopefully be resolved by the extensive genetic analysis of the cY region currently under way (D. Wulff, personal communication) and by the biochemical analysis of leftward transcription initiated near *ori* (see Fig. 1; J. Roberts, personal communication; M. Rosenberg, personal communication).

The two-promoter hypothesis is illustrated by a comparison of the properties of $\lambda\ prm116$ and $\lambda\ cY$. These mutants, neither of which lysogenizes efficiently alone, lysogenize well after mixed infection. The resulting lysogens carry $\lambda\ cY$, either alone or accompanying $\lambda\ prm116$;

TABLE 1

Repressor Levels after Infection[a]

	Phage 2	
Phage 1	None	λcY (2 mutants)
λ	100	
λcY (2 mutants)	3.0	3.9
λcII (2 mutants)	0.85	229
$\lambda cIII$ (2 mutants)	16.2	108
λcI amber	0.88	5.7
λ prophage	6.6	

[a] The cells were harvested 30 min after infection with six virus particles per cell. Results are expressed as the percent of the level of repressor accumulated in cells infected by wild-type λ (1230 μg repressor/g protein). Repressor levels were determined by measuring the ability of protein in a crude extract of infected cells to compete with a known amount of radioactively labeled pure repressor for binding to antirepressor antibody. (From Reichardt and Kaiser, 1971, with permission of authors and of *Proceedings of the National Academy of Sciences of the USA.*)

they never carry λ prm116 alone (Yen and Gussin, 1973). On single infection of sensitive cells cY mutants produce little repressor antigen, while the normal initial high-level synthesis of repressor is seen with λ prm116 (Gussin $et\ al.$, 1975). The converse is true when lysogens are infected. Here, λ prm116 makes no repressor, while the maintenance level of repressor synthesis by a λ cY prophage is normal (Gussin $et\ al.$, 1975).

One important biochemical prediction of the two-promoter model has been tested. The cro gene lies between P_R and P_{re}. Transcription of the sense strand of this gene derives from P_R. When P_{re} is activated, the antisense strand should also be transcribed (see Fig. 2). This prediction is confirmed by the following experiment (Spiegelman $et\ al.$, 1972). Radioactively labeled P_R-derived cro RNA, isolated from an induced lysogen deleted for P_{re}, was annealed to unlabeled RNA extracted from infected cells. Some of the label became resistant to RNase after annealing; such resistance is characteristic of double-stranded RNA (Geiduschek and Grau, 1970). Annealing with unlabeled RNA extracted from cells infected with cII, cIII, or cY mutants, or from an established lysogen, provided little or no protection. P_L-derived message in all cases remained RNase sensitive. Thus, protection appears to be due to the presence of P_{re}-derived cro antisense message in the infected cells.

The mechanism of cIII action appears to differ from that of cII. This was first suggested by the properties of λ imm21, a hybrid between λ and the related temperate phage, 21. Lambda imm21 has the same cIII gene as λ, but structurally different cI and cII genes and a non-homologous P_{re} (see Fig. 2) (Liedke-Kulke and Kaiser, 1967; Davis and Davidson, 1968). This suggests that while cII product may act directly at or near P_{re}, the activation of P_{re} by cIII may be indirect.

Both λ cIII and λ imm21 cIII show significant residual lysogenization in the absence of cIII function. Unlike λ cII, which makes very little repressor, cIII mutants can make as little as 10% or as much as 50% of the normal repressor after infection of the same bacterial strain grown in different media (Table 1). Two types of host mutation influence this cIII-independent lysogenization. The first, hfl, results in nearly 100% lysogenization by λ or λ cIII (Belfort and Wulff, 1971; 1973a and b). Activation of P_{re} is still required in hfl mutants, since cII mutants lysogenize poorly. It has been suggested that the product of the hfl^+ gene interferes with activation of P_{re} by cII, and that this interference is antagonized by cIII product (Belfort and Wulff, 1974).

The second type of host mutation affecting lysogenization by λ cIII causes a deficiency in the cyclic AMP pathway of gene activation. In such strains, λ cIII is completely unable to lysogenize, while the efficiency of lysogenization by λ is only slightly reduced (Belfort and Wulff, 1974; Grodzicker *et al.*, 1972). Presumably, *c*III product and some product under cyclic AMP control can partially substitute for each other and therefore might act in the same way to stimulate P_{re} transcription. In fact, they both appear to act by inactivating *hfl* product, since *hfl* mutants are lysogenized well in the absence of both *c*III product and the cyclic AMP pathway.

The establishment of immunity also requires the presence of a functional *N* gene. Lambda N^- synthesizes only 7% as much repressor protein as λ N^+ (Reichardt, 1975*b*). As a consequence, infection of sensitive cells with *N* mutants leads not to the formation of immune lysogens, but to the appearance of carrier cells bearing λ N^- plasmids (Signer, 1969). Since *N*-function is required for the synthesis of most phage proteins, e.g., cII/cIII, as well as for normal phage DNA replication, the requirement for *N* product could be indirect. Although there is a direct relationship between gene copies and frequency of lysogenization (Brooks, 1965; Kourilsky, 1973; see below), it is clear that replication is not an absolute requirement for repressor synthesis. Repressor is formed in the absence of *O*-product, or when λ replication is prevented by the host mutation, *gro*P (Reichardt and Kaiser, 1971; Reichardt, 1975*b*). That the *N* product requirement is indeed indirect is shown by the following experiment. A sensitive cell was coinfected with λ N^-c17, which synthesizes cII product constitutively by an *N*-independent promoter, and λ *imm*21cII$^-$. The latter phage provided cIII, but not *N* product. The result, which was the production of normal levels of λ repressor, indicates that *N* acts through cII and cIII in promoting repressor synthesis (Reichardt, 1975*b*).

2.2.2. *c*II and *c*III Products Retard Late Gene Expression and DNA Synthesis

It has been known for some time (McMacken *et al.*, 1970) that *c*II and *c*III mutants synthesize at least two λ late proteins, endolysin and tail antigen, sooner than their wild-type counterparts. The delay in late gene expression promoted by *c*II and *c*III products appears to be mediated through P_{re} since it is not seen in *c*Y mutants (Court *et al.*, 1975). Spiegelman *et al.* (1972) have suggested that leftward transcription from P_{re} directly depresses rightward transcription from P_R and that this limits the expression of *Q*, whose product is required for late gene expression. Inhibition of late gene expression might be important for lysogenization. Lysogeny is favored by increasing the multiplicity of infection, which appears to affect late functions more than cI synthesis (Court *et al.*, 1975).

The gene analogous to cII in the λ-related phage P 22 causes a depression of the rate of phage DNA synthesis starting about 6 min after infection (Smith and Levine, 1964). If the phage can make repressor and the conditions of infection favor lysogenization, this depression is followed by permanent repression; if the phage cannot make repressor, the depression is transient. If the cII gene of P 22 acts like its λ analog, as suggested by the close genetic and functional homology between the two phages (Botstein and Herskowitz, 1974; Gough and Tokuno, 1975), the primary action of both cII-products may be to promote leftward transcription from P_{re}. This, in turn, may depress DNA synthesis by impeding the transcription of the replication genes (genes O and P in λ; see Fig. 1; Gough and Tokuno, 1975). The relation between the transient depression of phage DNA synthesis and lysogeny by P 22 is unclear, although recent evidence suggest that there may be a causal connection in this phage (B. Steinberg and M. Gough, personal communication) as well as in λ (P. Kourilsky, personal communication).

2.2.3. The Role of *cro*

2.2.3a. The Discovery of the *cro* Gene

Lysogens bearing the defective prophage λ $N^-cIts857\ O^-$ or P^-, like those bearing λ $N^-cIts857\ P_R{}^-$ (see previous section), survive at elevated temperatures. When cultures of each of these lysogens were shifted to 42°C, the capacity to rapidly regain immunity to superinfection upon cooling gradually disappeared as preexisting repressor was diluted by cell division and not replaced (Eisen *et al.*, 1968*a* and *b*; Calef and Neubauer, 1968). However, shifting a culture that had been growing for many generations at 42°C down to 32°C revealed an important difference between the $P_R{}^-$ and the O^- or P^- lysogens: The former slowly returned to the immune state while the latter never regained immunity or *rex* function. This long-term loss of immunity is not due to a mutation as it is induced in 100% of the population during a few generations of growth at high temperature (Neubauer and Calef, 1970). Since $P_R{}^-$ phages express neither gene O nor gene P, these results suggest the existence of another gene transcribed from P_R whose product blocks the synthesis of repressor (Calef and Neubauer, 1968; Neubauer and Calef, 1970). We shall call this gene *cro*; other workers have referred to it as *tof, fed,* or *Ai.* Since the O^- and P^- lysogens are also N^-, the synthesis of *cro* must be N-independent. Superinfecting phage are also affected by this function: $\lambda\ c^+$ makes a clear plaque on nonimmune $\lambda\ N^-$ *c*I*ts*857 O^- lysogen, a phenomenon known as "antiimmunity" or "chanelling" (Neubauer and Calef, 1970; Calef *et al.,* 1971). The heteroimmune phage λ *imm*434 is not directed toward the lytic pathway in these lysogens, suggesting that the function acts only in the immunity region.

, *cro* mutants were isolated by selecting the rare cells in an antiimmune culture of a $\lambda\ N^-$ *c*I*ts*857 O^- lysogen that had regained immunity at low temperature (Eisen *et al.,* 1970; Calef *et al.,* 1971). Lambda N^- *c*I*ts*857 *cro*$^-$ O^- lysogens behaved like $\lambda\ N^-$ *c*I*ts*857 P_R^- lysogens: All the cells slowly regained immunity when shifted from high to low temperature. In addition, the *cro*$^-$ strain did not channel superinfecting λ into the lytic pathway. *cro* mutations are located in the region between P_R and P_{re}, under the *imm*434 region (see Fig. 2). This region is transcribed in the absence of N product (Kourilsky *et al.,* 1968). Therefore, the location of *cro* is consistant with its N-independent expression. Lambda *imm*434 has a *cro* homolog that is active on λ *imm*434 but not on λ (Pero, 1970).

2.2.3b. The Behavior of *cro* Mutants

Lambda *c*I*ts*857 *cro*$^-$ fails to form a plaque at 32°C because every infecting phage represses its lytic functions (Eisen *et al.,* 1970). The high frequency of repression is associated with a continuous elevated rate of repressor synthesis after infection. This contrasts with wild-type phage, which switch from the P_{re}-dependent, high-level rate of repressor synthesis to the low-level P_{rm}-dependent rate at about 10 min after infection (Reichardt and Kaiser, 1971; Reichardt, 1975*b*). It appears, therefore, that the accumulation of *cro* product is responsible for this shut-off of P_{re}.

How does *cro* product act to slow repressor synthesis? The following experiments show that *cro* acts by blocking the synthesis of *c*II and *c*III products. Lambda *imm*434 has inherited the *cro* and *c*I genes of phage 434 (Pero, 1970; Kaiser and Jacob, 1957) and the *c*II and *c*III genes of λ (Kaiser and Jacob, 1957). Echols *et al.* (1973) and Reichardt (1975*b*) have shown that λ *imm*434 can promote abundant synthesis of repressor by a coinfecting λ in an antiimmune λ lysogen. Therefore, in a cell containing *cro* product, λ *imm*434 but not λ can express genes needed for λ repressor synthesis. These genes must include *c*II and *c*III because neither λ *imm*434 *c*II⁻ nor λ *imm*434 *c*III⁻ promote λ repressor synthesis as well as λ *imm*434 *c*⁺. Reichardt (1975*b*) has shown that the *cro* product of phage 434 acts in an analogous way to that of λ: It represses the expression of genes *c*II and *c*III by λ *imm*434, but not by λ. The effect of *c*II and *c*III in overcoming the *cro* inhibition of repressor synthesis from P_{re} is reflected in an increase in the frequency of λ lysogens in the antiimmune host (Echols *et al.*, 1973; Galland *et al.*, 1973).

Since blocking the synthesis of *c*II and *c*III products leads to a rapid fall-off in the rate of repressor synthesis, it follows that these products are functionally unstable. This has been demonstrated (Reichardt, 1975*b*) by sequentially infecting cells, first with λ *imm*434 and then with λ *c*III⁻ or λ *c*II⁻, and then measuring the level of λ repressor. The ability of λ *imm*434 to complement λ *c*II⁻ or λ *c*III⁻ for λ repressor synthesis is lost 20 min after infection. In contrast, λ *imm*434 *cro*⁻ retains the ability to complement the λ mutants, presumably because it cannot shut off the expression of its own *c*II and *c*III genes. These data indicate that: (1) *cro* product shuts off *c*II and *c*III expression, (2) this is followed by the rapid inactivation of *c*II and *c*III products, and (3) this leads to loss of activity of P_{re}.

2.2.4. The Sites of *cro* Action

It is likely that *cro* diminishes transcription from P_L and P_R, probably by binding to sites within O_L and O_R. First, λ is insensitive to the action of the *cro* product of phage 434, and vice versa (Pero, 1970). This suggests that sites within the *imm*434 region (see Fig. 2) are required for the specific action of *cro*. Second, *cro* product reduces the expession of genes transcribed from P_L. This can be seen by measurement of mRNA synthesis (Kumar *et al.*, 1970; Kourilsky, 1971), by assays of λ exonuclease (the product of the *exo* gene; Radding, 1964; Pero, 1970), and by assays of bacterial enzymes when the corresponding genes are fused to P_L (Franklin, 1971; Adhya *et al.*, 1974). The expression of gene *N* is also depressed by *cro* (Franklin, 1971; Greenblatt, 1973; Reichardt, 1975*b*; S. Adhya, personal communication). Third, *v*2, a mutation of O_L, reduces *cro* repression of λ exonuclease synthesis (Sly *et al.*, 1971) and *c*III expression (Echols *et al.*, 1973; Reichardt, 1975*b*). This finding is the strongest evidence that *cro* acts by specifically binding to DNA. Fourth, *cro* represses *c*II expression in a deletion mutant that lacks the left part of the immunity region but retains an intact O_R and P_R (Reichardt, 1975*b*).

Most of the above evidence is circumstantial. Proof of the hypothesis that *cro* represses transcription by binding specifically to O_L and O_R requires isolation of *cro* product and demonstration of its activity *in vitro* as already described for repressor. Such evidence has very recently been obtained (Folkmanis, A., Takeda, Y., Simuth, J., Gussin, G., and Echols, H., in press).

Although *cro* action mimics repressor action in many respects, some important differences exist. (a) The binding of repressor to O_R stimulates *c*I expression from P_{rm} (see Section 2.1.10); *cro* product binding, if it occurs, has the opposite effect. This is most clearly shown by the properties of a λ N^- *c*I*ts*857 lysogen deleted for *cro*, P_{re}, *c*II, and *O*. Because of its *cro* defect, this lysogen did not become anti-immune after prolonged growth at high temperature. However, it did become antiimmune when provided with *cro* product from a second *cro*$^+$ prophage (Castellazzi *et al.*, 1972*c*). Since the missing DNA included P_{re} and *c*II, *cro* presumably can act to prevent the expression of *c*I from P_{rm}. This inference was confirmed by measurement of repressor antigen in λ *N*c*I*ts857 *c*II$^-$*O*$^-$ *cro*$^+$ and *cro*$^-$ lysogens that were maintained at high temperature: the *cro*$^+$ lysogen contained less than 1%, while the *cro*$^-$ lysogen contained about 10%, the level of repressor antigen found in an immune lysogen (Reichardt, 1975*a*). In fact, *cro* reduces *c*I expression from P_{rm} even in the presence of active repressor: Superinfection with λ O_R^-, which synthesizes *cro* product

TABLE 2

Control Elements[a]

Site	cI	cII, cIII[b]	cro
		Gene product	
O_L,P_L	↓	0	↓
O_R,P_R	↓	0	↓
O_R,P_{rm}	↑,↓[c]	0	↓
P_{re}	0	↑	0

[a] The symbol "↑" indicates that the gene product in line 1 directly stimulates gene expression from the site given in column 1, "↓" indicates depression, and "0" indicates no direct effect.

[b] It appears likely that *c*III product may act indirectly, perhaps by antagonizing the action of a host protein (see text).

[c] At low repressor-to-operator ratios, repressor stimulates *c*I expression from P_{rm}; at high ratios, it depresses such expression (see text).

constitutively (Reichardt, 1975*b*), converts immune cells into the antiimmune state by blocking the synthesis of repressor from P_{rm} (Sly *et al.*, 1971; Calef *et al.*, 1971; Reichardt, 1975*a*). (b) *Cro* represses *c*II expression not only in wild-type phage, but also in mutants that have a much reduced affinity for repressor at O_R (Reichardt, 1975*b*); therefore, the O_R sequence (if any) recognized by *cro* is not identical to the one recognized by repressor. Alternatively, the extent of repression necessary for the *c*I and *cro* functions is different.

Our current ideas on the direct effects of the *c*I, *c*II, *c*III, and *cro* products on gene expression from sites P_L, P_R, P_{rm}, and P_{re} are given in Table 2. These elementary controls can be combined to give complicated control circuits that allow phage gene expression to be modulated with great flexibility in response to external stimuli (see Thomas, 1971).

2.2.5. The Influence of Nonhereditary Factors on the Establishment of Lysogeny

2.2.5a. The Number of Infecting Phage

The efficiency of lysogenization increases with the number per cell of infecting phage particles (Boyd, 1951; Lieb, 1953). Not surprisingly, the increase is larger for mutants that are defective in replication than for wild-type phage (Brooks, 1965; Levine and Schott, 1971; Kourilsky, 1973); thus, replication can partially compensate for a low initial number of virus genomes. The inability of a single infecting virus genome to lysogenize efficiently could be due to a failure to establish repression or to a failure to integrate into the host chromosome. In fact, both explanations are probably correct. Kourilsky (1971) compared lysogenization of a sensitive cell to lysogenization of a homoimmune cell at virus–cell ratios less than one. Gene products required for integration in the immune cell were provided by brief heating of the *c*I*ts* lysogen after superinfection. Kourilsky found that a phage with a replication defect gave low lysogeny after single infection of a sensitive cell; the presence of preexisting repressor increased the lysogenization frequency to that obtained by single infection of a sensitive cell with a replication-proficient phage. Kourilsky concluded that an increase in the number of genomes increased the probability of establishing repression. Recently, Reichardt (1975*b*) measured the dependence of repressor synthesis on the number per cell of infecting replication-defective phage particles. He found that the repressor level increased with the square of the number of infecting phages and concluded that the synthesis of repressor was limited by the number of copies of two different genes. Appropriate complementation tests suggested that these genes are cI and $cIII$.

R. Weisberg (unpublished experiments) has also measured the ability of a superinfecting phage to lysogenize a homoimmune lysogen. Again, gene products required for integration of the superinfecting phage were provided by brief heating of the $cIts$ lysogen. He found that at a virus:cell ratio of 2, fewer than 10% of the cells were lysogenized while at a virus:cell ratio of 20, more than 50% of the cells were lysogenized. He concluded that increasing the number of phage genomes increases the probability that one will recombine with the host chromosome.

2.2.5b. The Temperature

Elevated temperature reduces the frequency of lysogens (Lieb, 1953). This is primarily due to decreased integration (Lieb, 1953; Zichichi and Kellenberger, 1963; Guarneros and Echols, 1973). Reichardt (1975b) has found a twofold decrease in the amount of repressor made at 42°C relative to 37°; this seems to have little if any influence on lysogenization frequency.

2.2.5c. Culture Conditions

It is a common observation that exponentially growing cells in broth are lysogenized less efficiently by λ than a culture that has been briefly incubated in $MgSO_4$ solution (see, for example, Kourilsky, 1973). The molecular basis of this effect is unknown. Echols *et al.* (1975) have measured the dependence of lysogenization frequency and repressor synthesis on cell growth rate which was varied by varying the carbon source. They found little change in either lysogenization frequency or repressor synthesis over a two- to threefold range of growth rates.

3. INSERTION AND EXCISION

3.1. Site-Specific Recombination and the Campbell Model

As we discussed in the introduction to this chapter, temperate bacteriophage such as λ are capable of long-term coexistence with their hosts. Lambda accomplishes this by repressing its lytic functions and inserting its DNA into the host chromosome. The inserted viral DNA is replicated, as part of the host chromosome, entirely by the host replication machinery. Integration thus provides a reliable method for transmitting the viral genome to daughter cells.

In 1962 Campbell (1962) proposed a model to explain the association of the viral and host DNA. He asserted that λ DNA circularizes by end-joining after infection, and that this circular molecule undergoes recombination with the host chromosome at specific sites on the viral and the host DNAs (the attachment sites, or *att*'s; see Fig. 5). As a result of the recombination, the viral DNA is linearly inserted in the chromosomal DNA. Excision of the viral DNA, which occurs after prophage induction, reverses this process. Recombination between the two prophage *att*'s regenerates a circular phage genome. Although autonomous phage DNA synthesis and the expression of late viral genes does not require prophage excision, the production of normal phage particles does (Little and Gottesman, 1971). Normally, prophage excision is followed by the full lytic cycle, with cell death and the release of several hundred λ virions. Under special circumstances, however, the lytic cycle can be aborted, and excision leads to "curing": the restoration of an intact bacterial chromosome free of inserted prophage.

The Campbell model was originally proposed to explain the observation that the order of three genetic markers appeared to differ between the phage and the prophage genetic maps (Calef and Licciardello, 1960). The apparent change in gene sequence upon lysogenization stimulated the formation of various baroque models for prophage attachment. Campbell, however, saw that the map alteration could be easily explained as a cyclic permutation of the phage gene order. The permutation results from recombination between a circular phage genome and the bacterial chromosome at a site different from the ends of the linear DNA of the phage virion.

Transducing phage, which arise occasionally from integrated prophage, and whose genetic constitution appeared entirely inexplicable by other theories, are nicely explained by the Campbell model. Transducing phage have replaced phage genes with bacterial DNA near the center of their genomes. One end of the substitution is fixed; the other varies among transducing phage lines of independent origin. Campbell suggested that these phage arise by abnormal excision of the integrated permuted prophage. A recombination between points in the bacterial and prophage genomes generates a phage in which bacterial genes replace phage genes in the observed order (see Fig. 5).

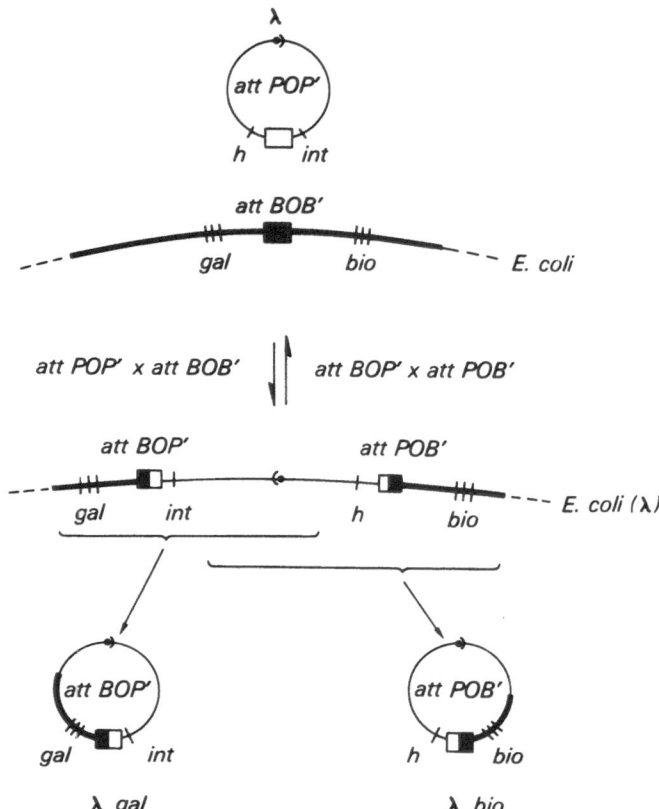

Fig. 5. The Campbell model. Campbell (1962) hypothesized that a circular lambda DNA molecule undergoes a recombination event with the host chromosome at a particular site on the phage DNA (here called *att* POP′) and at a particular site on the bacterial DNA (here called *att* BOB′). The recombination event linearly inserts the lambda DNA into the host chromosome. Reversal of this recombination (called excision) leads to reconstitution of the host chromosome and reconstitution of the circular phage genome (*att* BOP′ × *att*POB′). If, however, the excision mechanism does not operate properly, bacterial genes adjacent to the integrated phage genome may be excised with the phage DNA, while some phage genes may be left behind. This leads to the formation of transducing phage, carrying bacterial DNA for nearby genes, such as *gal* and *bio*.

Some predictions of the Campbell model are:

1.   Viral chromosomes circularize.
2.   The prophage is an insertion into the bacterial chromosome.
3.   Specific sites on the host chromosome (*att* BOB′) and on the

circular phage chromosome (*att* POP´) recombine during insertion.

4.   Insertion results in a cyclic permutation of the phage map.

These predictions are now confirmed by experimental evidence, reviewed below.

3.1.1.   Circularization of the λ Chromosome

λ DNA, as isolated from mature phage, consists, as stated previously, of a single linear molecule of 31×10^6 daltons (see Davidson and Szybalski, 1971). This molecule has the property of forming circular molecules or end-to-end aggregates under annealing conditions (Hershey *et al.*, 1963). These circles (called Hershey circles) can be returned to the linear state by heating and quick cooling. Circle formation is due to the presence on the ends of the mature phage DNA molecule of single-stranded segments of DNA which are complementary to each other. They anneal to form a circular molecule (Hershey and Burgi, 1965).

λ DNA extracted from cells directly after infection appears to be in the circular form as well (Dove and Weigle, 1965; Young and Sinsheimer, 1964; Bode and Kaiser, 1965*a*): Apparently, λ DNA circularizes immediately after injection.

3.1.2.   Linear Insertion of the Viral DNA into the Host Chromosome

3.1.2a.   Genetic Evidence

Prophage insertion increases the genetic distance between markers on either side of *att* BOB´. This is shown by the following experiments. P 1 is a generalized transducing phage, cutting and packaging random pieces of bacterial DNA about 100 kilobase pairs in length (see Stent, 1971). The frequency with which two genes can be cotransduced by P 1 decreases with increasing distance between the markers. In a non-lysogen, the bacterial genes *gal* and *bio* (which span *att* BOB´) are cotransduced more frequently than in a lysogen, suggesting that the prophage insertion has separated the markers (Rothman, 1965). Similar results were obtained by P 1 analysis of φ80 lysogeny (Signer, 1966).

Both prophage λ and prophage φ80 are located near bacterial genes for which deletions can be easily selected. These deletions often extend into the prophage genome, suggesting that the prophage DNA is directly adjacent to and continuous with the bacterial genome (see below; Franklin *et al.*, 1965; Adhya *et al.*, 1968).

3.1.2b. Physical Evidence

Physical studies of the association of phage and host DNA have also supported the idea of a linear insertion of phage DNA in the host DNA. Early experiments by Jacob and Wollman (1956*a*) and later ones by Menninger *et al.* (1968) confirmed a physical association of some sort between prophage and host chromosome.

Freifelder and Meselson (1970) measured the x-ray sensitivity of F′DNA covalent circles, using an F′-factor that contained *att* BOB′. X-ray sensitivity of covalent circles is directly proportional to the length of the DNA; a break anywhere in the circle converts it to a noncovalent form. It was found that the rate of conversion to a nonco-valent form of F′-DNA bearing an inserted prophage was more rapid than the rate of conversion of covalent F′-DNA from a nonlysogenic cell. The increase in sensitivity corresponded to the increase in DNA length expected for the linear insertion of the prophage DNA into the sex factor.

Direct visual proof of the linear insertion of λ DNA into host DNA comes from the work of Sharp *et al.* (1972). They heteroduplexed a strand of F′*gal* carrying a λ prophage with the complementary strand of a vegetative λ. The resulting structure demonstrates that (1) the λ is linearly integrated into the F′ DNA, and (2) the prophage is a cyclic permutation of the vegetative λ genome.

3.1.3.   A Specific Site on the Host Chromosome and the Viral Chromosome Is Involved in the Integration Reaction

The integration of λ prophage in normal cells always occurs between the bacterial *gal* and *bio* operons. Deletions of this region reduce the frequency of λ integration 200-fold (Shimada *et al.*, 1972). Partial deletions of the λ attachment site (e.g., λ *b2*) also significantly lower integration frequencies (Kellenberger *et al.*, 1961). These data suggest that almost all integration involves these regions.

More precise localization of *att* POP′ and *att* BOB′ comes from the analysis of DNA heteroduplexes between transducing and normal λ phage. According to Campbell's model (Fig. 5), the substitution of bac-terial DNA in transducing phage begins at the site of recombination between the λ and the host chromosomes. In heteroduplexes between the DNAs of wild-type and transducing λ, therefore, a region of non-homology should appear at that site and extend a variable distance toward either end of the vegetative DNA. Any variability in the site of recombination between host and phage DNA would result in two variable endpoints in the heteroduplexes. This is not found; the region of nonhomology, and thus the attachment site, always has a fixed endpoint, 0.574 λ lengths from the left cohesive end (Hradecna and Szybalski, 1969; Davis and Parkinson, 1971).

Similar reasoning pertains to *att* BOB´. If *att* BOB´ were not fixed, then heteroduplexes between like transducing phage (e.g., two independent *gal* transducing phage) would show a region of nonhomology at 0.574 λ lengths. This is not seen (Davis and Parkinson, 1971).

3.1.4. The Prophage Map Is a Cyclic Permutation of the Vegetative Map

The initial conclusion that the prophage and phage maps differed was based on the analysis of very few markers. As further markers became available (Campbell, 1963), it became clear that the prophage genome was a unique cyclic permutation of the vegetative genome.

While ordering of genes in vegetative crosses poses no particular problem, mapping a number of prophage genes required more subtle techniques. Spurning the laborious Hfr × F crosses originally used in prophage mapping, (Calef and Licciardello, 1960), Rothman (1965) used P 1 grown on a *gal*⁺ lysogen to transduce a *gal*⁻ lysogen to *gal*⁺. The prophage in the two strains differed in their genetic constitution. Scoring the prophage genotype in the *gal*⁺ P 1 transductants, Rothman was able to obtain an unequivocal linear map of prophage λ. The map confirmed Campbell's model; it was a cyclic permutation of the vegetative map. For reasons as yet unexplained, a similar analysis based upon P 1 cotransduction of *bio*⁺ and prophage markers failed to resolve the prophage markers (Rothman, 1965).

The second technique involved the isolation and characterization of deletions that extended from regions surrounding the prophage variable distances into the prophage genome. Marker rescue experiments then give a "deletion map" which is unequivocal. The most complete map, generated from *chl*D and *chl*A into prophage λ, clearly shows that a unique cyclic gene permutation occurs on prophage integration (Adhya *et al.*, 1968) (see Fig. 6). A similar analysis of prophage ϕ80—

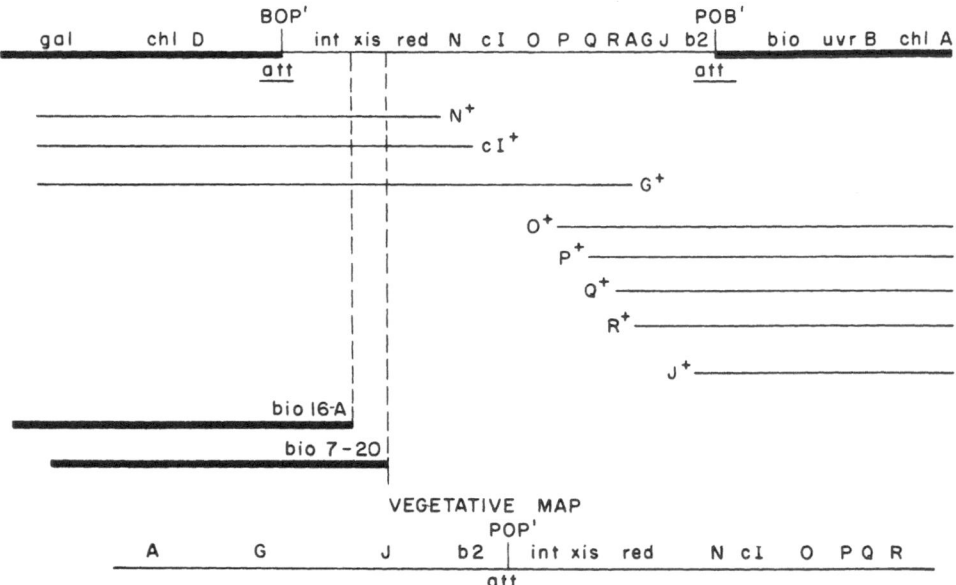

Fig. 6. Deletions and substitutions of the prophage map. The thin lines indicate the phage markers missing and the first phage marker present in various deletions of λ prophage generated as chlorate resistant (*chl*A or *chl*D) derivatives of a lysogen (Adhya *et al.*, 1968). These deletions served to order the genes shown on the prophage map and indicated clearly that the prophage map is a circular permutation of the vegetative map. The bold lines labeled *bio* 16-A and *bio* 7-20 represent 2 *bio* transducing phage which have replaced the indicated lambda genes with bacterial DNA. These two *bio* substitutions define the *xis* gene, as discussed in the text.

which is located near the *trp* operon—led to the same conclusion (Franklin *et al.*, 1965).

3.2. The Genetics of Site-Specific Recombination

3.2.1. *int*

The analysis of transducing phage was fruitful in another way. Signer and Beckwith (1966) constructed a φ80 phage transducing *lac* by inserting F$_{ts}$-*lac* near the φ80 attachment site at *trp*: φ80 *lac* was capable of integrating either at the φ80 attachment site or at *lac*. Lysogens carrying the prophage at one locus differed from lysogens carrying it at the other; when the phage was inserted at its attachment site, it could be cured by heteroimmune superinfection. When located at *lac*, the prophage was incurable. This observation led Signer and Beckwith to postulate the existence of a special recombination system (carried by the superinfecting phage) necessary for integration into and excision from the attachment site, a prediction that proved remarkably prescient.

The isolation of λ phage mutants unable to integrate was achieved simultaneously in several laboratories. Smith and Levine (1967) first reported such mutants in the temperate Salmonella phage P 22. Similar λ mutants were then described by Zissler (1967), Gottesman and Yarmolinsky (1968), and Echols, Gingery, and Moore (1968). These mutants (λ int^-) made normal turbid plaques but did not form stable lysogens. The mutations were recessive, i.e., coinfection of a cell with an int mutant and a wild-type λ could yield a lysogen carrying mutant and not helper phage. Such lysogens were of two classes. The first class, which bore a single λ int^- prophage, gave very few viable phage on induction, suggesting that int-function is also necessary for the excision reaction. The second class, which gave normal phage yields on induction, were found to contain two tandem int^- prophage. In these lysogens the requirement for int in excision was circumvented. Instead the phage function normally responsible for the formation of the λ cohesive ends could generate plaque-forming phage by cutting at the two tandem cos sites (see Yarmolinsky, 1971). Recently, large numbers of int mutations have been isolated by an ingenious technique (L. Enquist and R. Weisberg, manuscript in preparation; see below). These mutants fall into one complementation group and map immediately to the right of att POP´.

A protein present in λ int^+ infected cells but not in λ int^-_{am} infected cells has been identified and partially purified by differential radiochemical labeling (Ausubel, 1974; Nash, 1974). The labeled protein binds to DNA with and without attachment sites.

3.2.2. *xis*

Although the failure of int mutants to integrate or excise suggested an identical mechanism for the two reactions, such notions proved oversimplified. It was soon observed that some int mutants were capable of curing a heteroimmune int^+ prophage, i.e., leaving survivors that contained no prophage. Surprisingly, even a bio transducing phage, λ bio16-A, partially deleted for the int gene, was capable of curing on heteroimmune superinfection (Kaiser and Masuda, 1970; Weisberg, 1970). Since int function is required for curing (Gottesman and Yar-

molinsky, 1968), it must be provided by the prophage; in fact, *int*⁻
prophage are not cured by *int*⁻ superinfecting phage. Low levels of *int*-
function are known to be synthesized constitutively (see below, Section
3.5); whether the superinfecting phage raises these levels is not, as yet,
known. The superinfecting phage must, however, provide another function required for curing since a different *bio* transducing phage, λ *bio*7-
20, missing slightly more phage DNA than λ *bio*16-A, was completely
incapable of heteroimmune curing (Kaiser and Masuda, 1970) (see Fig.
6). The postulated function, Xis,* would be required for curing, or more
generally, for excision but not integration.

Point mutants in *xis* were isolated and characterized by Guarneros
and Echols (1970). The *xis* gene maps just to the right of *int*, and, as
predicted, *xis* mutants integrate but do not excise. The specificity of the
Xis requirement strongly suggests that integration and excision are different reactions.

A large set of *int* and *xis* mutations has been isolated and mapped
against deletions. From these data, as well as electron-microscopic
measurement of the size of the deletions, the space available for the *int*
and *xis* genes has a coding capacity equivalent to about 50,000 daltons
of protein (about 1500 nucleotide pairs) (Enquist and Weisberg,
manuscript in preparation).

3.2.3. A Temperature-Sensitive Function

λ mutations defective in integration but not in excision, termed
hen, have been reported (Echols *et al.*, 1974). Mapping studies
necessary to establish *hen* as a cistron separate from *int* and *xis* have
not been reported. No other phage mutants affecting integration or
excision have been reported, nor have host mutants with defects in either
reaction been isolated. Nevertheless, the participation of some function
other than Int or Xis appears likely on the basis of genetic, and now,
biochemical data (see Section 3.4). It was observed that the formation
of lysogens was strongly reduced at high temperature, whereas
prophage excision was relatively thermoinsensitive (Campbell and
Killen, 1967). This result suggests that either Xis stabilizes a temperature-sensitive Int-protein, or a temperature-sensitive element is required
for integration but not for excision (Echols *et al.*, 1974). This second
theory has a certain appeal in that it restores symmetry to site-specific
recombination. Two proteins would be required for integration, and

---

* The gene products of *int* and *xis* are represented as Int and Xis.

two for excision. The biochemical data discussed below in fact indicate that a second protein, thermolabile and host-derived, is required for integration.

3.2.4. The Attachment Sites: The Specific Elements

The isolation and properties of *int* and *xis* mutants indicated that the attachment sites of the phage and bacterium were not simply regions of homology. The requirement for Int and Xis in site-specific recombination shows that the general recombination functions of *E. coli* (Rec A) or λ (Red) will not, alone, promote integration or excision (Signer *et al.*, 1969). Nor are Rec or Red required: λ *red⁻* integrated and excised normally in a *rec*A host (Weil and Signer, 1968), conditions under which general recombination does not take place. Neither *int* nor *xis* functions themselves promote recombination outside the attachment sites (Weil and Signer, 1968). Taken together, these data suggest that the attachment sites contain specific elements recognized by the products of the *int* and *xis* genes.

In 1969, F. Guerrini suggested that the bacterial and phage attachment regions contained different specific recognition elements. This suggestion was confirmed by an analysis of the integration and excision of transducing phage (Guerrini, 1969; Signer *et al.*, 1969; Weisberg and Gottesman, 1969). Transducing phage carry an attachment site that contains elements of both phage and bacterial attachment sites (see Fig. 5). The integration-excision pattern of these transducing phage were unlike that of either parent, indicating that *att* POP′ and *att* BOB′ could not be identical. These experiments are considered in more detail below.

It had long been known that the transducing phage λ *gal* (called λ *dg* in early papers) lysogenizes poorly. This defect is overcome if the cell is coinfected with wild-type λ. In contrast to the λ *int⁻* mutants, the helper phage does not complement the λ *gal*; instead the lysogens contain both the transducing phage and the helper. Guerrini (1969) explained this phenomenon in terms of the structure of the attachment sites. In his model the phage attachment site contained specific recognition elements, P and P′, which flank the cross-over point (or region) "O." Different specific elements (B and B′) span the cross-over point in the bacterial attachment site. Site-specific recombination between phage and bacterium generates a prophage whose attachment sites are hybrids and therefore different from those of either parent (see Fig. 5). The attachment site of λ *gal*, *att*BOP′, recombines poorly with *att*BOB´. Wild-type λ helps the integration of λ *gal* by providing a different set of attachment sites rather than by providing a missing function. This is best demonstrated by the fact that prophage λ promotes the efficient integration of a heteroimmune λ *gal*. In this case, as seen in Fig. 5, the incoming λ *gal* genome is presented with *att*BOP´ and *att*POB´ prophage attachment sites for recombination, instead of *att*BOB´.

Site-specific recombination can be demonstrated in crosses between two lytically growing phage. Here too, the difference between *att*'s is reflected in the recombination frequency (Signer *et al.*, 1969). In all but one case there is good agreement between the ability of a pair of *att*'s to recombine in a lytic cross and the frequency with which integration can occur via recombination between those sites. The one exception, *att*BOP´ × *att*POB´, is efficient in vegetative crosses or in prophage excision but is quite inactive if integration is measured. This was first stressed by Guerinni who demonstrated that heteroimmune λ *gal* always integrated to the left of a prophage (*att*BOP´ × *att*BOP´) rather than to the right (*att*BOP´ × *att*POB´). This phenomenon became known as Guerinni's paradox.

The solution to the paradox came from the experiments of Weisberg and Gottesman (1971) and from Echols and Court (1971). The latter grew λ *gal* lytically on a heteroimmune lysogen and found in the lysate recombinants between the superinfecting λ *gal* and the prophage which must have involved *att*BOP´ × *att*POB´ recombination. Why then are lysogens not recovered in which the transducing phage has integrated via *att*BOP´ × *att*POB´? The answer is suggested by the work of Weisberg and Gottesman (1971), who found that while Int activity is stable, Xis activity is not. Recall that Xis is required for *att*BOP´ × *att*POB´, but not for *att*BOP´ × *att*BOP´ (Echols, 1970) nor for recombination between the products of *att*BOP´ × *att*POB´, i.e., *att*POP´ × *att*BOB´. Assuming that an equilibrium exists between the integrated and excised state of λ *gal* at the right-hand prophage *att*, it is apparent that the loss of Xis will result in a shift toward the excised state. At the left-hand prophage *att*, no such equilibrium shift occurs. The net result is the recovery of lysogens in which λ *gal* is situated only to the left of the resident prophage. The instability of Xis may play a role in ensuring the high efficiency of normal λ integration. Here, the integration reaction requires only Int, which is stable, whereas excision requires, in addition, the unstable Xis protein. This would tend to favor the integrated over the excised state, after the expression of the *int* and *xis* genes was repressed.

The functional requirements for recombination between a particular pair of *att*'s has a characteristic temperature dependence. Guarneros and Echols (1973) showed that at elevated temperature a number of different crosses became *xis*-dependent or *xis*-stimulated. It is possible that in the case of some *att*'s the thermolabile factor replaces *xis* at low temperature.

We have discussed the evidence for specific elements in the *att*'s. Two questions immediately arise: How big are the elements and how specific are they?

3.2.4a. Size of the *att* Elements

There is very little direct information on the size of the *att*'s. The fact that there is but one primary λ integration site suggests a minimum number of about 12 nucleotide pairs for *att*. Recognition elements do not extend further than 2000 nucleotide pairs to the left of the crossover point in the phage chromosome, since a deletion which begins at that location and extends to the left does not affect site-specific recombination (Davis and Parkinson, 1971). Other data come from the analysis of *int*-promoted deletions which have one end point at 0.574 fractional λ lengths and the other at variable distances to the right or left. The phage, of which the archetype is λ b2, have site-specific recombination patterns unlike normal or transducing phage. It is believed that these phage carry specific elements, referred to collectively as Δ if they are to the left of O and Δ′ if they are to the right, which resemble but are not identical to P, P′, B or B′. One phage has been isolated which appears like λ b2 (ΔOP′) by heteroduplex analysis in the electron microscope but behaves in crosses like λ *gal* (BOP′). If this phage carries B, B cannot be larger than about 50 nucleotide pairs, which is the limit of detectability of a segment of single-stranded DNA surrounded by duplex (Davis and Parkinson, 1971).

Similarly, deletions have been found in P′ which are too small to be visible by heteroduplex mapping but which extend into the *int* gene (L. Enquist and R. Weisberg, manuscript in preparation; M. Shulman and K. Mizuuchi, personal communication). If the deletion begins at the cross-over point (as most deletions of this sort do), either P′ is less than 50 nucleotides in length or the *int* gene overlaps with P′.

3.2.4b. Specificity of Site-Specific Recombinants

It is now known that Int and Xis can recognize a rather large number of different sequences. The efficiency of recombination at these sequences is, however, lower than at the normal *att* sites. This conclusion is based on the work of Shimada *et al.* (1972, 1973, 1975) who found that *E. coli* deleted for *att*BOB′ is lysogenized at about 0.5% the frequency of a normal host. This residual integration requires the *int* gene of the phage and occurs by a cross-over at the normal *att* on the phage chromosome. There are certain sites on the bacterial chromosome which are preferentially utilized, and some of these are within bacterial genes. When a bacterial gene has been inactivated by prophage integration, prophage excision can restore its function. This indicates the fidelity of site-specific recombination; evidently nucleotides are neither lost nor gained through a cycle of integration-excision.

Prophage excision from the abnormal sites, like normal excision, requires Int and Xis. This suggests that these sites resemble *att*BOB′ rather than *att*POP′, *att*BOP′, or *att*POB′. Transducing phage can be generated from the abnormal lysogens, and they have recombination characteristics like those of the *int*-promoted deletion phage λ ΔOP′ or λ POΔ′. They differ from one another in site-specific recombination efficiency. None of those tested appear to contain the normal specific elements P, P′, B, or B′ (Shimada *et al.*, 1975). Taken together, these data indicate some latitude in the specificity of *int-xis* promoted recombination.

As noted above, it is possible to insert λ into a gene. If one selects for the loss of a given function and, at the same time, for λ lysogeny in an *att*BOB′ deleted host, it is possible to direct the integration at a particular locus. Even under these conditions, integration is not random, and repeated insertions at the same locus are observed. In the case of the arabinose operon, there are three integration sites within a region of 2100 nucleotide pairs (Boulter and Lee, 1975). At this frequency, the minimum recognition element could be as small as five nucleotide pairs.

3.2.5. The Attachment Sites: The Common Core

We have presented evidence indicating that the attachment sites carry specific recognition elements. We will now discuss the evidence that they carry as well a recognition element "O" which is not specific. This element would be a sequence common to all the attachment sites, including the secondary attachment sites. It may, in fact, be the sole remaining recognition element in the more rarely utilized of these abnormal sites.

A common homologous core in the attachment sites was first proposed by Signer *et al.* (1969), Kaiser and Wu (1968), and Campbell *et al.* (1969) based on the properties of the cohesive ends of λ DNA. Site-specific recombination between *att*'s would occur in three steps: (1) recognition of the *att*'s by site-specific recombination functions; (2) endonucleolytic cleavage bracketing and on opposite strands of each core (this introduces staggered single-strand breaks); and (3) crossing-over in the same way that the cohesive ends of λ DNA open and close.

For some time no evidence supporting this model could be found. It was already known that the putative homologous sequence must be too small to serve as a substrate for general recombination. Electron-microscopic studies of heteroduplexes, in fact, placed an upper limit on the size of such homology (Hradecna and Szybalski, 1969; Davis and Parkinson, 1971). Heteroduplexes between λ and λ *bio* and between λ and λ *gal* demonstrated that in each case the region of nonhomology began at 0.574 fractional length. Furthermore, heteroduplexes between λ *bio* and a λ *gal* DNA showed no homology at all in the *att* region. On the basis of these experiments, it was concluded that if any region of homology exists between the phage and bacterial *att*'s it must be less than 50 nucleotide pairs in length, the lower limit of detection of homologous regions by this technique (Westmoreland *et al.*, 1969).

Data supporting the notion of a common core come from the work of Shulman and Gottesman (1971 and 1973). These workers attempted to isolate point mutations in the prophage *att*'s. Mutations were found which, on the basis of density measurements and examination by electron microscopy, did not appear to be insertions or deletions. The mutations reduced site-specific recombination about 100-fold. Although originally isolated in a particular prophage *att*, they could be crossed into the bacterial, phage, or reciprocal prophage *att*. They thus appeared to reside in a recognition element common to all *att*'s, or "O." In crosses between wild-type and mutant *att*'s, occasional hetero-diploids could be isolated. This provides support for the recombination model of Signer (Signer *et al.*, 1969) and Kaiser and Wu (1968). However, the segregation pattern of the recombinants is not 1:1 for reasons not as yet understood.

As yet no point mutations have been found in any of the specific elements; perhaps such mutations would not interfere sufficiently with *att* function to survive the selection procedure.

3.3. Effects of Macromolecular Synthesis on Site-Specific Recombination

It has been proposed that DNA must be replicating or transcribed to undergo recombination. Replicating forks have been suggested to be the initiators of recombination, while replication of recombination intermediates would serve to complete the recombination process. Stahl and his co-workers (Stahl and Stahl, 1971) have studied the interrelationships between λ generalized and site-specific recombination on the one hand, and replication on the other. A role for transcription in recombination has been suggested by Drexler (1972) and by Dove and his co-workers (Davies *et al.*, 1972; Inokuchi *et al.*, 1973).

It is impossible to dissociate the role of phage replication, transcription, and packaging in the formation of virions from their possible role in recombination. Therefore, to study directly the role of macromolecular synthesis on recombination, it is necessary to measure recombinant DNA molecules rather than recombinant phage. Two basic systems have been used.

Freifelder and his co-workers (Freifelder *et al.*, 1971) utilize an episome, F′*gal att*BOB′, into which λ can insert. The F′ and F′ (λ) can be distinguished on alkaline sucrose gradients, where they sediment as covalent DNA circles of characteristic size (Freifelder and Kirschner, 1971). Integration can be detected by infecting nonlysogenic cells carrying the episome and monitoring the loss of F′ and the appearance of F′ (λ) covalently closed DNA circles (Inokuchi *et al.*, 1973; Freifelder *et al.*, 1973). Similarly, excision can be detected as loss of an F′ (λ) covalent circle and the appearance of F′ circles.

Another assay designed to investigate recombination independently of virion formation utilizes λ phage carrying two attachment sites called λ*att*². The first of these was isolated in 1971 by Shulman and Gottesman (1971) as a rare transducing phage carrying both *att*BOP′ and *att*POB′ (see Fig. 7). Because the transducing phage carries two *att*'s, it can only be isolated under *int*⁻ or *xis*⁻ conditions. Exposure to Int and Xis results in intramolecular recombination between the *att*'s with the loss of the bacterial DNA lying between them and a consequent reduction of genome size by 7×10^6 daltons. As a result, the phage product of recombination is, unlike the parental λ*att*², resistant to chelating agents such as pyrophosphate or EDTA, which selectively inactivate λ virions with normal or large genomes. Another λ*att*² phage, carrying *att*BOB′ and *att*POP′, was isolated by Nash (1974). This phage undergoes intramolecular site-specific recombination which is analogous to integration; it is dependent on Int but not Xis. Such recombination converts an EDTA-sensitive phage into an EDTA-resistant one. The conversion follows the loss of bacterial DNA lying between the two *att*'s. Recombination proceeds preferentially by intramolecular recombination.

Recombinant DNA produced after infection by these phages is detected as follows (Nash, 1975*a*; Gottesman and Gottesman, 1975*a*):

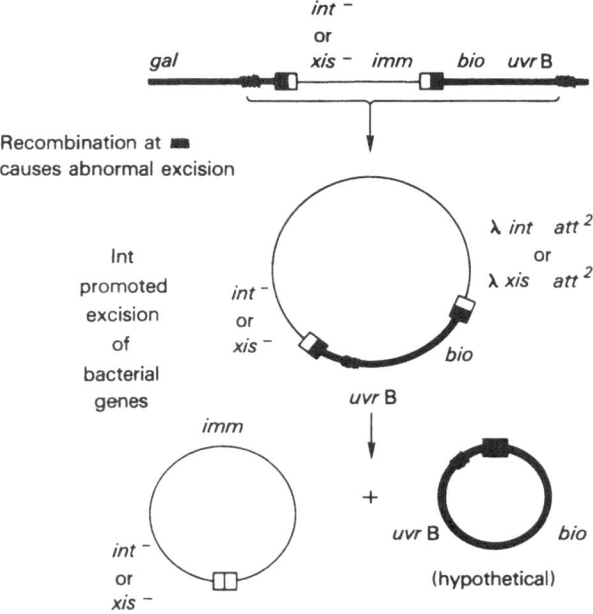

Fig. 7. The formation and excision of λ att^2 (BOP′ × POB′). The abnormal transducing phage, λ att^2, isolated by Shulman and Gottesman (1971), arose by an excision outside both the *att* BOP′ and *att* POB′ attachment sites. The resulting larger-than-normal phage carries both prophage attachment sites and the *bio* and *uvr* B genes of the host. The two attachment sites recombine intramolecularly in the presence of Int and Xis to produce a smaller phage which has lost the bacterial DNA and now contains only a POP′ attachment site.

(1) Lysogenic cells are induced: the prophage synthesizes the proteins required for recombination. (2) The cells are then infected with λatt^2 and incubated to permit recombination. (3) The DNA is then extracted from the cells and added to spheroplasts. The spheroplasts produce bursts of phage, which are then scored for parental (EDTA-sensitive) and recombinant (EDTA-resistant) phage types. The substrate phage λatt^2 is int^- and/or xis^-, while the spheroplasts contain no prophage. The assay thus measures the site-specific recombination of λatt^2 at the DNA level, since the function of the spheroplasts is solely to produce virions from the added DNA.

3.3.1. Lack of Involvement of Replication in Site-Specific Recombination

Destruction of covalently closed F′ (λ) circles occurs after induction of a lysogen according to Freifelder *et al.* (1973) and Freifelder and

Kirschner (1971). When the prophage was replication defective because of O or P mutations, loss of F′ (λ) circles required *int* and *xis* functions. They thus conclude that excision can occur (or at least begin) in the absence of replication. This agrees with earlier results of Shulman and Gottesman (1971), who observed recombination of λ*att*² in the presence of λ repressor, and of Nash and Robertson (1971), who showed that *int*-promoted recombinants can be found which have undergone little or no replication. Later experiments of Nash (1975*a*) and Gottesman and Gottesman (1975*a*), in which λ*att*² recombination was measured in the presence of both repressor and nalidixic acid, an inhibitor of DNA replication, have confirmed that both integration and excision proceed efficiently in the absence of extensive replication.

In this respect, site-specific recombination may differ from *red*-promoted recombination, where significant levels of DNA synthesis appear to be associated with the recombination event (Stahl and Stahl, 1971).

Although it is apparent that normal, O- and P-promoted replication is not essential for site-specific recombination, the participation of a small amount of repair replication, resistant to nalidixic acid and repressor, cannot be excluded by these *in vivo* experiments.

3.3.2. Lack of Involvement of Transcription in Site-Specific Recombination

The need for transcriptional activation has been examined for both integration (Nash, 1975*a*) and excision (Gottesman and Gottesman, 1975*a*) with the λ*att*²-transfection system described above. In these experiments, site-specific recombination of the two λ*att*²'s was analyzed in the presence of repressor and rifampicin, both inhibitors of transcription. In the case of the integration reaction, ongoing transcription did not enhance integrative recombination; on the contrary, the kinetics of the reaction were considerably faster in the presence of transcription inhibitors than in their absence (Nash, 1975*a*). For excisive recombination, no effect of transcription on the reaction was observed.

Earlier experiments implicating transcription in prophage excision used an N^- prophage and a superinfecting phage carrying N_λ or N_{21} (Davies *et al.*, 1972). After N_λ superinfection, the N^- prophage excised well. With the N_{21} helper, little prophage excision was observed, although both phage could provide Int and Xis. Inokuchi *et al.* (1973) confirmed the requirement in a biophysical assay by measuring the disappearance of covalent circles of F′ (λN^-). It was postulated that N_λ was required so that *att*BOP′ could be transcribed. Later work showed that the apparent stimulation of site-specific recombination by transcription was observed only when the *b2* region was expressed (Lehman, 1974; Gottesman and Court, unpublished data). They concluded that the *b2* region encoded a function inhibitory to recombination (*sif*) which was antagonized by *att* transcription (Lehman, personal communication). One can reconcile the conflicting data about the role of transcription in site-specific recombination by arguing that *b2* products have not been expressed under the conditions of Nash and Gottesman and Gottesman. It should be noted, however, that excision normally precedes the expression of the *b2* region. This would suggest that the role of Sif and of transcription in site-specific recombination may be ancillary.

3.4. *In Vitro* Site-Specific Recombination

In the last two years, two types of *in vitro* systems capable of carrying out site-specific recombination have been developed; these will lead, undoubtedly, to an understanding of the biochemical basis of site-specific recombination.

Syvanen (1974), studying *in vitro* packaging of λ DNA, found that covalently closed circular λ monomers cannot be packaged unless they first recombine with concatemeric DNA that is present in the extracts. The recombination-dependent phage formation that he observed appeared to involve the *int* system: (1) As expected for integrative recombination, it was temperature sensitive; general recombination is relatively thermoresistant. (2) Genetic analysis of the phage showed that recombination had taken place in the *J–c*I region, presumably at *att* POP'. Syvanen's assay has the advantage of detecting very low levels of recombination, since only monomers that have recombined will form virions. On the other hand, it suffers from its complexity: It has thus far not been possible to separate the maturation reaction from the recombination reaction.

The second system, developed by Nash (1975*b*) and by Gottesman and Gottesman (1975*b*), utilizes the conversion of the λ*att*^{2} DNAs to smaller molecules by the *int* or the *int-xis* functions. In this assay, λ*att*^{2} DNA is added to an extract prepared as described by Kaiser and Masuda (1972) from induced λ lysogens. After incubation, the DNA is extracted and added to spheroplasts. The phage released by the spheroplasts are titered in the presence or absence of chelating agents to determine the amount of recombination that took place in the extract. The systems have the following characteristics: (1) They are highly efficient: as much as 50% of the input DNA molecules recombine during the course of the reaction. (2) The still-crude systems require Mg^{2+}, spermidine, and ATP. (3) The λatt^2 DNA must be either in the form of open or covalently closed circles. (4) Nash's system, which measures integrative recombination, is thermolabile; Gottesman and Gottesman's system, which assays excisive recombination, is not. The former, as expected, required, among phage functions, only Int; the latter requires Int and Xis.

Recently Nash (H. Nash, personal communication) has shown that, in addition to Int, at least two host factors are required for integrative recombination in his extract system. Both Int and a host factor are thermolabile.

3.5. Control of *int* and *xis* Expression

The *int* and *xis* genes form part of the *N* operon and are under the control of the P_L promoter and the O_L operator (Signer, 1970; Gottesman and Weisberg, 1971). In addition, there is a second, low-level promoter for *int*, probably located within the *xis* gene. The existence of this promoter was detected by Shimada *et al.* (1973) and by Shimada and Campbell (1974a) who found a low-level constitutive expression of *trp*AB in a lysogen where λ had inserted in *trp*C. The source of the transcription of *trp*AB lay in the prophage. Mutations with higher levels of *trp*AB expression were obtained, and some of these were located in the prophage *xis* gene (Shimada and Campbell, 1974a and b). Whether these mutations, called *int*C, define the site of the *int* promoter is problematical. They all have in common, however, the property of rendering the prophage *xis*⁻ (Shimada and Campbell, 1974b; Enquist and Weisberg, in preparation).

The importance of the *int* promoter in lysogeny is not clear at present. A low level expression of *int* probably accounts for the relative instability of prophage bracketed by POP′ and BOB′, where Int alone can promote excision (Weisberg and Gottesman, 1971; Echols and Court, 1971; Echols, 1975). Very recent evidence suggests that the *c*II and *c*III functions can increase *int* expression from the *int* promoter (A. Oppenheim, D. Court, H. Nash, and S. Adhya, personal communications).

3.6. An Overview of Integration-Excision

After infection of a sensitive cell, λ*int* and *xis* products are synthesized in large amounts under the control of the P_L promoter.

The activity of this promoter is soon diminished by the action of *cro* and repressor. What follows is a decrease in the transcription of the phage *att* and the rapid decay of Xis. It is likely that integration does not occur until these processes are well under way, since transcription interferes with integrative recombination (Nash, 1975*a*). Furthermore, a direct inhibition of integrative recombination by Xis has been observed by Nash (Nash, 1975*a* and *b*). Integration occurs when Int and the host recombination function(s) recognize the elements of both *att*'s and promote recombination between them. The *att*'s do not appear to react separately, since Freifelder found no loss of F´ (λ) circles in cells induced carrying the integration-deficient λ*b2* prophage (Freifelder *et al.*, 1973). The absence of Xis, as well as possibly the continuous low-level expression of Int, assure that the prophage, once integrated, remains so.

After prophage induction, excision probably occurs quite rapidly. Little is known about why Xis is required for excision, but it appears to act in concert with Int, rather than sequentially, to promote excisive recombination (Gottesman and Gottesman, unpublished observations).

Our picture of site-specific recombination is obviously fragmentary. It is certain, however, to be filled in rapidly as the *in vitro* systems progress. In addition, there is reason to expect that the *att*'s, both normal and abnormal, will soon be sequenced.

ACKNOWLEDGMENTS

We gratefully acknowledge the helpful criticism of Drs. Sankar Adhya, Raymond Devoret, Lynn Enquist, Tom Maniatis, and Nat Sternberg. Drs. D. Court, H. Echols, M. Gough, G. Gussin, P. Kourilsky, B. Meyer, H. Nash, A. Oppenheim, V. Pirotta, M. Ptashne, and L. Reichardt kindly provided unpublished manuscripts and information. S. G. thanks Dr. D. Botstein for support. Ms. Martha Harshman speedily and skillfully typed the manuscript.

4. REFERENCES

Adhya, S., Cleary, P., and Campbell, A., 1968, A deletion analysis of prophage lambda and adjacent genetic regions, *Proc. Nat. Acad. Sci. USA* **61,** 956.
Adhya, S., Gottesman, M., and de Crombrugghe, B., 1974, Release of polarity in *Escherichia coli* by Gene N of Phage λ: Termination and antitermination of transcription, *Proc. Natl. Acad. Sci. USA* **71,** 2534.
Allet, B., Roberts, R., Gesteland, R., and Solem, R., 1974, Class of promoter sites for *Escherichia coli* DNA-dependent RNA polymerase, *Nature* **249,** 217.

Astrachan, L., and Miller, J., 1972, Regulation of λ rex expression after infection of *Escherichia coli* K by lambda bacteriophage, *J. Virol.* **9**, 510.

Attardi, G., Naono, S., Rouviere, J., Jacob, F., and Gros, F., 1963, Production of messenger RNA and regulation of protein synthesis, *Cold Spring Harbor Symp. Quant. Biol.* **28**, 363.

Ausubel, F., 1974, Radiochemical purification of bacteriophage λ integrase, *Nature* **247**, 152.

Bailone, A., Blanco, M., and Devoret, R., 1975, *E. coli* K12 inf: A mutant deficient in prophage induction and cell filamentation, *Mol. Gen. Genet.* **136**, 291.

Belfort, M., and Wulff, D., 1971, A mutant of *Escherichia coli* that is lysogenized with high frequency, *in* "The Bacteriophage λ" (A. D. Hershey, ed.), p. 739, Cold Spring Harbor, New York.

Belfort, M., and Wulff, D., 1973*a*, An analysis of the processes of infection and induction of *E. coli* mutant hfl-1 by bacteriophage lambda, *Virology* **55**, 183.

Belfort, M., and Wulff, D., 1973*b*, Genetic and biochemical investigation of the *Escherichia coli* mutant hfl-1 which is lysogenized at high frequency by bacteriophage lambda, *J. Bacteriol.* **115**, 299.

Belfort, M., and Wulff, D., 1974, The roles of the lambda cIII gene and the *Escherichia coli* catabolite gene activation system in the establishment of lysogeny by bacteriophage lambda, *Proc. Nat. Acad. Sci. USA* **71**, 779.

Benzer, S., and Champe, S., 1962, A change from sense to nonsense in the genetic code, *Proc. Natl. Acad. Sci. USA* **48**, 1114.

Bertani, G., 1956, The role of phage in bacterial genetics, *Brookhaven Symp. Biol.* **8**, 50.

Blattner, F., and Dahlberg, J., 1972, RNA synthesis startpoints in bacteriophage λ. Are the promoter and operator transcribed? *Nature (London) New Biol.* **237**, 227.

Blattner, F., Dahlberg, J., Boettiger, J., Fiandt, M., and Szybalski, W., 1972, Distance from a promoter mutation to an RNA synthesis startpoint on bacteriophage DNA, *Nature (London) New Biol.* **237**, 232.

Bode, V., and Kaiser, A., 1965*a*, Changes in the structure and activity of DNA in a superinfected immune bacterium, *J. Mol. Biol.* **14**, 399.

Bode, V., and Kaiser, A., 1965*b*, Repression of the c_{11} and c_{111} cistrons of phage lambda in a lysogenic bacterium, *Virology* **25**, 111.

Borek, F., and Ryan, A., 1958, The transfer of irradiation-elicited induction in a lysogenic organism, *Proc. Nat. Acad. Sci. USA* **44**, 374.

Botstein, D., and Herskowitz, I., 1974, Properties of hybrids between *Salmonella* phage P 22 and coliphage λ, *Nature* **251**, 584.

Botstein, D., Lew, K., Jarvik, V., and Swanson, C., Jr., 1975, Role of antirepressor in the bipartite control of repression and immunity by bacteriophage P 22, *J. Mol. Biol.* **91**, 439.

Boulter, J., and Lee, N., 1975, Isolation of specialized transducing bacteriophage lambda carrying genes of the L-arabinose operon of *Escherichia coli* B/r, *J. Bacteriol.* **123**, 1043.

Boyd, J., 1951, "Excessive dose" phenomenon in virus infections, *Nature,* **167**, 1061.

Brachet, P., and Thomas, R., 1969, Mapping and functional analysis of y and cII mutants, *Mutat. Res.* **7**, 257.

Brack, C., and Pirotta, V., 1975, Electron-microscopic study of the repressor of phage, and its interaction with operator DNA, *J. Mol. Biol.* **96**, 139.

Brooks, K., 1965, Studies in the physiological genetics of some suppressor-sensitive mutants of bacteriophage λ, *Virology* **26**, 489.

Brooks, K., and Clark, A., 1967, Behavior of lambda bacteriophage in a recombination-deficient strain of *Escherichia coli, J. Virol.* **1**, 283.

Calef, E., and Licciardello, G., 1960, Recombination experiments on prophage-host relationships, *Virology,* **12**, 81.

Calef, E., and Neubauer, Z., 1968, Active and inactive states of the cI gene in some λ defective phages, *Cold Spring Harbor Symp. Quant. Biol.* **33**, 765.

Calef, E., Avitabile, A., del Giudice, L., Marchelli, C., Menna, T., Neubauer, Z., and Soller, A., 1971, The genetics of the anti-immune phenotype of defective lambda, *in* "The Bacteriophage λ" (A. D. Hershey, ed.), p. 609, Cold Spring Harbor, New York.

Campbell, A., 1961, Sensitive mutants of bacteriophage λ, *Virology,* **14**, 22.

Campbell, A., 1962, The episomes, *Adv. Genet.* **11**, 101.

Campbell, A., 1963, Segregants from lysogenic heterogenotes carrying recombinant lambda prophages, *Virology,* **20**, 344.

Campbell, A., 1971, Genetic structure, Chapter 2 *in* "The Bacteriophage λ" (A. D. Hershey, ed.), p. 13, Cold Spring Harbor, New York.

Campbell, A., and del Campillo-Campbell, A., 1963, Mutant of lambda bacteriophage producing a thermolabile endolysin, *J. Bacteriol.* **85**, 1202.

Campbell, A., and Killen, K., 1967, Effect of temperature on prophage attachment and detachment during heteroimmune superinfection, *Virology* **33**, 749.

Campbell, A., Adhya, S., and Killen, K., 1969, The concept of prophage, *in* "Ciba Foundation Symposium on Episomes and Plasmids" (G. E. W. Wolstenholme and M. O'Connor, eds.), p. 13, J. A. Churchill, London.

Castellazzi, M., George, J., and Buttin, G., 1972a, Prophage induction and cell division in *E. coli.* I. Further characterization of the thermosensitive mutation tif-1 whose expression mimics the effect of UV irradiation, *Mol. Gen. Genet.* **119**, 153.

Castellazzi, M., George, J., and Buttin, G., 1972b, Prophage induction and cell division in *E. coli, Mol Gen. Genet.* **119**, 139.

Castellazzi, M., Brachet, P., and Eisen, H., 1972c, Isolation and characterization of deletions in bacteriophage λ residing as prophage in *E. coli* K12, *Mol. Gen. Genet.* **117**, 211–218.

Chadwick, P., Pirotta, V., Steinberg, R., Hopkins, N., and Ptashne, M., 1970, The λ and 434 phage repressors, *Cold Spring Harbor Symp. Quant. Biol.* **28**, 363.

Court, D., Green, L., and Echols, H., 1975, Positive and negative regulation by the cII and cIII gene products of bacteriophage λ, *Virology* **63**, 484.

Dambly, C., and Couturier, M., 1971, A minor Q-independent pathway for the expression of the late genes in bacteriophage λ, *Mol. Gen. Genet.* **113**, 244.

Dambly, C., Couturier, M., and Thomas, R., 1968, Control of development in temperate bacteriophages, *J. Mol. Biol.* **32**, 67.

Davidson, N., and Szybalski, W., 1971, Physical and chemical characteristics of Lambda DNA, Chapter 3 *in* "The Bacteriophage λ" (A. D. Hershey), p. 45, Cold Spring Harbor, New York.

Davies, R., Dove, W., Inokuchi, H., Lehman, J., and Roehrdanz, R., 1972, Regulation of λ prophage excision by the transcriptional state of DNA, *Nature (London)* New Biol. **238**, 43.

Davis, R., and Davidson, N., 1968, Electron-microscope visualization of deletion mutations, *Proc. Nat. Acad. Sci. USA* **60**, 243.

Davis, R., and Parkinson, J., 1971, Deletion mutants of bacteriophage lambda. III. Physical structure of attφ, *J. Mol. Biol.* **56**, 403.

Devoret, R., and George, J., 1967, Induction indirecte du prophage λ par le rayonnement ultraviolet, *Mutat. Res.* **4,** 713.

Dickson, R. C., Abelson, J., Barnes, W. M., and Reznikoff, W. S., 1975, Genetic regulation: the *lac* control region, *Science* **187,** 27.

Dottin, R., Cutler, L., and Pearson, M., 1975, Repression and autogenous stimulation *in vitro* by bacteriophage lambda repressor, *Proc. Nat. Acad. Sci. USA* **72,** 804.

Dove, W., 1966, Action of the lambda chromosome. I. Control of functions late in bacteriophage development, *J. Mol. Biol.* **19,** 187.

Dove, W., Hargrove, E., Ohashi, M., Haugli, F., and Guha, A., 1969, Replicator activation in lambda, *Jpn J. Genet. 44 Suppl.* **1,** 11.

Dove, W. F., and Weigle, J. J., 1965, Intracellular state of the chromosome of bacteriophage lambda. I. The eclipse of infectivity of the bacteriophage DNA, *J. Mol. Biol.* **12,** 620.

Drexler, H., 1972, Transduction of Gal$^+$ by coliphage T 1, *J. Virol.* **9,** 280.

Echols, H., 1970, Integrative and excisive recombination by bacteriophage λ: Evidence for an excision-specific recombination protein, *J. Mol. Biol.* **47,** 575.

Echols, H., 1972, Developmental pathways for the temperate phage: Lysis vs lysogeny, *Annu. Rev. Genet.* **6,** 157.

Echols, H., 1975, Constitutive integrative recombination by bacteriophage λ, *Virology* **64,** 557.

Echols, H., and Court, D., 1971, The role of helper phage in *gal* transduction *in* "The Bacteriophage λ" (A. D. Hershey, ed.), p. 701, Cold Spring Harbor, New York.

Echols, H., and Green, L., 1971, Establishment and maintenance of repression by bacteriophage λ: The role of the *c*I, *c*II, and *c*III proteins, *Proc. Nat. Acad. Sci. USA* **68,** 2190.

Echols, H., Gingery, P., and Moore, L., 1968, Integrative recombination function of bacteriophage λ: Evidence for a site-specific recombination enzyme, *J. Mol. Biol.* **34,** 251.

Echols, H., Green, L., Oppenheim, A. B., Oppenheim, A., and Honigman, A., 1973, The role of the cro gene in bacteriophage λ development, *J. Mol. Biol.* **80,** 203.

Echols, H., Chung, S., and Green, L., 1974, Site-specific recombination: Genes and regulation, *in* "Mechanisms in Recombination" (R. E. Grell, ed.), p. 69, Plenum Press, New York.

Echols, H., Green, K., Kudrna, P., and Edlin, G., 1975, Regulation of phage λ development with the growth rate of host cells: A homeostatic mechanism, *Virology* **66,** 344.

Eisen, H., Pereira da Silva, L., and Jacob, F., 1968a, Genetique cellulaire, *C. R. Acad. Sci.* **266,** 1176.

Eisen, H., Pereira da Silva, L., and Jacob, F., 1968b, The regulation and mechanism of DNA synthesis in bacteriophage lambda, *Cold Spring Harbor Symp. Quant. Biol.* **33,** 755.

Eisen, H., Brachet, P., Pereira da Silva, L., and Jacob, F., 1970, Regulation of repressor expression in λ, *Proc. Nat. Acad. Sci. USA* **66,** 855.

Fiandt, M., Hradecna, Z., Lozeron, H., and Szybalski, W., 1971, Electron micrographic mapping of deletions, insertions, inversions, and homologies in the DNAs of coliphages lambda and phi 80, *in* "The Bacteriophage λ" (A. D. Hershey, ed.), p. 329, Cold Spring Harbor, New York.

Franklin, N., 1971, The N operon of lambda: Extent and regulation as observed in fusions to the tryptophan operon of *Escherichia coli, in* "The Bacteriophage λ" (A. D. Hershey, ed.), p. 621, Cold Spring Harbor, New York.

Franklin, N., Dove, W., and Yanofsky, C., 1965, The linear insertion of a prophage into the chromosome of *E. coli* shown by deletion mapping, *Biochem. Biophys. Res. Commun.* **18,** 910.

Freifelder, D., and Levine, 1973, Requirement for transcription in the neighborhood of the phage attachment region for lysogenization of *Escherichia coli* by bacteriophage λ, *J. Mol. Biol.* **74,** 729.

Freifelder, D., and Kirschner, I., 1971, A phage endonuclease controlled by genes O and P, *Virology* **44,** 223.

Freifelder, D., and Meselson, M., 1970, Topological relationship of prophage λ to the bacterial chromosome in lysogenic cells, *Proc. Nat. Acad. Sci. USA* **65,** 200.

Freifelder, D., Folkmanis, A., and Kirschner, I., 1971, Studies on *Escherichia coli* sex factors; Evidence that covalent circles exist within cells and the general problem of isolation of covalent circles, *J. Bacteriol.* **105,** 722.

Freifelder, D., Kirschner, I., Goldstein, R., and Baran, N., 1973, Physical study of prophage excision and curing of λ prophage from lysogenic *Escherichia coli, J. Mol. Biol.* **74,** 703.

Fuerst, C., and Siminovitch, L., 1965, Characterization of an unusual defective lysogenic strain of *Escherichia coli* K12(λ), *Virology* **27,** 449.

Galland, P., Bassi, P., and Calef, E., 1973, On the mode of antirepressor action in anti-immune cells, *Mol. Gen. Genet.* **125,** 231.

Geiduschek, E. P., and Grau, D., 1970, T 4 anti-messenger, "Lepetit Colloquium on Biology and Medicine," Volume 1, p. 190.

Gellert, M., 1967, Formation of covalent circles of lambda DNA by *E. coli* extracts, *Proc. Nat. Acad. Sci. USA* **57,** 148.

George, J., 1966, Correlation entre la disparition de l'induction indirecte du prophage λ et la restriction du DNA transmis à la bactérie receptrice inductible dans un croisement F+ × F− hétérospecifique, *C. R. Acad. Sci.* **262,** 1805.

Gesteland, R., and Kahn, C., 1972, Synthesis of bacteriophage λ proteins *in vitro, Nature (London) New Biol.* **240.** 3.

Gilbert, W., and Maxam, A., 1973, The nucleotide sequence of the *lac* operator, *Proc. Nat. Acad. Sci. USA* **70,** 3581.

Gingery, R., and Echols, H., 1967, Mutants of bacteriophage λ unable to integrate into the host chromosome, *Proc. Nat. Acad. Sci. USA* **58,** 1507.

Goldthwait, D., and Jacob, F., 1964, Genetique biochimique. Sur le mécanisme de l'induction du développement du prophage chez les bactéries lysogènes, *C. R. Acad. Sci.* **259,** 661.

Gottesman, S., and Gottesman, M. E., 1975a, Elements involved in site-specific recombination in bacteriophage lambda, *J. Mol. Biol.* **91,** 489.

Gottesman, S., and Gottesman, M. E., 1975b, Excision of prophage λ in a cell-free system, *Proc. Nat. Acad. Sci. USA* **72:** 2188.

Gottesman, M. E., and Weisberg, R., 1971, Prophage insertion and excision, Chapter 6 *in* "The Bacteriophage λ" (A. D. Hershey, ed.), Cold Spring Harbor, New York, p. 113.

Gottesman, M. E., and Yarmolinsky, M., 1968, Integration-negative mutants of bacteriophage lambda, *J. Mol. Biol.* **31,** 487.

Gough, M., and Tokuno, S., 1975, Further structural and functional analogies between the repressor regions of phages P 22 and λ, *Mol. Gen. Genet.* **138,** 71.

Green, M., 1966, Inactivation of the prophage lambda repressor without induction, *J. Mol. Biol.* **16,** 134.

Greenblatt, J., 1973, Regulation of the expression of the N gene of bacteriophage lambda, *Proc. Nal. Acad. Sci. USA* **70**, 421.

Grodzicker, T., Arditti, R., and Eisen, H., 1972, Establishment of repression by lambdoid phage in catabolite activator protein and adenylate cyclase mutants of *Escherichia coli*, *Proc. Nat. Acad. Sci. USA* **69**, 366.

Guarneros, G., and Echols, H., 1970, New mutants of bacteriophage λ with a specific defect in excision from the host chromosome, *J. Mol. Biol.* **47**, 565.

Guarneros, G., and Echols, H., 1973, Thermal asymmetry of site-specific recombination by bacteriophage λ, *Virology* **52**, 30.

Guerrini, F., 1969, On the asymmetry of λ integration sites, *J. Mol. Biol.* **46**, 523.

Gussin, G., and Peterson, V., 1972, Isolation and properties of rex⁻ mutants of bacteriophage lambda, *J. Virology* **10**, 760.

Gussin, G., Yen, K., and Reichardt, L., 1975, Repressor synthesis *in vivo* after infection of *E. coli* by a p_{rm}^- mutant of bacteriophage λ, *Virology* **63**, 273.

Hayes, S., and Szybalski, W., 1973, Control of short leftward transcripts from the immunity and ori regions in induced coliphage lambda, *Mol. Gen. Genet.* **126**, 275.

Heinemann, S., and Spiegelman, W., 1970, Control of transcription of the repressor gene in bacteriophage lambda, *Proc. Nat. Acad. Sci. USA* **67**, 1122.

Hershey, A. D., and Burgi, E., 1965, Complementary structure of interacting sites at the ends of lambda DNA molecules, *Proc. Nat. Acad. Sci. USA* **53**, 325.

Hershey, A. D., Burgi, E., and Ingraham, L., 1963, Cohesion of DNA molecules isolated from phage lambda, *Proc. Nat. Acad. Sci.* **49**, 748.

Herskowitz, I., 1973, Control of gene expression in bacteriophage lambda, *Annu. Rev. Gen.* **7**, 289.

Herskowitz, I., and Signer, E., 1970, A site essential for expression of all late genes in bacteriophage λ, *J. Mol. Biol.* **47**, 545.

Hertman, I., and Luria, S., 1967, Transduction studies on the role of the rec⁺ gene in the ultraviolet induction of prophage, *Biol.* **23**, 117.

Honigman, A., Oppenheim, A., and Oppenheim, A. B., 1975, A pleiotropic regulatory mutation in λ bacteriophage, *Mol. Gen. Genet.* **138**, 85.

Hopkins, N., and Ptashne, M., 1971, Genetics of Virulence, *in* "The Bacteriophage λ" (A. D. Hershey, ed.), p. 571, Cold Spring Harbor, New York.

Horiuchi, T., and Inokuchi, H., 1967, Temperature-sensitive regulation system of prophage lambda induction, *J. Mol. Biol.* **23**, 217.

Howard, B., 1967, Phage lambda mutants deficient in rII exclusion, *Science* **158**, 1588.

Hradecna, Z., and Szybalski, W., 1969, Electron micrographic maps of deletions and substitutions in the genomes of transducing coliphages λdg and λbio, *Virology* **38**, 473.

Inokuchi, H., Dove, W. E., and Freifelder, D., 1973, Physical studies of RNA involvement in bacteriophage λ DNA replication and prophage excision, *J. Mol. Biol.* **74**, 721.

Isaacs, L., Echols, H., and Sly, W., 1965, Control of lambda messenger RNA by the C_1-immunity region, *J. Mol. Biol.* **13**, 963.

Jacob, F., and Campbell, A., 1959, Sur le système de répression assurant l'immunité chez les bactéries lysogènes, *C. R. Acad. Sci.* **248**, 3219.

Jacob, F., and Fuerst, C., 1958, The mechanism of lysis by phage studies with defective lysogenic bacteria, *J. Gen. Microbiol.* **18**, 518.

Jacob, F., and Monod, J., 1961, Genetic regulatory mechanisms in the synthesis of proteins, *J. Mol. Biol.* **3**, 318.

Jacob, F., and Wollman, E., 1954, Etude génétique d'un bactériophage temperé d'*Escherichia coli*. I. Le systéme génétique du bactériophage λ, *Ann. Inst. Pasteur* **87**, 653.

Jacob, F., and Wollman, E., 1965*a*, Recherches sur les processus de conjugaison et de recombinaison chez *Escherichia coli* I. L'induction par conjugaison ou induction zygotique, *Ann. Inst. Pasteur* **91**, 486.

Jacob, F., and Wollman, E., 1956*b*, Recherches sur les bacteries lysogenes defectives. I. Déterminisme génétique de la morphogenése chez un bactériophage tempére, *Ann. Inst. Pasteur* **90**, 282.

Jacob, F., Fuerst, C., and Wollman, E., 1957, Recherches sur les bactéries lysogènes deféctives. II. Les types physiologiques liés aux mutations du prophage, *Ann. Inst. Pasteur* **93**, 724.

Jacob, F., Sussman, R., and Monod, J., 1962, Sur la nature du répresseur assurant l'immunité des bactéries lysogènes, *C. R. Acad. Sci.* **254**, 4214.

Kaiser, A. D., 1957, Mutations in a temperate bacteriophage affecting its ability to lysogenize *Escherichia coli*, *Virology* **3**, 42.

Kaiser, A. D., and Jacob, F., 1957, Recombination between related temperature bacteriophages and the genetic control of immunity and prophage localization, *Virology* **4**, 509.

Kaiser, A. D., and Masuda, T., 1970, Evidence for a prophage excision gene in λ, *J. Mol. Biol.* **47**, 557.

Kaiser, A. D., and Masuda, T., 1972, In vitro assembly of bacteriophage lambda heads, *Proc. Nat. Acad. Sci. USA* **70**, 260.

Kaiser, A., and Wu, P., 1968, Structure and function of DNA cohesive ends, *Cold Spring Harbor Symp. Quant. Biol.* **33**, 729.

Kellenberger, G., Zichichi, M. L., and Weigle, J., 1961, A mutation affecting the DNA content of bacteriophage lambda and its lysogenizing properties, *J. Mol. Biol.* **3**, 399.

Korn, D., and Weissbach, A., 1964, The effect of lysogenic induction on the deoxyribonucleases of *Escherichia coli* K12 (λ). II. The kinetics of formation of a new exonuclease and its relation to phage development *Virology* **22**, 91.

Kourilsky, O., 1971, Lysogenization by bacteriophage lambda and the regulation of lambda repressor synthesis. *Virology* **45**, 853.

Kourilsky, P., 1973, Lysogenization by bacteriophage lambda. I. Multiple infection and the lysogenic response, *Mol. Gen. Genet.* **122**, 183.

Kourilsky, P., Marcaud, L., Sheldrick, P., Luzzati, D., and Gros, F., 1968, Studies on the messenger RNA of bacteriophage λ. I. Various species synthesized early after induction of the prophage, *Proc. Nat. Acad. Sci. USA* **61**, 1013.

Kourilsky, P., Bourguignon, M., Bouquet, M., and Gros, F., 1970, Early transcription controls after induction of prophage λ, *Cold Spring Harbor Symp. Quant. Biol.* **35**, 305.

Kourilsky, P., Bourguignon, M., and Gros, F., 1971, Kinetics of viral transcription after induction of prophage, *in* "The Bacteriophage λ" (A. D. Hershey, ed.), p. 647, Cold Spring Harbor, New York.

Kumar, S., Calef, E., and Szybalski, W., 1970, Regulation of the transcription of *Escherichia coli* phage λ by its early genes N and tof, *Cold Spring Harbor Symp. Quant. Biol.* **35**, 331.

Lehman, J., 1974, λ Site-specific recombination: Local transcription and an inhibitor specified by the b2 region, *Mol. Gen. Genet.* **130**, 333.

Levine, M., and Schott, C., 1971, Mutations of phage P 22 affecting phage DNA synthesis and lysogenization, *J. Mol. Biol.* **62**, 53.

Levine, M., Truesdell, S., Ramakrishnan, T., and Bronson, M., 1975, Dual control of lysogeny by bacteriophage P 22: An antirepressor locus and its controlling elements, *J. Mol. Biol.* **91**, 421.

Lieb, M., 1953, The establishment of lysogenicity in *Escherichia coli, J. Bacteriol.* **65**, 642.

Lieb, M., 1966, Studies of heat-inducible λ phage. I. Order of genetic sites and properties of mutant prophages, *J. Mol. Biol.* **16**, 149.

Liedke-Kulke, M., and Kaiser, A. D., 1967, The c-region of coliphage 21, *Virology* **32**, 475.

Little, J. W., and Gottesman, M., 1971, Defective lambda particles whose DNA carries only a single cohesive end, *in* "The Bacteriophage λ" (A. D. Hershey, ed.), p. 371, Cold Spring Harbor, New York.

Luzzati, D., 1970, Regulation of λ exonuclease synthesis: Role of the N gene product and λ repressor, *J. Mol. Biol.* **49**, 515.

Maniatis, T., and Ptashne, M., 1973a, Structure of the λ operators, *Nature (London)* **246**, 133.

Maniatis, T., and Ptashne, M., 1973b, Multiple repressor binding at the operators in bacteriophage λ, *Proc. Nat. Acad. Sci. USA* **70**, 1531.

Maniatis, T., Ptashne, M., and Maurer, R., 1973, Control elements in the DNA of bacteriophage λ, *Cold Spring Harbor Symp. Quant. Biol.* **38**, 857.

Maniatis, T., Ptashne, M., Barrell, B., and Donelson, J., 1974, Sequence of a repressor-binding site in the DNA of bacteriophage λ, *Nature (London)* **250**, 394.

Maniatis, T., Jeffrey, A., and Kleid, D., 1975a, Nucleotide sequence of the rightward operator of phage λ, *Proc. Nat. Acad. Sci.* **72**, 1184.

Maniatis, T., Ptashne, M., Backman, K., Kleid, D., Flashman, S., Jeffrey A., and Maurer, R., 1975b, Recognition sequences of repressor and polymerase in the operators of bacteriophage lambda, *Cell* **5**, 109.

Maurer, R., Maniatis, T., and Ptashne, M., 1974, Promotors are in the operators in phage lambda, *Nature (London)* **249**, 221.

McMacken, R., Mantei, N., Butler, B., Joyner, A., and Echols, H., 1970, Effect of mutations in the cII and cIII genes of bacteriophage λ on macromolecular synthesis in infected cells, *J. Mol. Biol.* **49**, 639–655.

Menninger, J., Wright, M., Menninger, L., and Meselson, M., 1968, Attachment and detachment of bacteriophage DNA in lysogenization and induction, *J. Mol. Biol.* **32**, 631.

Meyer, B., Kleid, D., and Ptashne, M., 1975, λ Repressor turns off transcription of its own gene, *Proc. Nat. Acad. Sci. USA* **72**, 4785.

Naono, S., and Gros, F., 1966, On the mechanism of transcription of the lambda genome during induction of lysogenic bacteria, *Cold Spring Harbor Symp. Quant. Biol.* **31**, 363.

Nash, H., 1974a, λ att B-att P, a λ derivative containing both sites involved in integrative recombination, *Virology* **57**, 207.

Nash, H. A., 1974b, Purification of bacteriophage λ int protein, *Nature (London)* **247**, 543.

Nash, H., 1975a, Integrative recombination in bacteriophage lambda: Analysis of recombinant DNA, *J. Mol. Biol.* **91**, 501.

Nash, H., 1975b, Integrative recombination of bacteriophage lambda DNA *in vitro, Proc. Nat. Acad. Sci. USA* **72**, 1072.

Nash, H. A., and Robertson, C. A., 1971, On the mechanism of int-promoted recombination, *Virology* **44**, 446.

Neubauer, Z., and Calef, E., 1970, Immunity phase-shift in defective lysogens: Non-mutational hereditary change of early regulation of λ prophage, *J. Mol. Biol.* **51**, 1.

Ordal, G., 1971, Supervirulent mutants and the structure of operator and promoter, *in* "The Bacteriophage λ" (A. D. Hershey, ed.), p. 565, Cold Spring Harbor, New York.

Ordal, G., and Kaiser, A. D., 1973, Mutations in the right operator of bacteriophage lambda: Evidence for operator-promoter interpenetration, *J. Mol. Biol.* **79**, 709.

Parkinson, J. S., and Huskey, R. J., 1971, Deletion mutants of bacteriophage lambda. 1. Isolation and initial characterization, *J. Mol. Biol.* **56**, 369.

Pereira da Silva, L., and Jacob, F., 1968, Etude génétique d'une mutation modifiant la sensibilité à l'immunité chez le bactériophage lambda, *Ann. Inst. Pasteur (Paris)* **115**, 145.

Pero, J., 1970, Location of the phage λ gene responsible for turning off λ-exonuclease synthesis, *Virology* **40**, 65.

Pirotta, V., 1975, Sequence of the O_r operator of phage λ, *Nature (London)* **254**, 114.

Pirotta, V., and Ptashne, M., 1969, Isolation of the 434 phage repressor, *Nature (London)* **222**, 541.

Pirotta, V., Chadwick, P., and Ptashne, M., 1970, Active form of two coliphage repressors, *Nature (London)* **227**, 41.

Pirotta, V., Ptashne, M., Chadwick, P., and Steinberg, R., 1971, Isolation of repressors, *in* "Procedures in Nucleic Acid Research" (G. L. Cantoni and R. D. Davies, ed.), Harper and Row, New York and London.

Ptashne, M., 1967a, Isolation of the λ phage repressor, *Proc. Nat. Acad. Sci. USA* **57**, 306.

Ptashne, M., 1967b, The λ phage repressor binds specifically to λ DNA, *Nature (London)* **214**, 232.

Ptashne, M., 1971, Repressor and its action, Chapter 11 *in* "The Bacteriophage λ" (A. D. Hershey), p. 221, Cold Spring Harbor, New York.

Ptashne, M., and Hopkins, N., 1968, The operators controlled by the λ phage repressor, *Proc. Nat. Acad. Sci.* **60**, 1282.

Radding, C., 1964a, Nuclease activity in defective lysogens of phage λ, *Biochem. Biophys. Res. Commun.* **15**, 8.

Radding, C., 1964b, Nuclease activity in defective lysogens of phage λ. 11. A hyperactive mutant, *Proc. Nat. Acad. Sci. USA* **52**, 965.

Radman, M., 1975, SOS Repair Hypothesis: Phenomenology of an inducible DNA repair which is accompanied by mutagenesis, *in* "Molecular Mechanisms for the Repair of DNA" (P. C. Hanawalt and R. B. Setlow), p. 355, Plenum Press, New York.

Reichardt, L., 1975a, Control of bacteriophage lambda repressor synthesis: Regulation of the maintenance pathway by the cro and cI products, *J. Mol. Biol.* **93**, 289.

Reichardt, L., 1975b, Control of bacteriophage lambda repressor synthesis after phage infection: The role of the N, cII, cIII, and cro products, *J. Mol. Biol.* **93**, 267.

Reichardt, L., and Kaiser, A. D., 1971, Control of λ repressor synthesis, *Proc. Nat. Acad. Sci. USA* **68**, 2185.

Riggs, A., and Bourgeois, S., 1968, On the assay, isolation and characterization of the *lac* repressor, *J. Mol. Biol.* **34**, 361.

Roberts, J., 1969, Termination factor for RNA synthesis, *Nature (London)* **224**, 1168.

Roberts, J., 1975, Transcription termination and late control in phage lambda, *Proc. Nat. Acad. Sci. USA*, **72**, 3300.

Roberts, J., and Roberts, C., 1975, Proteolytic cleavage of bacteriophage lambda repressor in induction, *Proc. Nat. Acad. Sci.* **72**, 147.

Rosner, J. L., 1967, Transfer of materials in bacterial conjugation: Their nature and role in indirect lysogenic induction, Ph.D. dissertion, Yale University.

Rosner, J., Adelberg, E., and Yarmolinsky, M., 1967, An upper limit on β-galactosidase transfer in bacterial conjugation, *J. Bacteriol.* **94**, 1623.

Rosner, J. L., Kass, L., and Yarmolinsky, M., 1968, Parallel behavior of F and P 1 in causing indirect induction of lysogenic bacteria, *Cold Spring Harbor Symp. Quant. Biol.* **33**, 785.

Rothman, J. L., 1965, Transduction studies on the relation between prophage and host chromosome, *J. Mol. Biol.* **12**, 892.

Sharp, P. A., Hsu, M., and Davidson, N., 1972, Note on the structure of prophage λ, *J. Mol. Biol.* **71**, 499.

Shimada, K., and Campbell, A., 1974a, Int-constitutive mutants of bacteriophage lambda, *Proc. Nat. Acad. Sci. USA* **71**, 237.

Shimada, K., and Campbell, A., 1974b, Lysogenization and curing by int-constitutive mutants of phage λ, *Virology* **60**, 157.

Shimada, K., Weisberg, R. A., and Gottesman, M. E., 1972, Prophage lambda at unusual chromosomal locations. I. Location of the secondary attachment sites and the properties of the lysogens, *J. Mol. Biol.* **63**, 483.

Shimada, K., Weisberg, R. A., and Gottesman, M. E., 1973, Prophage lambda at unusual chromosomal locations. II. Mutations induced by bacteriophage lambda in *Escherichia coli* K 12, *J. Mol. Biol.* **80**, 297.

Shimada, K., Weisberg, R. A., and Gottesman, M. E., 1975, Prophage lambda at unusual chromosomal locations. III. The components of the secondary attachment sites, *J. Mol. Biol.* **93**, 415.

Shulman, M., and Gottesman, M., 1971, Lambda att²: A transducing phage capable of intramolecular int-xis promoted recombination, *in* "The Bacteriophage λ" (A. D. Hershey, ed.), p. 477, Cold Spring Harbor, New York.

Shulman, M., and Gottesman, M. E., 1973, Attachment site mutants of bacteriophage lambda, *J. Mol. Biol.* **81**, 461.

Signer, E., 1966, Interaction of prophages at the att_{80} site with the chromosome of *Escherichia coli*, *J. Mol. Biol.* **15**, 243.

Signer, E., 1969, Plasmid formation: A new mode of lysogeny by phage λ, *Nature (London)*, **223**, 158.

Signer, E. R., and Beckwith, J. R., 1966, Transposition of the lac region of *Escherichia coli*, III. The mechanism of attachment of bacteriophage ϕ80 to the bacterial chromosome, *J. Mol. Biol.* **22**, 33.

Signer, E. R., Weil, J., and Kimball, P. C. 1969, Recombination in bacteriophage λ. III. Studies on the nature of the prophage attachment region, *J. Mol. Biol.* **46**, 543.

Silver, S., 1963, The transfer of material during mating in *Escherichia coli*. Transfer of DNA and upper limits on the transfer of RNA and protein, *J. Mol. Biol.* **6**, 349.

Sly, W., Rabideau, K., and Kolber, A., 1971, The mechanisms of lambda virulence. II. Regulatory mutations in classical virulence, *in* "The Bacteriophage λ" (A. D. Hershey), p. 575, Cold Spring Harbor, New York.

Smith, H. O., and Levine, M., 1964, Two sequential repressions of DNA synthesis in the establishment of lysogeny by phage P 22 and its mutants, *Proc. Nat. Acad. Sci. USA* **52**, 356.

Smith, H. O., and Levine, M., 1967, A phage P 22 gene controlling integration of prophage, *Virology* **31**, 207.

Spiegelman, W. G., Reichardt, L., Yaniv, M., Heineman, S., Kaiser, A. D., and Eisen, H., 1972, Bidirectional transcription and the regulation of phage λ repressor synthesis, *Proc. Nat. Acad. Sci. USA* **69**, 3156.

Stahl, M. M., and Stahl, F. W., 1971, DNA synthesis associated with recombination. 1. Recombination in a DNA-negative host, *in* "The Bacteriophage λ" (A. D. Hershey), p. 431, Cold Spring Harbor, New York.

Steinberg, R., and Ptashne, M., 1971, *In vitro* repression of RNA synthesis by purified λ phage repressor, *Nature (London) New Biol.* **230**, 76.

Stent, G. S., 1971, "Molecular Genetics," W. H. Freeman and Co., San Francisco, California.

Sussman, R., and Jacob, F., 1962, Génétique physiologique. Sur le mécanisme de l'induction du développement du prophage chez les bactéries lysogènes, *C. R. Acad. Sci.* **254**, 1517.

Syvanen, M., 1974, *In vitro* genetic recombination of bacteriophage λ, *Proc. Nat. Acad. Sci. USA* **71**, 2496.

Szybalski, E., and Szybalski, W., 1974, Physical mapping of the att-N region of coliphage lambda: Apparent oversaturation of coding capacity in the gam-ral segment. *Biochimie* **56**, 1497.

Taylor, K., Hradecna, Z., and Szybalski, W., 1967, Asymmetric distribution of the transcribing regions on the complementary strands of coliphage λ DNA, *Proc. Nat. Acad. Sci. USA* **57**, 1618.

Thirion, J., and Hofnung, M., 1972, On some genetic aspects of phage λ resistance in *E. coli* K12, *Genetics* **71**, 207.

Thomas, R., 1966, Control of development in temperate bacteriophages, I. Induction of prophage genes following heteroimmune superinfection, *J. Mol. Biol.* **22**, 79.

Thomas, R., 1970, Control of development in temperate bacteriophages. III. Which prophage genes are and which are not trans-activable in the presence of immunity? *J. Mol. Biol.* **49**, 393.

Thomas, R., 1971, Control Circuits, Chapter 10 *in* "The Bacteriophage λ" (A. D. Hershey), pp. 211, Cold Spring Harbor, New York.

Thomas, R., and Bertani, L. F., 1964, On the control of the replication of temperate bacteriophages superinfecting immune hosts, *Virology* **24**, 241.

Thomas, R., and Lambert, 1962, On the occurrence of bacterial mutations permitting lysogenization by clear variants of temperate bacteriophages, *J. Mol. Biol.* **5**, 373.

Tomizawa, J., and Ogawa, T., 1967, Effect of ultraviolet irradiation on bacteriophage lambda immunity, *J. Mol. Biol.* **23**, 247.

Tomizawa, J., and Ogawa, T., 1968, Replication of phage lambda DNA, *Cold Spring Harbor Symp. Quant. Biol.* **33**, 533.

Walz, A., and Pirotta, J., 1975, Sequence of the P_R promoter of phage λ, *Nature* **254**, 118.

Weil, J., and Signer, E. R., 1968, Recombination in bacteriophage λ. II. Site-specific recombination promoted by the integration system, *J. Mol. Biol.* **34**, 273.

Weisberg, R. A., 1970, Requirements for curing of λ lysogens, *Virology* **41**, 195.

Weisberg, R., and Gallant, J., 1967, Dual function of the λ prophage repressor, *J. Mol. Biol.* **25**, 537.

Weisberg, R., and Gottesman, M. E., 1969, The integration and excision defect of bacteriophage λ dg, *J. Mol. Biol.* **46**, 565.

Weisberg, R. A., and Gottesman, M. E., 1971, The stability of int and xis functions, *in* "The Bacteriophage λ" (A. D. Hershey, ed.), p. 489, Cold Spring Harbor, New York.

Westmoreland, B., Szybalski, W., and Ris, H., 1969, Mapping of deletions and substitutions in heteroduplex DNA molecules of bacteriophage lambda by electron microscopy, *Science* **163,** 1343.

Witkin, E., 1967, The radiation sensitivity of *Escherichia coli* B: A hypothesis relating filament formation and prophage induction, *Proc. Nat. Acad. Sci. USA* **57,** 1275.

Witkin, E., and George, D., 1973, Ultraviolet mutagenesis in polA and uvrA polA derivatives of *Escherichia coli* B/r: Evidence for an inducible error-prone repair system, *Genetics* **73,** (Supplement), p. 91.

Wu, A., Ghosh, S., and Echols, H., 1972*a*, Repression by the cI protein of phage λ: Interaction with RNA polymerase, *J. Mol. Biol.* **67,** 423.

Wu, A., Ghosh, S., Echols, H., and Spiegelman, W. S., 1972*b*, Repression by the cI protein of phage λ: *In vitro* inhibition of RNA synthesis, *J. Mol. Biol.* **67,** 407.

Yarmolinsky, M. B., 1971, Modes of prophage insertion and excision, *in* "Advances in the Biosciences," Volume 8 (Workshop on mechanism and prospects of genetic exchange), p. 31, Pergamon Press, Berlin.

Yen, K., and Gussin, G., 1973, Genetic characterization of a prm⁻ mutant of bacteriophage λ, *Virology* **56,** 300.

Young, E. T., II, and Sinsheimer, R. L., 1964, Novel intracellular forms of lambda DNA, *J. Mol. Biol.* **10,** 562.

Zichichi, M., and Kellenberger, G., 1963, Two distinct functions in the lysogenization process: The repression of phage multiplication and the incorporation of the prophage in the bacterial genome, *Virology* **19,** 450.

Zissler, J., 1967, Integration-negative (int) mutants of phage λ, *Virology* **31,** 189.

CHAPTER 3

Defective Bacteriophages and Incomplete Prophages

A. Campbell

Department of Biological Sciences
Stanford University
Stanford, California 94305

1. DEFINITION AND SCOPE

A bacteriophage is termed defective if it is virtually unable to carry out a complete infectious cycle in known circumstances. As it is impossible to test infectivity on every host under all conditions, categorization is arbitrary and provisional. The significance of failure to infect is most readily interpretable for variants derived in a single step from an active phage. Operationally, ability to carry out a complete infectious cycle is frequently equated with plaque formation.

Where the only available method for viral propagation is by infection, a defective variant is a lethal mutant. For various reasons, mutations may confer either conditional or absolute lethality. Conditionally lethal mutants behave like defective phages under nonpermissive conditions.

We must distinguish between a defective virion and a defective virus. A virus particle may be noninfectious for many reasons. Its tail fibers may have broken off, it may have been neutralized by antiserum, or it may have failed to package any DNA within its protein coat. For our purposes, a defective virus comprises a viral genome, either extracellular or intracellular, which, although unable to execute a normal infectious cycle, can nevertheless be propagated in some manner. Such viral variants, from which genetic stocks can be made, will comprise the principal subject matter of this chapter.

We consider as defective phages genetic variants, such as λ *dv* (Section 2.1.2) and the minivariants of Qβ phage (Section 5), that are not packageable into virions as such. We exclude not only products of variation that is manifestly phenotypic, but also those genotypic changes that preclude faithful replication by known means. Virions that have been damaged by irradiation (Doermann *et al.*, 1955), transducing particles of phage λ the DNA of which lacks one of the single-stranded termini characteristic of the active virus (Section 2.4.1) and the small head forms of viruses like T 4 and P 1, in which shortened DNA molecules are packaged into aberrant heads (Section 2.4.2), are all genetically alive in the sense that markers can be rescued from them by recombination; yet in none of these cases can the DNA species in question reproduce as such. For comparative purposes, it will be convenient to discuss some objects of this sort, although they do not satisfy the strict definition of a defective phage.

Our definition specifies a viral genome. When the defective virus has been derived in the laboratory from a known phage, the meaning of "viral" is very clear. Naturally defective phages can present problems. Strictly speaking, the inability of an object to carry out a complete infectious cycle precludes its unambiguous categorization as "viral." The difficulties attending this question are substantive rather than trivial and will be discussed in the latter part of this chapter.

One manifestation of some naturally defective phages is the formation of noninfectious particles that resemble phages morphologically and sometimes in the ability to kill cells or to transfer bacterial genes. I might have called such objects defective phages, but have elected instead to reserve that term for their genetic determinants. Two good reviews on the properties of defective particles and their possible significance are those of Bradley (1967) and Garro and Marmur (1970).

2. METHODS OF PROPAGATION

2.1. Propagation as Prophages

The prophages of some temperate phages (such as λ) are inserted into the host chromosome, whereas others (such as P 1) replicate autonomous plasmids. Defective phage variants can usually be maintained in the same manner as their nondefective parents. This is an expected result, and requires little discussion.

Some defective variants are perpetuated in a different manner than their parents. An inserting phage like λ can engender derivatives that replicate as plasmids; conversely, a plasmid phage like P 1 has derivatives that insert. Such aberrations can potentially shed considerable insight into the mechanism of prophage maintenance, and will be treated in some detail.

2.1.1. As Inserted Elements

A defective prophage may be maintained as a DNA segment inserted into the bacterial chromosome or some other replicon. The viral DNA can then be passively replicated as part of that replicon.

The most common examples are defective mutants of temperate viruses which are themselves carried as inserted prophages (see Section 3). These include the coliphages λ, P 2, and Mu, *Salmonella* phage P 22, and the *Bacillus* phages φ105 and SPO 2 (Chow and Davidson, 1973).

2.1.1a. Coliphages P 1 Cryptic and P 1 *dtet*

Where the wild-type prophage is not ordinarily inserted into the chromosome, defective variants sometimes are. Phage P 1, for example, usually lysogenizes by plasmid formation, and thus the prophage is perpetuated as a separate, extrachromosomal element (see Section 2.2.2). However, a P 1 derivative (P 1 cryptic), which includes most but not all of the genes of P 1, is inserted into the bacterial chromosome close to the *proAB* and *lac* genes.

Such abnormal insertion by variants of phages that normally form plasmids (and the converse situation of plasmid formation by mutants of phages that ordinarily lysogenize by insertion, Section 2.2.2) can potentially illuminate our understanding of the normal process.

The origin of P 1 cryptic is complex. Interpretation of the published descriptions is probably best done by analogy with properties observed for P 22 *tet* (see Section 2.2.4) and for drug-resistance determinants associated with specific insertion sequences (see Section 3.2). As part of a study of the conjugal drug resistance plasmid R(TC.CM), Kondo and Mitsuhashi (1964) used P 1 to transduce the determinant for chloramphenicol resistance (*cam*) from a plasmid-bearing donor into a chloramphenicol-sensitive recipient. Some of the transductants proved to be lysogenic for a P 1 derivative (P 1 CM or P 1 *cam*) that had incorporated the *cam* gene from the R plasmid. This P 1 forms plaques and breeds true to type, although about 10% of the phage in a typical lysate no longer carry the *cam* genes (as tested by the chloramphenicol resistance of lysogenic cells from the center of the plaque). P 1 *cam* may have originated by transfer of the *cam* determinant from the R factor by a process of nonhomologous insertion, as seen with *tet* and *kan* determinants (see Section 3.2).

Later, chloramphenicol resistance was transferred by conjugation from F$^+$ *E. coli* lysogenic for P 1 *cam* into *Salmonella typhi* (Kondo and Mitsuhashi, 1966). Most of the Cam^R Salmonellae thus derived proved to be defective lysogens: They were specifically immune to superinfection by P 1, but they liberated no plaque-forming particles, either spontaneously or following UV induction. Some of these defective lysogens produced no particles capable of *cam* transduction, even upon superinfection with wild-type P 1. However, Scott (1970) detected transducing activity on *E. coli* K12 in the supernatant of one such defective P 1 *cam* lysogen, without superinfection. Among the Cam^R transductants thus obtained, some carried a defective P 1 prophage, P 1 cryptic, inserted close to *proAB* and *lac*.

The P 1 cryptic prophage does not confer superinfection immunity. Its presence is detected by the formation of recombinants between the prophage and superinfecting P 1 phages marked with appropriate mutations. At one time, defective lysogens were generally distinguished from nonlysogens by the fact that the former, though not overtly lysogenic, continued to make repressor and therefore retained superinfection immunity (Jacob *et al.*, 1957). Defective prophages not conferring immunity were therefore given the special name *cryptic* (cf. Section 3.3.2a).

Figure 1 shows the gene order of P 1 cryptic prophage (as determined by deletion mapping) and that of P 1 phage (Scott, 1970, 1973). Two features may be noted: (1) The *cam*^{R} determinant is the terminal marker on the prophage. (2) With respect to the P 1 map, *cam*^{R} is in the position formerly occupied by the genes (*am* 2.17–*am* 2.14) that are absent from P 1 cryptic. These facts suggest that both the deletion of phage genes and the insertion into the chromosome were consequences of a specific ability of the *cam* determinant to translocate itself to new locations (and hence to break and rejoin DNA at the site of its own insertion, see Section 3.2).

A second P 1 derivative, P 1 *dtet*, bears some resemblance to P 1 *cry* both in its properties and in its mode of origin. P 1 *dtet*, like P 1 *cam*, was isolated following P 1 transduction of a drug-resistance determinant (in this case for tetracycline) from an R factor into a

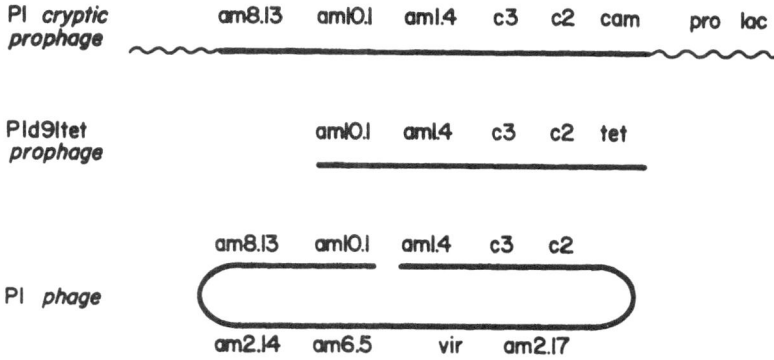

Fig. 1. Genetic maps of the defective prophage P 1 cryptic inserted into the bacterial chromosome (top), defective P 1 *d*91 *tet* (center), and wild-type P 1 phage (bottom). For ease of comparison, the linear P 1 map has been bent so as to align those genes common to P 1 and P 1 *cry*. Alleles of the mutations *am* 2.14, *am* 6.5, *vir*, and *am* 2.17 are absent from P 1 *cry* and P 1 *d* 91 *tet*. *am* 8.13 is present in P 1 *cry* but not in P 1 *d* 91 *tet*. The gene order of P 1 *cry* was deduced from deletion mapping, and that of P 1 by recombination frequencies. The gene order in P 1 *d*91 *tet* has not been determined. Presence of the indicated genes has been established by genetic marker rescue. Data from Scott, 1970, 1973, 1975.

nonlysogenic recipient. Unlike P 1 *cam*, P 1 *dtet* is a defective prophage with an extensive deletion of genes present in wild-type P 1. One terminus of the deletion is the same as in P 1 *cry* (except for the *cam* locus, which is not of viral origin, Fig. 1).

The exact genetic requirements for plasmid formation by P 1 or insertion by P 1 *cry* are not completely understood. Although lysogens of P 1 *dtet* are not immune to superinfection, the repressor gene *c*1 is present and expressed (because P 1 *dtet c*1 *ts* lysogens die at high temperature, Scott, 1975). The lack of superinfection immunity may result from the inability of P 1 *dtet* to regulate an antirepressor gene similar to that of phage P 22 (Botstein and Suskind, 1974).

2.1.1b. Staphylococcal Phage P 11 *de*

A defective prophage may be inserted either into the bacterial chromosome or into the DNA of a plasmid. Defective mutants of phage λ of the types of be described in Section 3, for example, can be derived from a prophage inserted either into the bacterial chromosome or into an F´ element that includes the λ attachment site. Where the genesis of a plasmid–phage complex entails loss of both viral and plasmid genes, it may be hard to distinguish a plasmid bearing an inserted defective phage from a defective phage that replicates as a plasmid. The critical difference lies in the genetic origin of the replication determinants used by the complex.

For example, an element P 11 *de* arose in *Staphylococcus aureus* by transduction of erythromycin resistance from a plasmid, called γ or *p* I258, by means of the generalized transducing phage P 11 (Novick, 1967). P 11 *de* multiplies as a plasmid, which, like γ itself, is incompatible with the presence of a second γ element in the same cell. This suggests that the γ replication machinery is used under these circumstances. The γ portion of P 11 *de* is incomplete, however, having lost some determinants for metal-ion resistance (cadmium and mercury) upon acquiring the P 11 genes. The P 11 moiety of P 11 *de* is manifested by recombination with superinfecting P 11 *ts* mutants to give P 11 *ts*⁺ recombinants. Cells carrying P 11 *de* are not immune to superinfecting P 11, and wild-type alleles of clear plaque mutants were not recovered by marker rescue. In fact, superinfection apparently induces extensive replication of the entire element as a phage. The phage replication origin is thus probably present but inactive in the absence of superinfection.

The erythromycin determinant of *p*I 258 can become inserted into the chromosome unaccompanied by most of the plasmid genome (Novick *et al.,* 1975). It thus seems to share the property of independent transposability with many plasmidborne drug-resistance determinants of the Enterobacteriaceae (see Section 3.2).

2.1.2. As Plasmids (with Specific Discussion of λ *dv*)

Whereas P 1 cryptic is inserted into the chromosome, the phage genome in most P 1 lysogens is maintained as a plasmid. Defective variants of P 1 can also be propagated as plasmids, provided that the genes for autonomous replication are intact.

All phage genomes can replicate autonomously. Generally, autonomous replication leads not to the perpetuation of the phage genome as a plasmid, but rather to destruction of the host cell and liberation of new virions. Plasmid formation requires controlled autonomous replication, unaccompanied by expression of viral genes that are lethal to the host. Certain phages, such as λ, though themselves incapable of plasmid formation, can generate mutants that are maintained as plasmids. The mutants are partially or totally defective. The special requirements for plasmid formation are comprehensible only with reference to the rather complex regulation of early genes in λ itself.

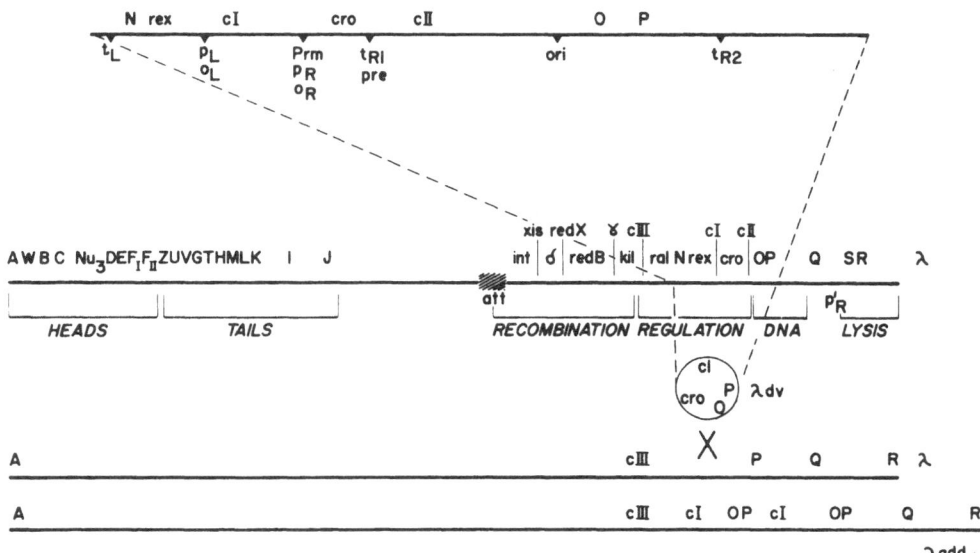

Fig. 2.   Top line: Expanded map of those genes of phage λ that may be present in λ *dv*. Genes (function) shown are N (extension of transcription starting at P_L and P_{R1} which otherwise would stop at t_L and t_{R1} or t_{R2}); *rex* (exclusion of T 4 *r* II), *cI* (repression of transcription starting at P_L and P_R, stimulation of leftward transcription from prm); *cro* (inhibition of transcription and from P_L, P_R, and prm), *cII* (stimulation of leftward transcription from pre), and O, P (DNA replication). Second line: Genetic map of λ showing some of its genes. Based in part of Szybalski, 1974. Third line: A circular λ *dv* monomer derived from λ. Fourth line: Homologous crossover between λ *dv* and λ. Fifth line: Addition phage produced by such homologous crossing over. (See Section 3.3.2.)

Autonomous replication of λ requires two phage-coded proteins, products of genes O and P (Fig. 2), which specifically promote replication initiated at the phage origin *ori* (Kaiser, 1971). Wild-type λ can follow two mutually exclusive pathways in which phage-directed replication is either rapid (as in infected cells about to lyse) or turned off almost completely (as in lysogenic bacteria).

In lysogenic cells, almost all viral genes are turned off, directly or indirectly, by the repressor protein coded by gene *cI* (Ptashne, 1971). This repressor combines with λ DNA at the operator sites o_L and o_R and prevents RNA polymerase from initiating transcription leftward from p_L or rightward from p_R. At the same time, repressor bound at or near o_R stimulates leftward transcription from a site (prm) very close to p_R (Eisen and Ptashne, 1971). The leftward transcript initiated at prm includes genes *cI* and *rex*. The *cI* product enhances its own rate of synthesis, while repressing genes whose expression is needed for virus production (see also Chapter 2 for detailed discussion).

An infected cell can be shunted into one pathway or the other by several viral regulatory products in addition to repressor itself (Echols, 1971). The first genes expressed are N (transcribed leftward from p_L) and *cro* (probably transcribed rightward from p_R). The N product interacts with RNA polymerase and/or promoters p_L and p_R so that transcription from these promoters becomes resistant to normal termination signals t_L, t_{R1}, and t_{R2} in the DNA template (Franklin, 1974; Adhya *et al.*, 1974). Thus, in the absence of N product, the transcripts from these promoters are very short, including (of known genes) only the messages for N on the left and *cro* on the right. Once N is present, longer transcripts are formed in both directions.

The DNA immediately to the right of *cro* includes, besides O, P, and *ori*, gene *cII*, whose product acts at site *pre* to activate leftward transcription of the *cI-rex* operon. The *cII* product is much more effective in the presence of the product of another gene, *cIII*, which lies to the left of the N gene. Genes *cII* and *cIII* thus cooperate to promote the establishment of repression. The *cro* product, which accumulates as lytic development proceeds, inhibits transcription from p_L, p_R, and prm (Echols, 1972; Takeda *et al.*, 1975; Reichardt, 1975a, b). It thus serves to regulate transcription both of *cI* and of the early genes in the lytic pathway.

In a general way, these several interactions can be understood as components of a switching mechanism that allows the infected cell to select either of the two possible pathways, leading toward lysis or lysogeny, respectively. A mathematically complete description in terms of all the relevant kinetic parameters is not yet possible. The whole complex of interactions renders the system responsive to the physical environment (e.g., the ion concentration and also the presence of glucose, as monitored by internal cyclic AMP formation; Grodzicker *et al.*, 1972), to viral population density (as monitored by the multiplicity of infection), and to the internal state of the system (e.g., whether or not there has been an opportunity to synthesize the proteins needed for inserting viral DNA into the chromosome; Echols, 1972).

For wild-type λ, the switch is not really bistable: Of the two alternative routes an infected cell may select, only the lysogenic pathway is stable in the sense that the system can perpetuate itself indefinitely in the same state. Choice of the lytic pathway, on the other hand, provokes a cascade of subsequent events that culminate in destruction of the cell and liberation of new virions. Thus, λ DNA does not establish itself as a plasmid like P 1 that replicates in harmony with the host.

Stable expression of lytic pathway genes has been observed under three circumstances: (a) In an inserted prophage those viral genes responsible for cell killing may have been deleted or inactivated by mutation, or may fail to be transcribed because of regulatory mutations. Cellular death can be caused by the product of a gene *kil* located to the left of *cIII* (Greer, 1975), or by *in situ* replication of viral DNA. Thus, if one mutation has inactivated either *kil* itself or the N gene (whose product is needed to extend transcription beyond t_L) and a second mutation destroys either O, P, or *ori*, then transcription from p_L and p_R can take place without killing the majority of the cells. Such cells can survive under conditions where repressor is permanently inactivated (by growth at high temperature when the prophage carries a *ts* mutation in the *cI* gene); in fact, they were originally selected as thermoresistant survivors of *cI ts* lysogens, and their genetic basis was later determined. Furthermore, because the *cro* product inhibits spontaneous transcription from prm, cells that have been heated to destroy all repressor remain derepressed for many generations even after they are shifted back to low temperature. (b) Although derepression of integrated N^- O^+ P^+ prophages can kill the cell, in some laboratories some such lysogens survive derepression. When repressor is inactivated by heat, these strains are converted to a new growth habit, in which the inserted provirus is amplified to many copies, probably by *in situ* replication; and the normal relationship between cellular growth and septum formation is deranged so that filaments form. This converted state has the same long-term stability as does the nonimmune state of N^- O^- or N^- P^- lysogens (Cross and Lieb, 1967; Lieb, 1972). It is not known why *in situ* replication is lethal in some circumstances and merely deleterious in others, but the underlying regulatory interactions are doubtless the same. (c) Where gene N has been damaged by mutation but O and P are still intact, λ can also replicate as a nonintegrated plasmid, without killing the host. Genes O and P are expressed at a low rate, whereas both the *kil* gene of the leftward operon and the late viral functions causing virion formation and cellular lysis are turned off completely.

Plasmid formation is observed with N^- mutants of λ (Signer, 1969), as well as with a variety of deletion mutants (λ *dv*'s) that have lost not only N but 70–90% of the viral genome as well (Matsubara and Kaiser, 1968: Fig. 2). For plasmid formation, genes O and P must be functional to mediate autonomous replication. Transcription of the *cro OP* operon also activates initiation of replication from *ori* (Dove *et al.*, 1971). The *cro* product is also needed, probably to reduce rightward transcription from p_R (Berg, 1974a). The *cro* gene thus prevents excessive replication by acting as a governor on the operon to which it belongs. Neither N product nor repressor is needed; λ dv's can be formed from cI mutants, and most λ dv's lack the N gene altogether (Berg and Kellenberger-Gujer, 1974). Whereas λ dv's derive their name ("defective virulent") from the presence of three operator mutations ($v2$, $v3$, vl) that render the parent phage insensitive to repression, none of these mutations is needed for plasmid formation (Matsubara, 1974); rather, the name implies that this is a type of defective variant of a virulent mutant that can be maintained as a prophage.

Natural plasmids fall into two classes with respect to replication control (Arai and Clowes, 1975). In some (such as F and P 1), control is *stringent*. A small number (about one per bacterial genome) of plasmid genomes are present in each cell. During each division cycle, each plasmid genome replicates exactly once. At cell division, daughter copies of the plasmid segregate from one another so that each daughter cell receives one copy of the plasmid, just as it receives one copy of the chromosome. The precision of this process is such that P 1 prophage is accidentally lost from the cell only once in 10^5 cell divisions (Rosner, 1972). Such a plasmid behaves like an extra chromosome of the host.

The other class of plasmids (which includes the colicin E 1 factor) exhibit *relaxed* control. Generally, several plasmid copies are maintained for each bacterial chromosome. During steady-state growth, each plasmid copy must, on the average, replicate once per division cycle; but a particular molecule may fail to replicate at all or may replicate more than once in a single cycle. Relaxed plasmids need not segregate in any regular manner at cell division. Because many copies are present, it is likely that each daughter cell will receive at least one. If we liken stringent plasmids to accessory chromosomes, relaxed plasmids are more analogous to cytoplasmic factors.

Those λ variants forming plasmids resemble relaxed plasmids at least in maintaining many copies at the steady state. A precise theory relating the concentrations of regulatory molecules, replication proteins, and DNA target sites to number of plasmid copies at the steady state is not yet available. λ dv carriers typically contain about 50 monomer equivalents of λ dv per cell (Matsubara, 1972). In rec^+ cultures, these are found as a mixture of monomers and higher oligomers. The monomer–oligomer transitions seem to be largely recombinational, because in $recA^-$ hosts each carrier line has one predominant molecular species (Kellenberger-Gujer *et al.*, 1974; Hobom and Hogness, 1974). Carriers lose λ dv, to become nonlysogens, at a low spontaneous rate (10^{-3} per generation or less). Loss is accelerated in stationary phase cultures and by treatment with acridine orange. Carriers of λ N^- plasmids seem to be more unstable (Signer, 1969).

λ phage fails to form plaques on λ *dv* carriers. This exclusion seems to have a dual basis: inhibition by *cro* product, and competition for replication proteins or other substances (Matsubara, 1972*b*). Among λ-related phages, those like λ *imm* 21 that share with λ neither the specificity of repression nor that of replication grow well on λ *dv* carriers. Those with a different *cro* product but the same replication genes (such as λ *imm* 434) are inhibited, but less so than λ itself. A high concentration of *cro* perhaps accounts also for the low rate of transcription of the *cI* operon (Kumar and Szybalski, 1970) and the failure of *N* gene expression by a superinfecting homoimmune phage (Matsubara, 1972*a*; Takeda *et al.*, 1975).

In summary, plasmid formation by λ contrasts with that by P 1 in two important respects. (1) First, P 1 has two alternative modes of autonomous replication—one used in the productive cycle of infection, the other in the plasmid state—and some sort of regulatory switch allowing stable commitment to one mode or the other. In λ, on the other hand, the normal alternatives are unrestrained autonomous replication or complete repression. The stable plasmid state results from commitment to the productive pathway in certain mutants that are unable to follow through with steps of that pathway that kills the cell. The basic requirement is a controlled rate of replication, which is achieved by the expression of genes *O* and *P* at a rate governed by the concentration of *cro* product and reduced by the absence of *N* protein. (2) Second, whereas P 1 replication is stringently coordinated with cell division, that of λ plasmids is relaxed.

The structure of λ *dv*'s and their possible mode of origin from λ DNA will be discussed in Section 3.3.2.

2.2. Propagation by Mixed Infection with Complete Phage

If a viral mutant cannot form plaques because of failure to elaborate some diffusible product(s), a cell infected with both mutant and wild type should liberate a mixture of the two—because the wild-type genome can supply the mutant as well as itself with substances the mutant lacks. A lysate produced by this method is a physical mixture that must be separated to study the defective variant itself. Thus, this is seldom the method of choice for simple stockkeeping. Most efforts have been concentrated on more tractable systems.   ·

When a defective mutant has been propagated as a prophage, it is sometimes of interest to know whether infectious but genetically defective virions can be formed by complementation with wild type, and what the physical and biological properties of these particles are. This is generally accomplished by superinfecting a derepressed defective lysogen with wild-type virus. In an early study of defective variants of λ, Appleyard (1956) demonstrated the existence in such lysates of particles that could infect and lysogenize a sensitive host to generate new defective lysogens. Jacob *et al.* (1957) extended these observations to many other mutants, and could monitor defective particles by their ability to kill cells, even where lysogenization in single infection was undetectable.

Whereas some of the biological properties of defective particles can be surmised by infecting appropriate hosts at various multiplicities with such a mixed lysate, for other purposes physical separation is desirable. This is readily achievable with a phage like λ that can give rise to viable variants with different DNA lengths and therefore different densities. The complementing phage can thus always be chosen to have a density far removed from the defective variant under study. With other phages such as P 22 or T 4, which use a headful packaging mechanism (Streisinger *et al.*, 1967), this option is not available.

2.3. Conditional Lethals

2.3.1 Conditionality Dependent on Nature of the Mutant Allele

A conditional mutation confers on its bearer a mutant phenotype under some (nonpermissive) conditions and not under other (permissive) ones. If the mutant phenotype is inviability (which in phage means inability to form a plaque), the mutation is conditionally lethal. Under nonpermissive conditions, a conditionally lethal phage mutant is a defective phage.

Several choices of permissive and nonpermissive conditions allow isolation of general classes of conditional mutants that can occur in almost any gene. Thus, systematic characterization of large numbers of such mutants yield a fairly exhaustive catalog of genes required for viability: (a) Supressible mutations inactivate gene function except in hosts carrying translational suppressor mutations. The commonest types are those in which mutation has generated in the messenger RNA a nonsense codon (UAA, UAG, or UGA), which a normal cell reads as a termination signal for translation. The suppressor mutation changes the anticodon of some transfer RNA molecules so that it now recognizes one of the nonsense codons as sense. Depending on the codon corrected, suppressible mutations of this type are termed ochre (UAA), amber (UAG), or azure (UGA). (b) Thermosensitive (*ts*) mutations allow gene function at low, but not high, temperature. Frequently, this is because the mutant protein is less thermostable than wild-type protein. (c) Cryosensitive (*cs*) mutations allow gene function at high, but not low, temperature. Sometimes this is because aggregation of mutant monomers fails at low temperature (Guthrie *et al.*, 1969).

In the first comprehensive study of defective mutants of phage λ, Jacob *et al.* (1957) analyzed a number of defective lysogens for the nature of the physiological block in phage development, as well as the approximate location of the responsible mutation on the genetic map. Subsequent studies of conditional lethals largely supplanted this line of endeavor because of the much greater facility with which conditional lethals can be handled in mapping, complementation, and production of pure virions. The results of such endeavors in phage λ are summarized in Fig. 2, where the capital letters designate genes needed for viability.

All mutations categorized above influence protein structure. Thus, they can provide representatives of the structural genes but do not include alterations in DNA target sites, such as promoters or replication origins. Such mutations have been obtained as unconditional lethals, propagated as prophages. An example is the t_{11} mutation in the major rightward promoter p_R of λ (Eisen *et al.*, 1966). This mutation eliminates rightward transcription initiated at p_R. Derepressed lysogens of λ t_{11} form none of the products of the *cII O P Q* operon and cannot complement superinfecting phages with point mutations in these genes (Brachet and Green, 1970).

2.3.2. Conditionality Dependent on Nature of the Mutated Function

In various special cases, a viral gene is essential for growth in certain circumstances, but not in others. Benzer's (1957) classical studies of T4*r*II mutations exploited the fact that these mutants do not form plaques on K (λ) hosts, whereas wild-type T 4 does. For reasons still not completely understood, the products of the two *r*II genes are needed for T 4 growth in the presence of the λ *rex* gene product. Similarly, in order to grow on *recA⁻* hosts, phage λ must have either the *redX* and *redB* genes, or else the γ gene (Fig. 2). Apparently a λ replication intermediate is sensitive to exonuclease V of the host (Skalka and Enquist, 1974). The intermediate can be circumvented by recombination (catalyzable either by the *rec* genes of the host or the *red* genes of the phage). In the absence of recombination, phage survives only if exonuclease V is inhibited by the γ protein.

The resemblance between these situations and either the first type of conditional lethal or unconditionally defective viruses may seem superficial, and their inclusion under the same heading needlessly formal. However, every real study of defective variants must be carried out under arbitrarily chosen conditions. Hence the variants obtained may well include some types that would be viable under other circumstances.

Moreover, among thermosensitive or cryosensitive mutants one expects not only missense mutants that render essential proteins conditionally unstable, but also mutations that inactivate completely any proteins that are needed for growth only under the nonpermissive conditions picked. A possible case of this type was discussed by Scott (1970). All the mutations found in the $c3$ gene of phage P 1 are lethal at 30°C and confer a clear plaque phenotype at 42°C. Whereas genes with products essential at only one temperature are probably rarer than those essential at all temperatures, a larger fraction of the mutations occurring at loci of the former type score as conditional lethals. Hence their representation in any mutant sampling will be disproportionately high.

2.3.3. Host Genes Required for Viral Development

One method for probing viral–host interactions is the isolation of host mutations that block viral development. The classical selections for resistance to virulent phages (Luria and Delbrück, 1943) yielded mainly cell lines deficient in the very early steps of surface interaction and adsorption. Study of viral mutants able to infect these host mutants has helped to define the viral genes whose products interact directly with the host envelope. Screening for bacterial colonies that are less damaged by virus than wild type (even though not fully resistant to the high virus concentrations used in earlier selections) can yield host mutants blocked at later stages in the infectious cycle. Phage mutations that reverse the block help define those phage proteins that interact with host gene products at the specific stage at which development is arrested. Insofar as the designation of particular host and phage genotypes is arbitrary, either host or phage can be considered defective where infection is abortive.

DNA replication of the single-stranded virulent phage ϕX174, the double-stranded temperate phage P 2, and certain other temperate phages is blocked by mutations (rep^-) mapping at 75 min on the E. coli chromosome (Calendar et al., 1970). The rep^- mutations have little effect on the uninfected host. Certain mutations (*groP*) in or near the *dnaB* gene of *E. coli* inhibit bacterial replication at high temperature and prevent replication of λ DNA even at low temperature. Certain mutations in the λ replication gene *P* allow λ to replicate on *groP* hosts (Georgopolous and Herskowitz, 1971).

Mutations in a gene (*groE*) mapping at about 84 min on the *E. coli* chromosome derange assembly in a variety of coliphages, including λ, T 4, T 5, φ80, and 186. λ head assembly is arrested in a *groE* host at the same stage as λ *B⁻* or λ *C⁻* mutants in wild-type cells (see Fig. 11). Mutations reversing the *groE* phenotype can take place in either the λ *B* gene or in the major structural gene *E* (Sternberg, 1973; Georgopolous *et al.*, 1973). The effects on T 4 assembly of similar (perhaps identical) host mutations are reversed by certain mutations in T 4 head gene 31 (Coppo *et al.*, 1973).

In hosts bearing mutations (called *nusA*), mapping at 61 min on the *E. coli* chromosome, wild-type λ behaves like an *N⁻* mutant (Friedman, 1971; D. I. Friedman and M. F. Baumann, private communication). Another mutation, called *groN* (Georgopolous, 1971), has a similar phenotype but has not yet been completely mapped. λ mutations reversing this phenotype lie either in the *N* gene itself or between genes *P* or *Q*. The latter region contains one of the sites of transcription termination in the absence of *N* product and is the locus of mutations that allow viral growth in the absence of *N*. Certain *E. coli-Salmonella* hybrids exhibit a phenotype similar to that of *nus* mutants (Baron *et al.*, 1970). Another host mutation (*ron*), apparently within the structural gene for the β-subunit of RNA polymerase, at 79 min on the *E. coli* map, has an effect similar to *nusA* and *groN*, but only on λ phage carrying certain natural or spontaneous variants of the *N* gene (Ghysen and Pironio, 1972). When coliphage P 2 infects *E. coli* cells bearing a recessive mutation (*gro₁₀₉*) at 64 min on the bacterial map, normal transcription of late genes fails (Sunshine and Sauer, 1975). The phenotype thus mimics that of a λ*Q* mutant on a wild-type host (Fig. 2). P 2 has no known gene analogous to the *Q* gene of λ. However, mutations of P 2 (called *ogr* for "overgrow") lying between *D* and *att* (Fig. 12) allow normal late function on *gro₁₀₉* and may represent such a gene.

2.4. Nonpropagable Variants

As indicated in Section 1, certain viral variants of genetic interest cannot be propagated as such because their linear DNA molecules lack the proper termini.

2.4.1. λ Particles Lacking One Cohesive End

The DNA of bacteriophage λ is double-stranded except for 12 bases at the 5′ ends of the two polynucleotide chains, which are complementary to each other. A large body of indirect evidence supports the general scheme of viral reproduction shown in Fig. 3. Phage DNA is arranged in the virion with the right (R) end in close proximity to the phage tail, and the left (A) end somewhere within the head (Chattoraj and Inman, 1974). Following injection, these two ends are joined by polynucleotide ligase to form a covalently closed ring. At least the first round of replication of these rings is by way of theta forms. At some later stage, the mode of replication changes to one of the rolling-circle type which spins out multigenomic DNA sequences of indefinite length. These concatemers constitute the preferred substrate for DNA packaging, the molecular ends of the virion DNA being generated by cutting successive cohesive site (*cos*) sequences in the concatemer. (The name

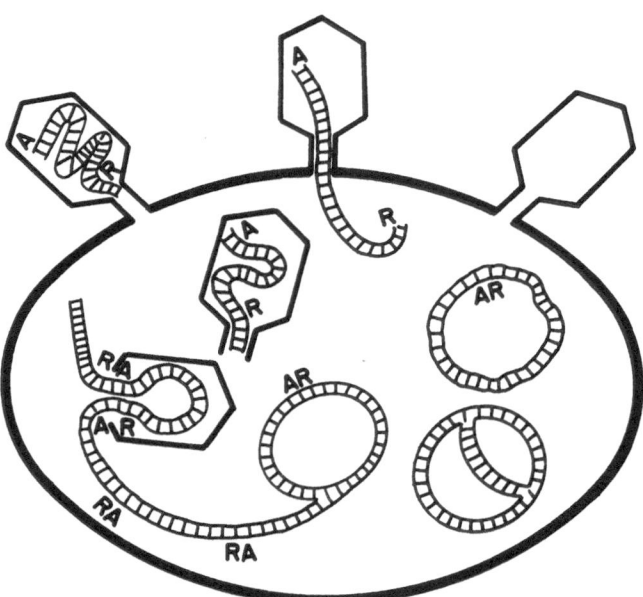

Fig. 3. Schematic diagram of λ infectious cycle. Successive stages (clockwise from 10 o'clock): Attachment of virion, injection of DNA, circularization by joining of cohesive ends, theta-form replication, rolling-circle replication, and packaging of new virions. For simplicity, the DNA cutting that accompanies packaging is shown as though both *cos* sites bracketing the packaged genome were juxtaposed near the site of subsequent tail attachment. The DNA fiber might instead pass through the head, entering and leaving at two different points.

cohensive site derives from the tendency of λ DNA to cohere because of intermolecular hydrogen bonding between complementary ends. The term *cos* refers to a specific base sequence that is recognized and cut during packaging to generate single-stranded ends.)

As both molecular ends are needed to form covalent circles, a molecule lacking one or both of the normal ends should be unable to replicate. By mechanisms not fully understood such molecules are occasionally packaged, however (Little and Gottesman, 1971). These are detectable where normal development is blocked and where the DNA thus packaged includes host genes not present in the typical virus particle. For example, derepression of bacteria lysogenic for λ mutants deficient in excision of phage DNA from the chromosome results in cellular lysis accompanied by production of very few infectious particles. These include some defective particles detectable by their ability to transduce bacterial genes close to the prophage insertion site. Many of these contain molecules with only one of the two cohesive ends (Fig. 4). Such defective particles have been called λ *doc* (for defective, one cohesive end).

Particles (*docR*) with only the right cohesive end include those viral genes between *int* and *R*, and also host genes such as *gal* that lie to the left of the λ insertion site. They can attach to bacterial cells and inject their DNA into them. This DNA can then recombine either with the bacterial *gal* operon or with a coinfecting, structurally normal λ. However, even with a coinfecting phage there is no way that the λ *docR* structure can be reproduced. Its ends cannot join, nor can it become inserted as such into any of the intracellular forms of λ DNA shown in Fig. 3.

The reciprocal class of defective particles (λ *docL*) have only the left molecular end and include viral genes *A* through *J* and bacterial genes such as *bio* that lie to the right of the prophage. Unlike λ *docR*, these particles cannot inject their DNA. In fact, they do not even have tails. Some DNA, presumably from the right end of the molecule, protrudes from the head and is digestible by DNase. DNase treatment of lysates containing *docL* particles produces heads that can now join to tails *in vitro* (N. Sternberg and R. Weisberg, private communication). Like λ *docR*, λ*docL* genomes cannot reproduce as such, even in the presence of helper phage.

Whereas detection of λ *doc* particles is facilitated by conditions where normal phage production is reduced, they probably occur in lysates of excision-proficient phages as well. Such lysates produce a class of transductants in which bacterial genes have replaced their homologs (as in λ *doc* transduction) rather than adding to the genome

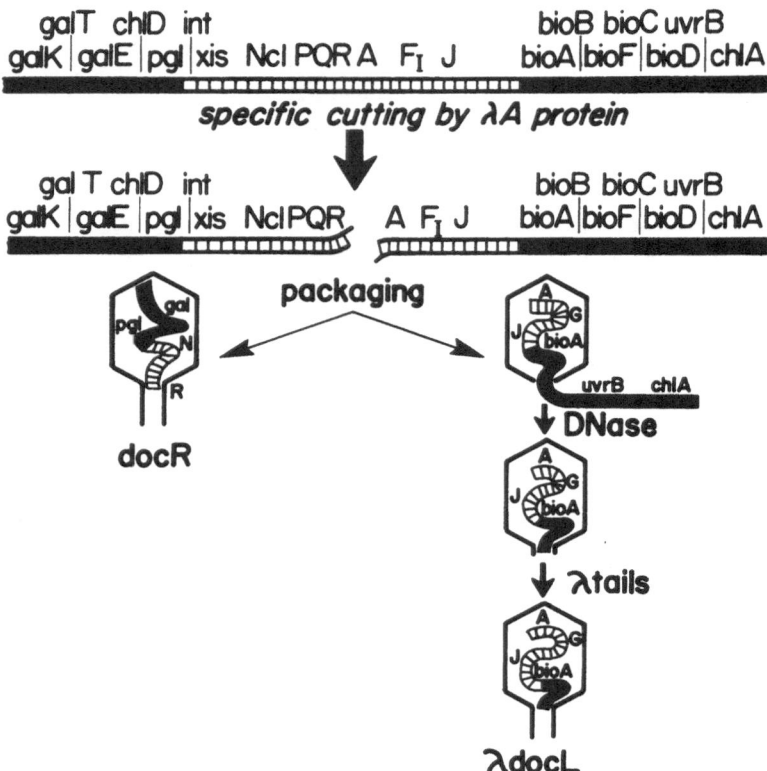

Fig. 4. Formation of λ *docL* and *docR* particles. DNA is packaged directly from the chromosome of the lysogenic bacterium. DNA cutting by the λ *A* protein (Wang and Kaiser, 1973) is shown as a separate step, although in reality it probably accompanies packaging (cf. Fig. 3 and Wang and Kaiser, 1973). It is not known whether a *docL* and a *docR* particle can both be formed from the same chromosome, as indicated here. In λ *docR* packaging, bacterial DNA is somehow broken to give a molecule of packageable size. λ *docL* molecules are too large to package. Excess DNA must be digested away before tails can be added.

in the manner of the specialized transducing phages to be discussed in Section 3.3.1 (Kayajanian, 1970).

2.4.2. Incomplete Genomes Created by Headful Packaging in T 4 and P 22

Incomplete genomes with a similar transient existence occur within smaller than normal particles that constitute a minor fraction of the virions formed in infection with phage T 4. Unlike λ, T 4 DNA lacks specific termini. It is apparently packaged from a concatenate precursor. The origin of the DNA of an individual virion may lie anywhere along the concatemer. Given a particular origin, the terminus is determined by the amount of DNA that fits into the phage head (Fig. 5). The average packaging length exceeds the genome length by about 2%, so that a typical molecule of packaged DNA contains the same base sequence repeated in double-stranded form at its two molecular

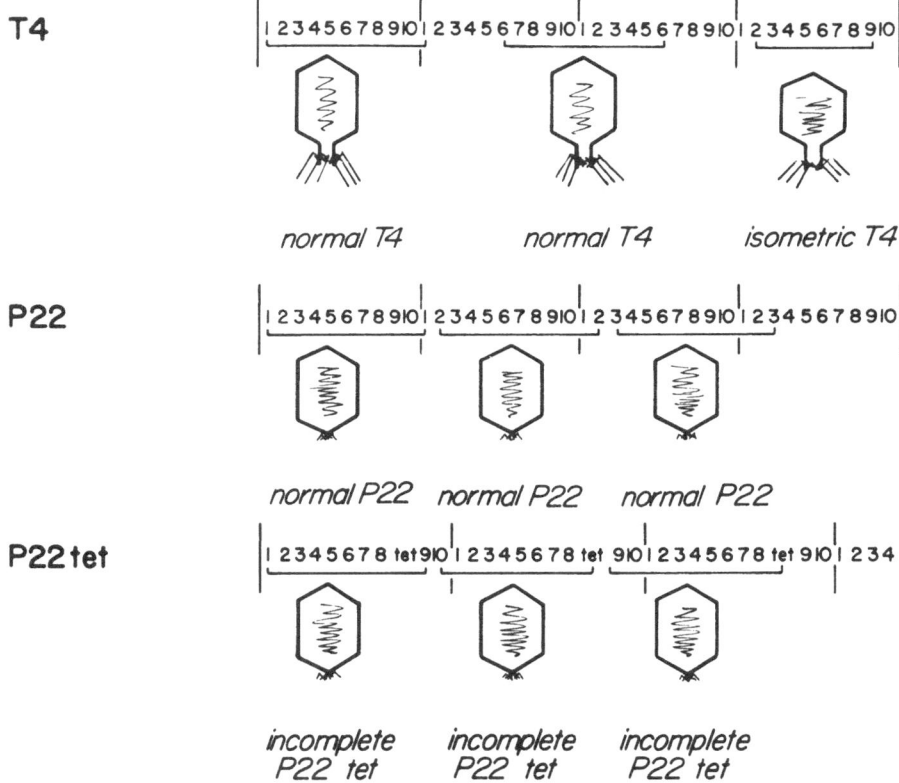

Fig. 5. Headful packaging in phages T 4 and P 22. T 4 heads are generally elongated and include more than one genome length of DNA. Shorter (e.g., isometric) heads contain less than a full genome and thus are defective. Headfuls appear to be chosen from random partners on the concatemer. P 22 packaging is similar, but in P 22 packaging starts from a fixed point, with subsequent headfuls becoming progressively out of phase with the original one. P 22 *tet* is a derivative of P 22 into which a tetracycline-resistance determinant has been inserted. The mechanics of packaging are the same as in P 22, but, because of the insertion, a headful is less than one genome length of P 22 *tet* DNA. Hence only defective particles with incomplete genomes are packaged.

ends. Whereas the typical T 4 capsid is a deformed icosahedron the linear dimension of which along the axis of the tail exceeds its lateral dimensions, some virions deviate from the normal length, apparently by adding or subtracting rows of capsomers, to give particles that are lengthened or shortened compared to wild type (Walker *et al.*, 1972). Certain mutations in gene 23, which determines the major capsid protein, augment the frequency of such abnormal particles (Doermann *et al.*, 1973). The DNA length packaged in the shortened heads is less than a complete genome. Thus, each such molecule lacks some genetic information, and all such molecules lack the terminal redundancy of normal T 4.

The role of terminal redundancy in viral development is uncertain. Recombination between the two ends may generate covalently closed rings, as in λ; or recombination between ends of linear replicas may generate concatemers. Some evidence indicates that incomplete genomes, unable to circularize, can nonetheless replicate extensively as such, provided that all gene products required in replication are available (Doermann, 1973). In any event, even in cells coinfected with wild-type phage, there is no reason to suppose, and no data to suggest, that an incomplete infecting particle ever engenders a clone of packaged progeny genomes with the genetic constitution of the parent. It is thus not feasible to prepare stocks, and the genotype of an individual can be surmised only from the recombinants it generates with a multiply marked coinfecting phage.

Virions in which the packaged length is less than a complete genome can arise in two ways: (1) As just discussed, genotypic or phenotypic variation may allow occasional packaging of DNA lengths shorter than the wild-type genome. (2) Alternatively, the packaged molecules may be of normal length, but the genome length may have been increased by addition mutations such as insertions or duplications. The latter situation obtains in lysogens of phage P 22 where a tetracycline-resistance determinant of plasmid origin has been inserted into the provirus (Tye *et al.*, 1974*a*). Here the entire lysate behaves in some respects like the small particle fraction of T 4. No individual virion can infect a cell and engender progeny; however, two or more virions can cooperate to do so (Chan *et al.*, 1972).

As in T 4, this cooperation does not allow cloning of the individual defective types, which must recombine to regenerate the original genome in order to replicate. Unlike T 4, the genome formed by recombination, being oversized, is still unpackageable. Thus, in each infectious cycle the whole spectrum of defective particles is regenerated. Plaques formed by such cooperative infection contain no virions that are individually capable of further propagation.

Like T 4, phage P 1 forms some small particles with shortened DNA molecules (Walker and Anderson, 1970). Packaging of segments from an oversized genome has been reported for a P 1 derivative that has acquired the *lac* operon of the host (Stodolsky, 1973).

Whereas the T 4 small particles comprise random genetic segments (Parma, 1969), the genotypic distribution of P 22 particles (as judged from electron microscopy of DNA heteroduplexes) is highly nonrandom, as though packaging always begins from a specific point on the genome, successive molecules being packaged processively from the concatemer (Tye *et al.*, 1974*b*; Fig. 5).

3. GENESIS OF DEFECTIVE VARIANTS FROM ACTIVE PHAGE

In terms of the DNA change from wild type, known defective mutations fall into four main categories: point mutations, insertions, deletions, and substitutions. Mutants of each type have their special utility.

3.1. Point Mutations

Point mutations have been induced by various mutagens in growing phages (to yield conditional lethals) or in prophages (yielding either conditional or absolute defectives). Mutants of the latter type (mostly induced by ultraviolet light) have been characterized extensively in phage λ by Fuerst and collaborators (Jacob *et al.*, 1957; Eisen *et al.*, 1966; Mount *et al.*, 1968). Complementation and mapping of absolute defectives have been accomplished by their interaction with conditional lethals. Deletion mapping is also possible, especially by use of transducing phages (Section 3.3; Campbell, 1968).

A point mutation behaves in crosses as though DNA had been altered at a single point, rather than throughout a larger segment. The known molecular changes that produce this behavior are substitution or deletion of a single base and addition of one or more bases. A point mutation alters the structure of at most a single gene. With some exceptions (polar effects), the direct consequences of a point mutation within a gene are limited to deranging the activity of that gene alone. They are thus the most useful type of mutation for discovering individual gene functions.

As the exact molecular nature of a mutational alteration is seldom known with certainty, the category of "point mutations" is operationally useful. Because of their special properties, insertions of some specific DNA sequences will be considered separately from other point mutations.

3.2. DNA Insertions

Several genetic elements can insert their DNA into other DNA molecules (such as the bacterial chromosome) in a large number of locations. Insertion within a gene inactivates that gene and frequently prevents transcription of genes distal to it in the same operon. Bacteriophage Mu is one such element. It has a DNA length of about 40,000 bases and appears able to insert between any two nucleotides of host DNA. Under special circumstances, insertion can be reversed to regenerate the original gene that was interrupted (A. I. Bukhari, private communication). Other insertion sequences (Is1, IS2, IS3) are much smaller (800–1400 bases), have no known autonomous phase, and are indigenous to *E. coli* (Starlinger and Saedler, 1972). Insertion of these elements is nonrandom, but a given element can insert at least in several sites within a single gene. The insertions revert readily. Excision is sometimes precise but may also remove bacterial DNA adjacent to the insertion site (Reif and Saedler, 1975). Neither insertion nor excision requires the recombination genes of the host.

Besides being able to insert and excise their own DNA, the IS sequences are involved in insertion and translocation of other DNA segments with which they become associated. These include not only the interactions between the bacterial chromosome and the conjugation factor F (see Section 3.4.2), but also the translocation of individual drug-resistance determinants among plasmids and between plasmids and chromosomes. For example, the tetracycline-resistance determinant of the R 6 plasmid (whose insertion into P 22 phage was discussed in Sec. 2.4) is flanked by two copies of the IS3 elements, in inverted orientation with respect to each other (Ptashne and Cohen, 1975). This segment IS3-tet-(IS3)′ is inserted and excised as a unit. The precise mechanism is not understood, but such *rec*-independent translocation of drug-resistance determinants associated with insertion sequences seems to be widespread. Inverted repetitions likewise are found at the site of specific insertions of one plasmid into another (Kopecko and Cohen, 1975).

Defective λ lysogens have been isolated in which Mu, IS1, or IS2 has inserted into the provirus (Toussaint, 1969; Fiandt *et al.,* 1972). Because these insertions are generally completely polar on distal genes, they are especially useful in probing the organization of the virus into operons. The polar effect of IS2 is confined to insertion in one orientation. Insertion in the opposite direction may in fact have a strong promoter effect on distal genes—as though transcription from a promoter within the element extended into adjacent DNA (Saedler *et al.*, 1974).

Besides inactivating gene expression, insertions can make the viral genome too large to package. In a phage with a headful packaging mechanism, this generates a collection of incomplete particles (see Section 2.4.2). Phage λ is normally packaged to include DNA between two *cos* sites, so that DNA from oversized genomes is rarely packaged in infectious form.

3.3. Deletions

3.3.1. Simple Deletions

Deletion mutations that remove portions of an inserted prophage are found among bacteria selected for mutational loss of genes close to the prophage. Phage φ80 of *E. coli* inserts close to a locus (*tonB*) that causes formation of the surface receptor for the virulent phage T 1. Loss of *tonB* thus renders a cell line T 1-resistant. Among T 1-resistant mutants, some have experienced deletions that include DNA adjacent to the *tonB* locus itself. Starting with bacteria lysogenic for φ80, some of the resulting *tonB* deletions penetrate the prophage to various extents (Franklin *et al.*, 1965).

Similar techniques have been applied to other prophages. Deletions entering λ prophage from the left can be selected as resistant to chlorate (deletion of *chlD*) or galactose (deletion of the *gal* promoter from *galT* or *galE* mutants that are galactosemic because they accumulate galactose-1-phosphate). Deletions entering from the right are selectable by chlorate resistance (deletion of *chlA*, Fig. 4). *E. coli* is sensitive not to chlorate itself but to its reduction product, chlorite, whose formation is catalyzed by nitrate reductase. Mutations in *chlD* and *chlA* reduce nitrate reductase activity.

Deletions internal to the prophage are selectable as thermoresistant variants of lysogens whose prophage carried a *cIts* mutation that renders repressor thermolabile. Raising the temperature induces virus production and cell death in a *cIts* lysogen. Thus, the only individuals that can originate colonies at high temperature are those from which all or part of the prophage has been removed or inactivated. As cell death is caused by the products of two different noncontiguous genetic regions (Section 2.1.2), only a deletion can inactivate both in a single event.

In a simple λ *cIts* lysogen, the most common deletion is loss of the entire prophage, catalyzed by the products of phage genes *int* and *xis*. Internal prophage deletions are readily selected only where whole prophage loss has been minimized by using *int* or *xis* mutants, prophages from which one terminus has been eliminated by prior deletion, or prophages at abnormal sites (see Section 3.4.2d) from which specific excision is rare.

Prophage deletions are useful in genetic mapping. Together with specialized transducing phages (see Section 3.4), they provide the most reliable methodology for ordering point mutations. A second use has been the detection of sites on DNA that interact with specific proteins. Even where no point mutations are available to identify an operator or promoter, for instance, it can be located among known genes by examining which deletions derange or prevent normal transcription. Herskowitz and Signer (1970) showed that transcription of all λ late genes (*S* through *J*) was prevented by deletions that removed the DNA between *Q* and *S*, and permitted by deletions leaving this region intact. Thus, late transcription depends on a site between these two genes. Late genes on a molecule lacking such a site are not transcribed. Likewise, Shimada and Campbell (1974) located a weak internal promoter for integrase by surveying internal prophage deletions for transcription starting from this promoter and extending into the neighboring bacterial DNA.

3.3.2. Deletions Accompanied by Rearrangements

3.3.2a. λ Cryptic

In lysogens, as in nonlysogens, deletions occur at a low frequency (10^{-5} to 10^{-7} per cell), and their termini are widely distributed, though not completely random. Lysogens of a particular λ variant (called cryptogen, *crg*), when exposed to very high doses of ultraviolet light, repeatedly generate deletions removing a particular segment of the prophage (Marchelli *et al.*, 1968). The deleted segment extends from the left prophage end, or nearby, to the *cos* site between genes *A* and *R* (Fischer-Fantuzzi and Calef, 1964). More rarely, λ *crg* lysogens also give rise to deletions that start near the left prophage end and extend through gene *P*, but leave genes *Q-J*.

Because these defective lysogens have lost the *cI* gene that codes for repressor, they are not immune to superinfection and hence were given the name λ cryptic (cf. Section 2.1.1a).

Whereas most λ cryptics have lost the same segment of viral

DNA, they do not always arise by simple deletion. Frequently a chromosomal rearrangement accompanies deletion, as indicated by the fact that the bacterial genes *gal* and *bio* that flank the λ prophage lose their ability to be cotransduced by phage P 1 when the cryptic is formed. In some isolates *gal* and *bio* remain linked. Bacterial conjugation experiments indicate that, where linkage has been lost, various rearrangements are present in different isolates (Zavada and Calef, 1969). The determinant for cryptic formation that distinguishes λ *cry* from λ maps between *att* and *cI* and is unexpressed in λ *cry imm* 434 recombinants (Adhya and Campbell, 1970). λ *cry* is also deficient in normal excision from the chromosome. The precise mechanism of cryptic formation remains to be elucidated. Some of the properties of λ *crg* are what we might predict for a phage in which an insertion sequence of the type described in Section 3.2 had inserted in the neighborhood of the *xis* gene.

3.3.2b. λ *dv*

λ *dv*'s are plasmids that include some of the regulatory and replication genes of λ. λ *dv*'s arise spontaneously from λ at a frequency, which has been measured for the λ derivative λ *gal8 v3 vI*, of 10^{-6}–10^{-7} (Berg, 1974*b*). Their frequency is increased by ultraviolet irradiation of the phage prior to infection. Various isolates include segments of different extents, as determined both by physical measurement and by genetic marker rescue.

The size distribution of λ *dv*'s obtained from parents of various genotypes is shown in Table 1. These sizes were deduced from the

TABLE 1

Size Distribution of λ *dv* Isolates

Phage parent	Size relative to λ[a]				
	<11%	11–14%	14–17%	17–27%	>27%
λ *gal8 v2 v3 vI*	50%	41%	4%	4%	2%
λ *gal8 (att-red)* del *v2 v3 vI nin5*	93%	0%	0%	7%	0%
λ *gal8 v2 v3 vI nin5*	16%	11%	11%	64%	0%

[a] The size measured is that of the fundamental unit that can add by homologous recombination to a λ chromosome, inferred from the ability to produce viable addition phages (Fig. 2, last line) in conjunction with tester λ stocks of various lengths. Data from Berg (1974*b*).

ability of the λ *dv* to add by recombination to various deletion derivatives of λ (Fig. 2, bottom line), yielding an addition phage of packageable length (i.e., less than about 109% the length of wild-type λ, see Section 3.4.2a). The high frequency of large λ *dv*'s obtained from λ *gal8 v2 v3 v1 nin5* (where *nin5* is a deletion of some DNA between genes *P* and *Q*) results from the fact that most of these isolates prove to have inverted repeat sequences (Chow *et al.*, 1974; Fig. 6). The two monomer components of a given inverted repeat are usually not of identical extent. λ*dv*'s with inverted repeats can form oligomers, the fundamental repeating unit of the oligomer being (as in insertion) the amphimer rather than the monomer.

The role of the *nin5* deletion in generating inverted repeats is unknown. Amphimers probably are formed directly from λ rather than by subsequent interaction between λ *dv* molecules because the unique sequences lie at one or both termini relative to the linear map and the junction between inverted repeats lies at the opposite segment: If "uvwxyz" and "u´v´w´x´y´z´" represent the basic sequences in the two complementary strands of a segment of λ (both read from left

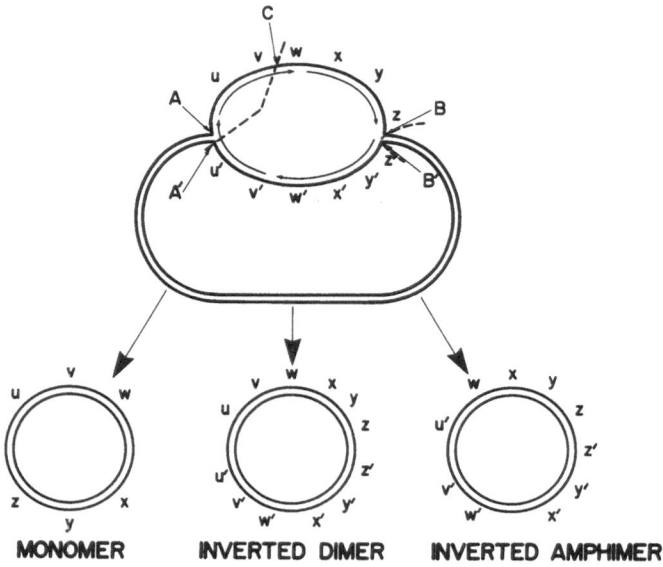

Fig. 6. Possible mechanism of λ *dv* formation from theta form of replicating DNA. DNA is assumed to be especially prone to breakage at replication forks. Breakage and rejoining of one parental chain and its complementary new strand at forks A and B generate a monomer; breakage of both parental strands (AA´, BB´) gives an inverted dimer, whereas imprecise breakage at A´C and BB´ could produce an inverted amphimer. Redrawn from Chow *et al.* (1974).

to right along the molecule), then a single strand of the amphimer may be "uvwxyzz´y´x´w´´" or "uvwxyzy´x´w´v´," but never "uvwxyzz´y´v´u´." As suggested by Chow *et al.* (1974), both amphimers and monomer *dv*'s could arise by breakage and rejoining of theta form intermediates in λ replication (Fig. 6).

One implication of this model is that most of the small circular fragments generated should include the replication origin and adjacent DNA. In alternative models, such fragments might arise at about equal frequency for all segments of the genome. The question is not directly testable by examination of λ *dv*'s, which must include the origin and replication genes to survive as plasmids. However, where either phage or host is recombination-proficient, circular fragments can add to complete chromosomes to generate duplication phages, and likewise can be formed by excision from such phages. Among spontaneous duplications observed in λ, not all include the replication origin (Busse and Baldwin, 1972; Emmons and Thomas, 1975). Their distribution is, in fact, fairly random. Neither inverted repeats nor the addition products of amphimeric *dv*'s have been isolated as spontaneous duplications; but they have not been specifically looked for in a phage that generates amphimers. It is likely that at least some monomeric *dv*'s arise by mechanisms not related to the location of the replication origin.

3.4. Substitutions and Specialized Transducing Phages

A specialized transducing phage is a viral variant in which some DNA segment not originally part of that virus' genome has become covalently joined to viral DNA, giving a molecular species that can replicate and be packaged into virions. Such phages generally arise rarely and are detected as prophages in transductants selected for acquisition of a dominant marker. Such transductants have acquired the donor phenotype not by replacement of the recipient marker by its homolog (as in general transduction) but rather by addition to the genome of a viral variant that includes the donor gene. Once acquired, the marker remains associated with the viral genome in subsequent growth and infection.

Specialized transducing phages arise by various mechanisms. Insertion of drug-resistance determinants into viral DNA, by a specific process determined by sequences associated with that determinant (see Section 3.2), can account for the origin of phages like P 1 *cam* (Section 2.1.1) and P 22 *tet* (Section 2.4). In these cases, extra DNA has been added to the viral genome, and no viral DNA has been lost in the process.

In other cases, such as the transductions of erythromycin resistance by Staphylococcal phage P 11 (Section 2.1.1b), lactose fermentation by coliphage P 1 (Luria *et al.*, 1960), or mandelate oxidation by Pseudomonas phage pf 16 (to give a defective transducing phage with the remarkable capacity to impart donor ability in bacterial conjugation; Chakrabarty and Gunsalus, 1969), host or plasmid genes have been acquired, and some viral DNA has been lost, but the mechanism by which this has come about is obscure.

3.4.1. Transducing Phages as Products of Abnormal Excision

One situation where replacement of a segment of viral DNA by host DNA occurs by a comprehensible mechanism is the acquisition of genes adjacent to an inserted prophage that sometimes transpires on induction of lysogenic bacteria.

Figure 7 shows the formation of a transducing variant of λ, in which phage DNA has been substituted by host DNA that includes the *gal* genes. The left column shows the normal λ cycle: (1) Following derepression of a lysogenic bacterium, λ prophage is excised by specific breaking and joining of DNA at the two prophage termini (*att*), catalyzed by the phage-coded enzymes integrase and excisionase. (2) Replication and packaging then proceed as in Fig. 3. (3) The virion infects another cell; DNA injects and circularizes. (4) In those cells giving the lysogenic response, viral DNA is inserted into the chromosome by exact reversal of step (1).

The right column shows the corresponding steps for the defective transducing phage λ *dgal*. (1) Starting from a normal λ lysogen, about 1 excision in 10^5 does not take place by specific recombination between *att*'s, but rather by breakage and joining within heterologous DNA. The excised DNA thus lacks the segment from the right prophage end (including gene *J*), which is replaced by a bacterial segment (not necessarily of the same length) including genes *gal*, *chlD*, and *pgl*. Steps (2) and (3) are the same as for normal λ. Step (2) requires the products of λ genes that are absent from λ *dgal*. These genes must therefore still be present in the cell. Insertion into the chromosome can take place either by recombination between *att*'s, or by recombination within homologous DNA. In either event, the product is a defective lysogen with a duplication of genes *gal-pgl*, which can generate λ *dgal* chromosomes by specific excision.

The viral life cycle is thus normal except for one step. Excision from the chromosome, usually an exact reversal of insertion, happens

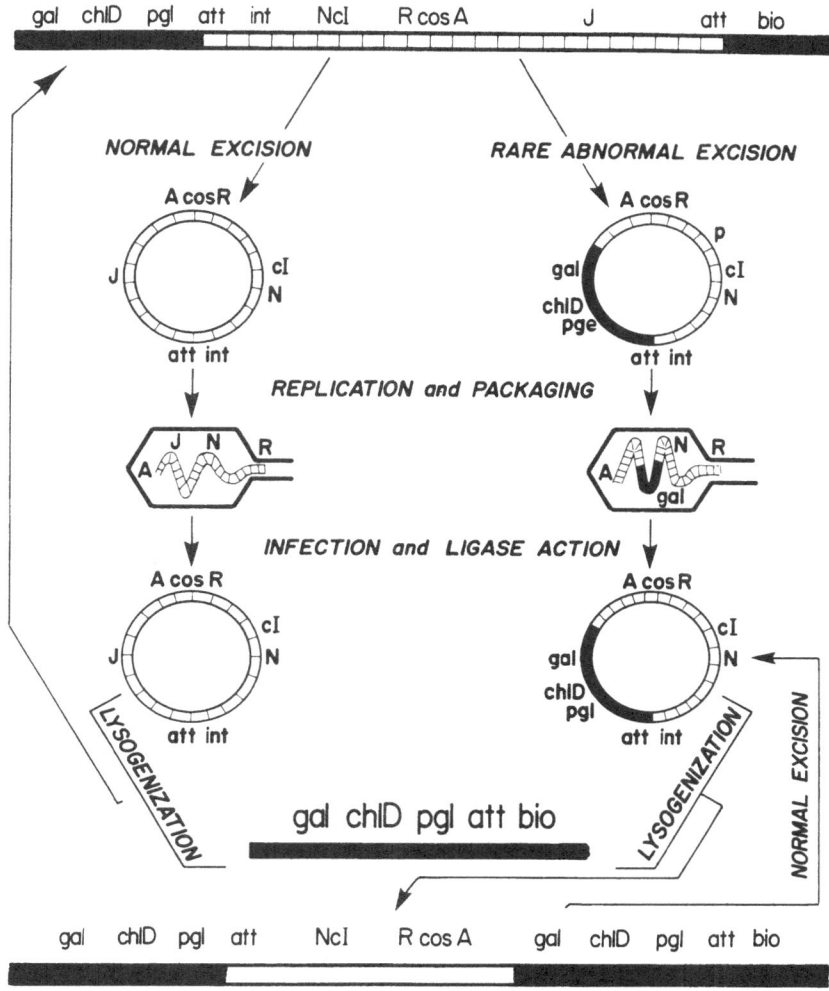

Fig. 7.   Formation of λ *gal* from a λ lysogen.

instead by breaking and joining within heterologous regions—the same kind of molecular event that generates deletion mutations (Franklin. 1971). Subsequent packaging of replicas of this DNA by cutting at *cos* sites (Fig. 3) produces molecules in which viral DNA at the right prophage end has been substituted by bacterial DNA adjacent to the left prophage end. The substitution cannot extent so far as to remove the *cos* site: otherwise the resulting molecule could not be packaged to yield an infectious particle.

If the viral DNA replaced by substitution includes any genes essential for phage development, the resulting virus is defective and cannot be propagated by infection as a pure stock. It can be replicated either as an inserted prophage, or in mixed infection with a complete λ phage.

The defective lysogens formed when transducing phages lysogenize appropriate hosts can be used in the same manner as deletion mutants for genetic mapping and studies of gene function. Their ease of propagation also makes feasible the precise physical localization of viral genes by examination of heteroduplexes formed between transducing and wild-type phage. They also offer a convenient means to amplify specific small segments of the bacterial genome for physical characterization.

In the particular case of λ *gal* phages, lysogenization by the transducing phage (Fig. 7, step 4, right column) is much less frequent than lysogenization by normal λ, because the "hybrid" *att* site (part bacterial, part phage) of the transducing phage does not recombine well with the bacterial *att* site. For that reason, λ *gal* phages lysogenize much better in the presence of a coinfecting normal λ, and the resulting lysogens are double lysogens carrying both λ and λ *gal* (Weisberg and Gottesman, 1969).

3.4.2. Methods for Moving Prophage Closer to Specific Genes

The scheme shown in Fig. 7 explains the origin of specialized transducing phages containing host genes next to the insertion site of the prophage. A specific phage would be of limited value if transduction were entirely restricted to contiguous DNA. Methods have thus been developed that extend the range of genes which can become substituted into the viral chromosome. The aims have been technical, and, wherever possible, the resulting transducing phages have been active rather than defective. Nevertheless, some issues of general principle have emerged that relate to the origin and properties of defective phages.

3.4.2a. Deletion Mutations and Chromosomal Rearrangements

A specialized transducing phage is derivable from a normal λ prophage as shown in Fig. 7 only if the length of DNA between the *cos* site of the prophage and the gene to be transduced is less than the maximum length that can be efficiently packaged into infectious particles. This length is about 1.09 times the size of wild-type λ (Weil *et al.*, 1975). Specialized transduction of more distant genes requires multiple events (see Section 3.4.3) and is consequently much rarer. The genes can be moved closer to the *cos* site of the prophage by chromosomal rearrangement.

Deletion of viral or host DNA in the vicinity of λ prophage has been used to construct transducing phages carrying genes that are otherwise slightly beyond reach (Campbell, 1971; Fig. 8). Because the deletion stocks must be isolated and cloned prior to induction, the deletions must not remove any host genes essential for bacterial growth. This limitation can be bypassed by use of partially diploid bacteria (such as F´ strains). This technique has been employed (H. Ozeki, private communication) in isolation of λ variants including the *supE* gene.

Small inversions are relatively common in *E. coli* (Berg and Curtiss, 1967). Insertional translocations have also been reported following heavy doses of x-irradiation (Jacob and Wollman, 1961). When one terminus of such an aberration lies close to a prophage site, genes normally distant from the prophage may become readily transducible.

3.4.2b. F´ Insertion

Bacterial genes can be brought close to a prophage by prior incorporation into F´ factors. F´ factors originate in a manner qualitatively

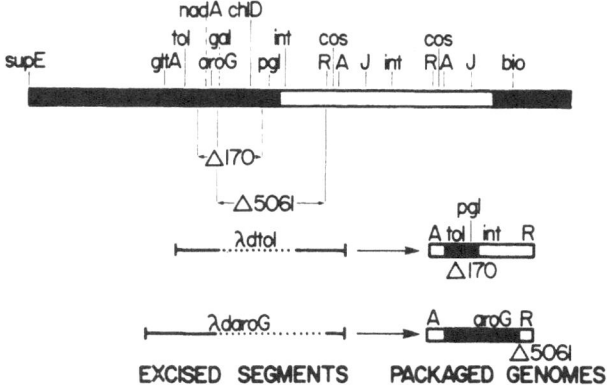

Fig. 8. Use of deletion mutations to move bacterial genes closer to λ *cos* site. To obtain phage carrying the *tol* genes, Bernstein *et al.* (1973) used a host (JP170) with a deletion removing genes *nadA-chlD*. To transduce *aroG*, Sato and Campbell (1970) used a deletion (Δ5061) that removes both bacterial and viral DNA. The bacterium shown is a double lysogen because a second phage is needed to supply replication functions with the 5061 deletion.

similar to transducing phages. In an Hfr strain, where the fertility factor F is inserted into the chromosome, occasional imprecise excision produces autonomous F plasmids that include a segment of bacterial DNA adjacent to the insertion site. F differs from λ in having no highly preferred insertion site on the chromosome. In various Hfr strains, F is located at different sites. When an F′ factor that has picked up genes from elsewhere on the chromosome inserts close to a prophage site, the genes may then come within reach for specialized transduction by that phage. This methodology has been extensively exploited by J. Beckwith and collaborators in deriving new types of transducing phages (Beckwith, 1970).

Whereas the mechanism by which F inserts into the chromosome is incompletely understood, recent evidence (Davidson *et al.*, 1975) indicates that F usually inserts by recombination between one of three special insertion sequences (see Section 3.2) within F, and corresponding sequences on the host chromosome. As some of these sequences recur at many chromosomal sites, and are themselves transposable, F can insert at various locations. The most frequent mode of F′ generation probably also takes place by interaction between special insertion sequences in F and adjacent bacterial DNA. Hence the bacterial segments that become incorporated into F′s were usually bounded by such special sequences in the wild-type chromosome.

3.4.2c.   Interaction between F′s

One special means of creating new linkages between usually distant bacterial genes has been the simultaneous introduction into the same cells of two different F′s carrying the DNAs to be fused, followed by selection for clones in which determinants from both F′s are perpetuated. For reasons not completely understood, two copies of F cannot be maintained in the same cell line. In the derivatives obtained, both segments of bacterial DNA have been joined to a single F. A prophage site on one F′ is thus brought close to genes on another F′ derived from a different part of the chromosome (Press *et al.*, 1971).

Some ways in which two F′ might interact to produce this result, based on models for F insertion and F′ formation of Davidson *et al.* (1975), are shown in Fig. 9. F actually has three kinds of special sequences ($\alpha\beta$, $\gamma\delta$, and $\epsilon\int$) at which insertion and excision can occur. $\epsilon\int$ is indistinguishable from IS2 (see Section 3.2). For simplicity, all events represented in Fig. 9 involve the same sequence (represented by a solid bar). F is shown inserting into the chromosome between episomal and

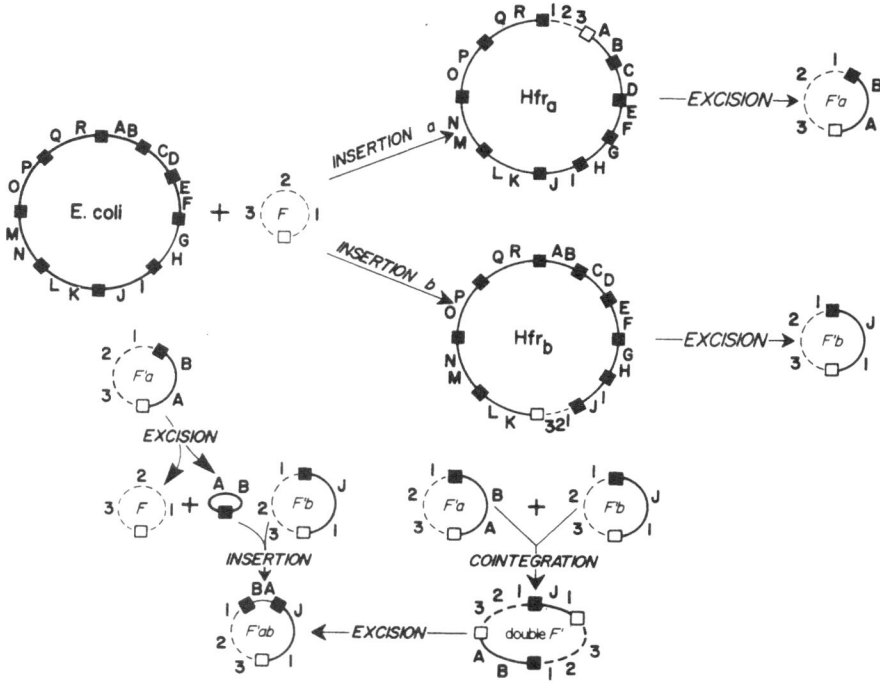

Fig. 9. Possible interactions between F′s that could bring together genes normally distant on the chromosome by recombination between special insertion sequences (designated by solid bar).

chromosomal $\gamma\delta$ sequences. Excision of F′s takes place by a similar process.

Introduction of two F′s (F_a' and F_b') into the same cell and selection for maintenance of bacterial genes from both sources yields cells harboring the recombinant F_{ab}'. This recombinant can arise by several routes, two of which are shown in Fig. 9. Direct evidence that formation of F_{ab}' species occurs by recombination between insertion sequences has recently been reported by S. Palchaudhuri, W. K. Maas, and E. Ohtsubo (private communication).

3.4.2d.   Secondary Insertion Sites

Lysogenic bacteria isolated from infection by λ of wild-type *E. coli* have always proven to carry the prophage inserted at a site between the *pgl* and the *bio* genes (Fig. 7). Infection of bacteria from which this preferred site has been eliminated by genetic deletion makes it possible to detect rare instances in which the virus has inserted elsewhere. By this means, lysogens can be isolated that carry λ at a variety of secondary sites. Insertion is not random, but recurs at specific places, where the base sequence is probably similar but not identical to the preferred site (Shimada *et al.*, 1975). Insertion and excision from these sites requires the virus-specific *int* and *xis* gene products (Shimada *et al.*, 1972). Some of these sites lie within known genes or operons; λ insertion has the predicted effect of disrupting both translation of the gene and transcription of the operon (Shimada *et al.*, 1973; Shimada and Campbell, 1974).

From each such abnormal lysogen, a family of transducing phages can be derived that carry genes on either side of the insertion site. Table 2 shows how transducing phages were used to demonstrate that three independent insertions in the *gal* operon all took place at the same site, within the "6b" segment of the *galT* gene. The results from transduction of bacterial mutations show that the three insertion sites are identical within the precision of that measurement. The fact that two transducing phages such as Y 256 and Y 253 can recombine to give a *galT*⁺ phage shows that, in any one host such as RW 614, insertion has

TABLE 2
Transducing Phages from Lysogens with λ Inserted in the *GalT* Gene[a]

| Phage | Host of origin | GalK | GalT | Point mutants located in deletion segment No. | | | | | | | Recombination to give *GalT*⁺ phage | |
				9	7	6a	6b	5	4	3	2	1	Y 256	Y 253
Y 256	RW614	+	+	+	+	−	−	−	−	−	−	−	+
Y 253	RW614	−	−	−	−	−	+	+	+	+	+	+	−
Y 252	RW616	+	+	+	+	−	−	−	−	−	−	−	+
Y 254	RW616	−	−	−	−	−	+	+	+	+	+	+	−
Y 257	RW617	+	+	+	+	−	−	−	−	−	−	−	+
Y 255	RW617	−	−	−	−	−	+	+	+	+	+	+	−

[a] Transducing phages were isolated from three independently derived lysogens in which λ had inserted into the *gal* operon. For each phage line studied, the bacterial loci present were scored by their ability to generate *Gal*⁺ transductants on infection of recipients carrying the mutations in question (which had previously been ordered by mapping against bacterial deletions). The transducing phages were also crossed with each other to see whether those carrying the left and right halves of the *gal* operon could recombine to give a complete *gal* operon, as assayed by formation of phage able to transduce a recipient with a mutation in the *gal* 6b segment. (Data from Shimada *et al.*, 1973.)

merely interrupted the *galT* gene without destroying any of it. Production of *gal*⁺ phage from cell pairs such as Y 256 and Y 254 bearing the left and right halves of the *galT* gene but derived from different lysogens proved that the insertion site was in fact identical in all three. These data support the inference that Fig. 7 is an adequate guide to the mode of origin of specialized transducing phage at secondary sites as well as at the preferred one.

3.4.3. Intensive (Brute Force) Selection for Multiple Events

The above methods require first isolating lysogens where the prophage is next to the gene of interest, then inducing phage development in them. Whereas the range of transducible genes is thus greatly augmented, it is not yet possible to derive specialized transducing phages for every gene of *E. coli* from available lysogenic stocks. One approach for the impatient investigator is to bypass this two-step procedure by simply inducing a large-enough number of lysogenic cells so that, within the population, the possibility of a double event—chromosomal aberration plus abnormal excision—becomes appreciable.

This tactic was first employed by Gottesman and Beckwith (1969), who isolated a defective λ derivative carrying the arabinose operon, starting from a lysogen that carried λ prophage at a position much closer to *ara* than the normal λ site but still too far away to generate directly a λ *ara* of packageable length. From the gene order of the transducing phage finally obtained, an inversion bringing the *ara* genes closer to the λ prophage apparently preceded or accompanied excision. The frequency of arabinose transductions per active phage particle in their original lysate was about 10^{-11}, as compared to 10^{-6} *gal* transductions per phage from a normal lysogen. However, the efficiency of transduction per physical λ *dara* particle was about 10^{-3}. Hence the ratio of λ *dara* to λ was about 10^{-8}, compared to 10^{-5} λ *gal*'s per λ.

Almost all subsequent applications have employed a similar procedure, in that the starting material was a lysogen in which the genes were close to the prophage, but slightly beyond reach for a single abnormal excision. The rationale for this particular methodology has never been explicitly defined: In the absence of specific information to the contrary, one might expect that, once a double event is required, it should be about as easy to obtain a λ *ara* from induction of a lysogen with λ at its normal location, between *gal* and *bio*, as from one where it is separated from the *ara* genes by a much shorter molecular distance. There are many possible reasons that the latter strategy *could* be more effective—if small inversions are more frequent or more viable than larger ones, for example. But these *ad hoc* explanations have no firm basis either in independent experimental evidence or as extensions of known principles.

There is also a paucity of published data on how important the distance between prophage and genes really is. When a double event is required, a large-scale preparation is usually needed to obtain a few successes. Similar large-scale negative controls from donors where the prophage is more distant have seldom been reported.

3.4.4. Secondary Alterations

By no means have all the transducing phage lines generated by such efforts been defective. On the contrary, for many purposes it is more convenient to use a nondefective, plaque-forming phage, which can be grown in large numbers physically uncontaminated with helper virus. These apparent advantages are sometimes compensated by an enhanced tendency to select for variants bearing additional, secondary changes.

Some examples with bacteriophage λ were described by Campbell (1971). By historical happenstance, the first specialized transduction to be thoroughly investigated was that of the galactose genes. The DNA length from the *gal* operon to phage gene *J* (the rightmost gene needed for lytic growth) is more than λ can package. Therefore, all the original λ *gal* stocks were defective and were of necessity maintained as prophages. Of many isolates examined genetically and/or physically all have obeyed the rules implied by Fig. 7: Viral DNA originally at the right prophage end has been replaced by bacterial DNA to the left of the prophage. The genome of the transducing phage is a connected segment of the lysogenic chromosome.

Somewhat later, it was discovered that λ could transduce the *bio* genes. The *bio* cluster is closer to λ than the *gal* operon is, and plaque-forming λ *bio* phages are readily obtained. Among the first seven λ *bio*'s analyzed in detail (Manly *et al.*, 1969), two constituted exceptions to the simple rule for transducing phage genomes, as though an additional genetic deletion had transpired early in the history of the phage before it was isolated as a stock. Additional examples have turned up in which plaque-forming variants, propagated lytically, have undergone deletion, insertion, substitution, or point mutation subsequent to their primary formation.

This suggests that viral variants are more likely to accumulate additional alterations when propagated lytically than when grown as prophages. The generality of this rule is not established, and the reason for it has not been experimentally demonstrated. However, it is a plausible expectation on general principles.

In the course of autonomous viral growth, the gross characteristics of the DNA itself—total length, GC content, secondary structure, etc.—very probably have at least some influence on rate of replication, time of gene expression, efficiency of packaging, and other aspects of the life cycle. A phage such as λ *bio* differs from λ in these properties, as well as in sublethal biochemical deficiencies due to loss of viral genes. Further changes may therefore frequently impart a selective advantage by counterbalancing the effect of the original alteration.

Replication of the bacterial chromosome is probably equally sensitive to DNA structure and function. However, the chromosome of a λ *gal* lysogen differs from that of a λ lysogen by a much smaller percentage of its total DNA than the chromosomes of λ and λ *gal* differ from each other. No bacterial genes have been deleted, and most of the λ genes are repressed and play no role in cellular processes. Therefore, the selective effect of the substitution is negligible when the phage is kept as an inserted prophage, but much larger when it is grown lytically.

3.4.5. How to Keep Stocks

An implication of the previous section is that viral mutants, especially structural variants such as transducing phages, are better maintained as prophages than as phages. This is especially important where the arrangement of host genes in the transducing phage is inferred to reflect their natural disposition in the cell line from which they were derived. The same considerations apply, perhaps with even greater force, to cloning of eukaryotic DNA segments joined enzymatically to prokaryotic vectors. The safest method may be to join foreign DNA to that of a transducing phage and select immediately for transductants in which the chimeric molecule has inserted into the bacterial chromosome.

3.5. Phenotypes of Defective Lysogens

The phenotype of a mutant phage depends on what genes have been deranged by mutation, and their role in normal viral development.

An extensive catalog of known mutations will not be attempted here. A description of the general results with mutants of phage λ illustrates the properties of laboratory mutants, which may be compared with some of the naturally defective viruses to be treated in Section 4.

Table 3 shows the phenotypes imparted by lethal mutations in single genes (cf. Fig. 2). The most defective are the *N* mutants, in which all the major processes in viral development are markedly reduced. Mutations in genes *O* and *P* curtail viral replication completely and reduce everything else to some extent. In *Q* mutants, DNA synthesis is normal, but all subsequent processes are delayed. In *S* and *R* mutants, only cell lysis is altered. Infectious virions are formed intracellularly but cannot escape to infect other cells. Mutations in head assembly genes allow normal lysis, with liberation of tails that can combine with heads *in vitro* to produce active phage. Likewise, tail mutants liberate active heads.

The results generate the scheme for normal λ development shown in Fig. 10, which has been verified and refined by additional molecular and genetic studies. Its main features are that the late steps all depend on the prior occurrence of early steps, but that the three major pathways of late functions (head and tail formation and lysis) are parallel rather than sequential.

Each late pathway can be further dissected. Table 4 shows the frequency of various head-related structures in lysates produced by induction of various mutant lysogens. From *in vitro* assembly studies, the isometric, empty petit λ particles are considered normal head

TABLE 3

Properties of Lethal Mutations in Genes of Coliphage λ[a]

Genotype	Viral DNA synthesis	Cellular lysis	Heads	Tails	Phages	Gene function
Wild type	+	+	+	+	+	
N^-	r	−	−	−	−	early reg.
O^-,P^-	−	r	r	r	r	dna
Q^-	+	r	r	r	r	late reg.
S^-,R^-	+	−	+	+	+	lysis
$A^-,W^-,B^-,C^-,Nu3^-,D^-,$ E^-,F_I^-,F_{II}^-	+	+	−	+	+	head
$Z^-,U^-,V^-,G^-,T^-,H^-,$ M^-,L^-,K^-,I^-J^-	+	+	+	−	+	tail

[a] The following symbols are used: + = indicated process occurs with same kinetics as wild type, or indicated component found in wild-type amount; − = process either does not take place or is drastically reduced: r = reduced or delayed relative to wild type.

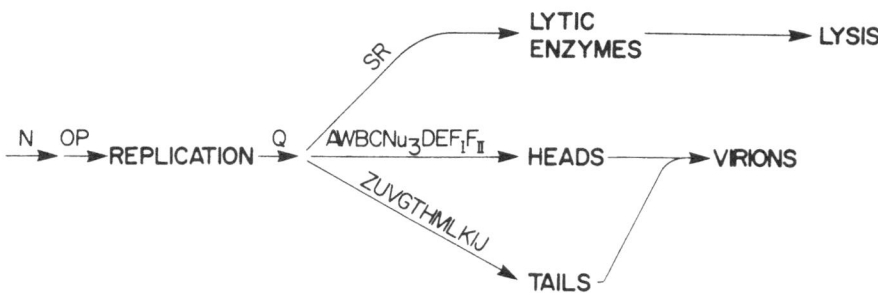

Fig. 10. Skeletal scheme for phage λ development, based on data in Table 3.

precursors, whereas the elongated, empty monsters and tubular forms, as well as empty heads and empty phages, are products of side reactions (Fig. 11). Those petit λ particles seen in B^-, C^-, or $Nu3^-$ lysates, though morphologically indistinguishable from normal petit λ, are chemically different (Hendrix and Casjens, 1975) and likewise come about by side reactions. Also, the full heads observed in W and F_{II} mutants, though resembling normal heads in appearance, cannot join to tails until the appropriate gene products are added.

Because λ genes are clustered according to function (Fig. 2), extensive deletions or substitutions can still leave intact some of the functions listed in Table 3. For example, all λ *dgal* strains show normal DNA synthesis and cellular lysis; and those that include all the head

TABLE 4

Percentages of Head-Related Structures in Head Mutants of λ^a

Genotype[b]	Phage	Empty phage	Full heads	Empty heads	Petit λ	Monsters	Tubular forms	Tails
Wild type	26	8	0.2	3.5	62	0.2	0	11
*Aam*32	0	0	0	0.1	99	0.4	0	22
*Wam*403	0	0	5.4	11	79	4.1	0	59
*Bam*10	0	0	0	0.1	79	8.8	12.5	60
*Cam*39	0	0	0	0.1	99	0.6	0	45
*Nu3t*78	0	0	0	1.4	92	4.6	0	50
*Dam*15	0	1.5	0	13.7	83	0.2	0	40
*Eam*4[c]	–	–	–	–	–	–	–	+
F_{II}*am*423	0	0	8.7	10.4	81	0	0	65

[a] Data from Murialdo and Siminovitch (1972).
[b] All mutants are amber mutants in nonpermissive hosts, except for *Nu3t*78, which is an absolute defective.
[c] *E* mutants lack the major head protein and form no head-related structures. Tails are formed in about the normal amount.

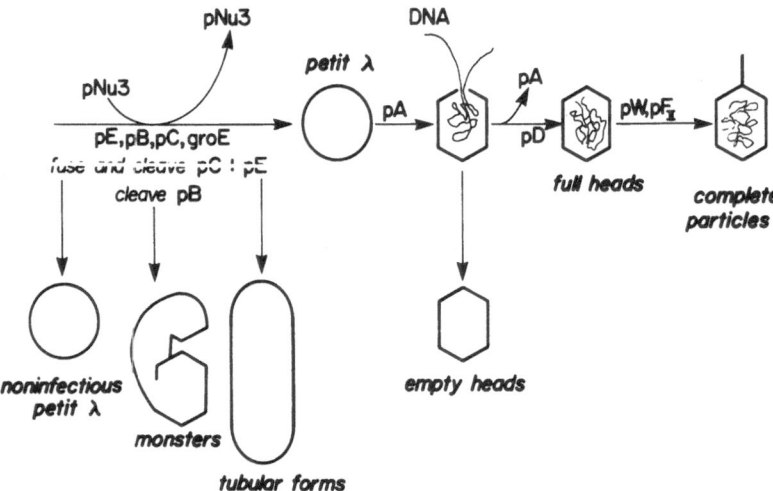

Fig. 11. Scheme for λ assembly based on *in vitro* assembly studies (Kaiser *et al.*, 1975), extended to show possible origin of various structures listed in Table 4. Protein products of viral genes are indicated as "pA" for gene *A* product, etc. The host gene *groE* is discussed in Section 2.3.3.

genes produce DNA-filled heads that can be completed by addition of tails *in vitro* (cf. Fig. 7).

4. NATURALLY DEFECTIVE PHAGES

Having described variants arising from known viruses within the laboratory, we turn to natural objects that resemble such variants. Operationally, these fall into two categories. On the one hand, many bacterial strains can produce particles that, though not demonstrably infectious, are morphologically similar to known viruses or components thereof. Other bacterial strains can, upon infection by known viruses, liberate occasional recombinants between the infecting virus and endogenous DNA that is functionally equivalent to it. For comparative purposes, satellite viruses that require a helper virus of a different species (rather than of the same species, as in Section 2.2) will also be considered here.

4.1. Viruslike Particles and Their Genetic Determinants

Objects resembling phages are very common and have been observed in the culture supernatants of many bacterial strains, especially after exposure to agents such as ultraviolet light or mitomycin C that can induce viral development in some lysogenic systems (Bradley, 1967; Garro and Marmur, 1970). These range from complete, DNA-containing particles to structures resembling various viral parts—empty heads, free tails, or mixtures thereof. Operationally, some of these are classified as bacteriocins because they can kill sensitive cells but fail to propagate therein.

Doubtless some such viruslike particles will prove to be complete viruses for which the proper host or the proper conditions remain undiscovered. Some may even multiply to produce infectious particles, but with such kinetics that plaques are not detectable on agar surfaces. Others may be mutants derivable from active viruses by a few simple steps of the kind described in Section 3. Still others may represent nonviral systems of DNA packaging, cell killing, or unknown interactions between bacterial cells.

These distinctions can easily become clouded by purely semantic problems, both conceptual and operational. The conceptual questions can be handled first.

As we will discuss in Section 6, one way of viewing a virus is as a collection of basic modules individually concerned with specific functions such as replication, virion formation, cellular lysis, insertion into host chromosome, etc. Our later discussion will focus on the potentially separable nature of these modules and the consequent possibilities of generating new combinations of preexisting modules. Independently of such specific considerations, we can ask whether morphological resemblance to known infectious agents is an adequate reason for considering an object to be virus-related.

The question of whether viruslike objects might represent evolutionary precursors to, rather than degeneration products of, viruses is not new (Bradley, 1967). Nevertheless, the highly specialized and distinctive morphology of viruses had led many investigators to dub such objects as "viruses" without qualification. The view that "Nothing else looks like a virus" is of course logically circular. Viruslike particles look like viruses. To classify them at the outset as virusrelated prejudges the question of whether some of them might in fact constitute cellular products where function and evolutionary origin are distinct from those of known infectious agents.

Independently of the evolutionary origin of defective viruses, we can ask whether they serve any function in their present condition and what that function is. At least two possibilities are easily imagined: First, some phagelike particles may be part of a host-determined system for gene transfer among bacteria. A generalized transducing phage like

P 22 serves this function as well—but as a byproduct of its own infectious spread. Most P 22 particles contain P 22 DNA; an occasional particle, perhaps by pure accident, happens to carry a segment of bacterial DNA. In contrast, a host-determined packaging system might be expected to show no preferential packaging of the determinants of packaging themselves and consequently no possibility of propagating those determinants in serial passage.

Second, some defective phages may serve to kill potential competitors of the bacteria that produce them. This seems especially plausible where the particles do not contain DNA. Operationally, such particles are classified as bacteriocins.

Bacteriocins are substances, usually protein in nature, that are elaborated by certain bacteria and are specifically toxic to other, closely related bacterial strains. Some are simple proteins such as colicin Ia (Konisky, 1975) of molecular weight 75,000 that deranges the function of the cell membrane, or colicin E3 (Boon, 1972) that causes the breakdown of ribosomal RNA into specific fragments. Others are much larger, resemble phage tails or whole virions, with or without DNA, and are considered to be defective phages. The genetic determinants for the simple bacteriocins are normally carried on plasmids rather than on the bacterial chromosome.

Cell killing by empty phage or phage tails resembles killing by some of the simple colicins in its gross pathology, but there is no firm evidence for homology between the two. An indirect argument against such homology that has been repeated in various forms, even quite recently, merits some discussion.

Lwoff (1953) wrote, "The hypothesis can be considered that bacteriocinogenic bacteria are related to lysogenic bacteria in the sense that they possess genetically the ability to form a protein corresponding to the tail's protein of a phage. But bacteriocins being lethal, the hypothetical corresponding phage, because it would kill, could not be temperate. Bacteriocinogeny, at least corresponding to the type of bacteriocins we know, therefore cannot be considered as a step in the phylogeny of lysogeny."

It is certainly true that, if every cell infected by a virus were to die as a result of the initial interaction at the cell envelope, there then could be no such thing as a lysogenic survivor. However, it is also true that, if every infected cell were to undergo the extensive, almost immediate damage wrought by most colicins, it would not only die but also be unable to produce and liberate phage. The same is true of the damage inflicted by empty phage particles (ghosts) of the T-even phages (Duckworth, 1970). The implication is that, although the phage tail may have a bacteriocinlike effect on the cell, this effect must be to some extent neutralized following infection. Infected cells are, in fact, resistant to added ghosts (Vallee *et al.*, 1972), and this resistance can be abolished by a mutation of the viral genome (Vallee and Cornett, 1972).

Just as infected cells can moderate the lethal damage of phage-tail protein, some bacteriocinogeny plasmids determine the production not only of the colicin itself but also of a second protein that prevents its toxicity. Thus, cells carrying the colicin E3 plasmid make, in addition to colicin E3, a specific inhibitor of the E3-induced cleavage of ribosomal RNA (Jakes *et al.*, 1974; Sidikaro and Nomura, 1974). There is thus no reason to believe that a lethal tail protein necessarily renders a phage virulent rather than temperate nor to postulate mixed populations of killing and nonkilling particles.

Some specific examples of bacterial strains that produce viruslike particles illustrate the possible functions mentioned here.

4.1.1. The Defective Prophages of *Bacillus subtilis*

Many strains of *Bacillus subtilis*, isolated independently from nature in different parts of the world, produce phagelike particles at a rate that is greatly enhanced by exposure to mitomycin C or ultraviolet light (Garro and Marmur, 1970). All are very similar in morphology. Three types have been distinguished on the basis of their specificity of cell killing: PBSX (or PBSH), PBSY, and PBSZ. PBSX lysates can kill strains carrying PBSY or PBSZ, but not another strain carrying PBSX. A similar defective phage (PBLB) is harbored by *Bacillus licheniformis*.

The hybridization kinetics of DNA extracted from virions with host DNA indicate that the entire host genome is packaged, with no specific sequence comprising a major fraction of the total. Infection of hosts prelabeled with heavy isotopes shows that 40% of the DNA present prior to mitomycin addition becomes packaged by PBSX. This comprises about 14% of the packaged DNA (Haas and Yoshikawa, 1969).

Following mitomycin C induction, seven PBSX proteins, all virion components, are synthesized and assembled into virions containing DNA fragments of molecular weight 8.35×10^6 (Thurm and Garro, 1975a). Transformation studies show that all genes are packaged. A gene (*purA*) close to the replication origin is overrepresented both in packaged and total DNA. Mitomycin C apparently induces multiple initiation at the origin, even where packaging is eliminated by mutation

(Thurm and Garro, 1975*b*). Another gene, *metC,* close to the PBSX prophage, is also preferentially replicated after induction.

Four mutations, two eliminating tail formation, one preventing head assembly, and one (selected as UV resistant) that eliminates both functions, as well as the preferential amplification of *metC* following induction, have all been mapped by transduction between the *metA* and *metC* genes (Thurm and Garro, 1975*b*). Thus, the known determinants of packaging are clustered rather than scattered on the bacterial chromosome. This cluster will be called the PBSX prophage. A fourth, unlinked mutation, *tsi,* which causes PBSX induction at high temperature, also induces several other unrelated phages and probably represents a host gene whose effect mimics that of UV or mitomycin.

If the preferential replication of the linked gene *metC* is caused by replication initiated within the prophage, the prophage must contain not only packaging genes but a replication origin as well. Furthermore, prophage DNA, like *metC* DNA, should be more abundant in the packaged DNA, than in the uninduced cell (though not enough to appear as a unique fraction in physical tests). If Thurm and Garro's UV-resistant mutant, *xin,* alters the repressor gene, then it must also be linked to the packaging genes; however, it could equally well affect an operator-promoter rather than a repressor.

About 40% of the bacterial DNA present prior to infection becomes packaged. By comparison, the generalized transducing phage P 22 packages only about 0.7%. However, P 22 mutants have been isolated in which this fraction is increased to 20–50% (Schmieger and Backhaus, 1973). Thus, the properties of PBSX are compatible with a one-step change from active phage. PBSX can transduce appropriate recipients under some circumstances (Haas and Yoshikawa, 1969) but not others (Garro and Marmur, 1970).

The main reason to consider PBSX as something other than the useless remains of a once-active phage is its presence in so many *B. subtilis* strains. The existence of three types, PBSX, PBSY, and PBSZ, each with its own killing specificity, verifies that the acquisition of this kind of defective phage was either ancient enough to allow subsequent evolution or else has frequently recurred.

4.1.2. The Denizens of *E. coli* Strain 15

E. coli 15 was the strain employed by Cohen and Barner (1954) for some of their pioneering studies on macromolecular synthesis and its regulation. The strain proved to be an unfortunate choice because it turned out to harbor several plasmids and prophages whose activities can be induced by various treatments once imagined to affect normal host processes. Early observations on a colicinlike activity in culture supernatants the production of which could be induced (with concomitant cell lysis) by mitomycin. UV. or thymine starvation provoked a careful examination of extrachromosomal DNA.

E. coli 15 carries three plasmids. of molecular lengths 32.9, 53.9. and 0.96-μm (Ikeda *et al.*, 1970). The 32.9-μm plasmid is at least 90% homologous with phage P 1 by *in vitro* hybridization. Like P 1. it is present in about one copy per bacterial genome. Lysogenization by P 1 *cam* (see Section 2.1.1a) can cause loss of the 32.9-μm plasmid. as though the two belonged to the same incompatibility group. Unlike P 1. however. the 32.9-μm plasmid is never packaged into virus particles. It thus can be considered together with the defective phages discussed in Section 4.3 as an object whose relation to viruses is detectable by genetic homology rather than by the manifestation of viruslike properties.

Strain 15 also liberates phagelike particles (called coliphage 15) with DNA-filled heads and long tails. The colicin activity of strain 15 cultures is due to the presence of phagelike particles and free tails. The majority of the heads are not attached to tails. Unlike *Bacillus* phage PBSX (Section 4.1.1). phage 15 contains a unique DNA species. 11.9 μm long. of molecular weight 27×10^6, cyclically permuted and terminally redundant (Lee *et al.*, 1970). At least some of the tailless particles contain the same kind of DNA as those with tails (Grady *et al.*, 1971). A second. short-tailed phage species whose DNA is not hybridizable to the long-tailed variety has also been reported (Ikeda *et al.*, 1970).

4.1.3. Defective Phages of *Vibrio cholerae*

Various bacteria that liberate structures resembling phage tails. frequently with killer activity. are described in older reviews (Bradley. 1967: Garro and Marmur. 1970). Recent results with *Vibrio cholerae* (Gerdes and Romig. 1975) will be described here.

Many classical *Vibrio* strains lyse following treatment with inducing agents. liberating structures resembling phage tails. A particular strain (NIH 41) is lysogenic for two nondefective phages. VcA-1 and VcA-2. VcA-2 is quite similar to previously described (Kappa type) phages. VcA-1 and VcA-2 are both temperate but differ in plaque morphology. host range. and immunity. The heads of the two phages are indistinguishable, but their tails differ in length. An induced lysate of a typical classical strain (569B) contained tails of both the VcA-1 and VcA-2 types, as well as occasional empty heads and, rarely, complete particles with contracted tails. Most classical strains, including 569B, are immune to both VcA-1 and VcA-2.

Taken together, these observations suggest that many *Vibrio* strains harbor defective phage genomes that retain most of the information of the active phage. The classical strains might well differ from active lysogens by a single mutation (either host or viral) affecting head assembly. As with the *Bacillus* phages (Section 4.1.1), the wide dispersion of the defective condition raises the question of its possible present function.

4.2. The Satellite Phage P 4

If host-determined packaging systems exist, some viruses might employ them for their own intercellular transfer. Productive infection by such viruses would necessarily be limited to those hosts carrying the necessary determinants.

Most bacteriophages have turned out to determine their own packaging. Viral and host proteins may interact (Section 2.3.3, Fig. 11), but most of the proteins that end up in virions are virus-coded. The various biases in selection of particular phages for study preclude our drawing any strong conclusion from this fact.

One phage that does not determine its own packaging but rather borrows the packaging machinery of another virus is coliphage P 4. P 4 forms plaques only on hosts lysogenic for a larger phage, P 2, or related phage. In nonlysogenic hosts, P 4 replicates and lysogenizes by insertion but forms no infectious particles. P 4 was isolated from the supernatant of a bacterial strain (K-235) that was lysogenic for both P 4 and a P 2-related phage, PK (Six and Klug, 1973).

The genome of P 4 is a double-stranded DNA weighing 7×10^6 daltons compared to 22×10^6 for P 2 (Barrett *et al.*, 1974). Furthermore, all 17 P 2 genes required for formation of infectious P 2 particles, and the one gene needed for lysis (Fig. 12), are also needed for P 4 development. If any one of these genes is inactivated by mutation, the resulting defective P 2 lysogen will not allow plaque formation by P 4. Less than 1% of P 4 DNA is homologous to P 2 DNA by *in vitro* hybridization (Lindqvist, 1974).

In extracts of P 4-infected P 2 lysogens, P 4 DNA, but not P 2 DNA, can be packaged as viable phage (Pruss *et al.*, 1974). Thus, the

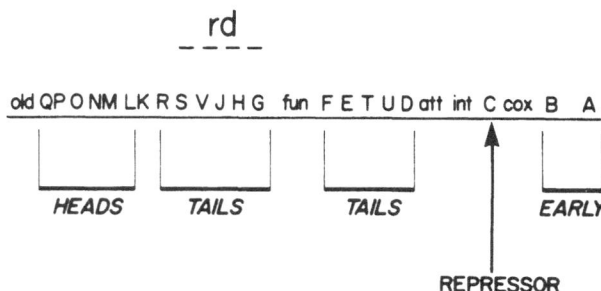

Fig. 12.   Genetic map of coliphage P 2, adapted from Barrett *et al.* (1974) and Bertani and Bertani (1971). Genes denoted by capital letters (except for gene C) are needed for plaque formation. *Old:* prevents lysogenization of *recB* hosts; *fun:* sensitivity of lysogens to fluorouracil; *int:* insertion during lysogenization; *cox:* excision; *rd:* round plaques, map location not precisely determined.

specificity of the packaging mechanism seems to be modified by the presence of P 4 proteins.

All known conditional lethal mutations of P 4 fall into one complementation group and prevent replication of P 4 DNA without impairing its ability to stimulate transcription of the late genes of P 2 prophage (Barrett *et al.*, 1974). The P 2 prophage itself is not replicated following P 4 replication, and the P 2 genes for replication and early regulation are not required by P 4.

Four P 4-specified proteins are formed following infection: Two of these appear soon after infection, whereas the others are synthesized later. Transcription of these late genes of P 4 is strongly delayed in the absence of P 2. Thus, not only does P 4 turn on the late genes of P 2, but P 2 reciprocates by turning on late genes of P 4. Certain P 2 prophage deletions abolish the ability to activate P 4, indicating that an unidentified gene in the early region of P 2 causes the activation. In these deletion strains, phage production by P 4 is delayed from its usual latent period of 50 to about 160 min. Visible plaques are not observed on these hosts, although the phage can potentially be propagated in serial passsage on them.

Whatever their molecular mechanisms, the regulatory and structural interactions of P 2 and P 4 suggest a long history of associated existence.

4.3.   Recombination between Prophage Fragments and Known Phages

Bacterial strains not known to elaborate phagelike particles have sometimes proven to harbor latent viral genes, as revealed by recombination between these genes and active phages. Such recombinants are distinguished from ordinary mutants of the active phage because they arise only on specific hosts and substitute whole segments of the viral genome with functional homologs.

4.3.1. An Immunity Region for Coliphage P 2 from *E. coli* B

When P 2 is grown for one cycle on the B strain of *E. coli*, about one progeny phage in a thousand differs from P 2 in its specificity of repression (Cohen, 1959). These variants were noticed originally as turbid plaque formers that appeared when a virulent mutant of P 2 was grown on B. Subsequent testing showed that the new phage could form plaques on P 2 lysogens, and likewise that P 2 could plate on lysogens of the new phage. P 2 and the new phage are therefore heteroimmune or "dismune." Infection of B with various mutants of P 2 available at that time showed that, except for the repressor gene and determinants closely linked to it, other determinants of the progeny with the new immunity usually, but not always, were derived from P 2. DNA heteroduplexing of one of these phage (called P 2 Hy^{1}Dis for "hybrid dismune") showed that it differed from wild-type P 2 by three separate substitutions located near the right end of the genetic map of Fig. 12 (Chattoraj and Inman, 1975). Probably the entire right end of P 2 Hy^{1}Dis is derived from an element in *E. coli* B that has alternating segments of homology and nonhomology with P 2 itself.

E. coli B does not spontaneously produce any phage that plate on *Shigella,* suggesting that a complete viral genome is absent. It is immune to infection of P 2 Hy^{1}Dis. Cured derivatives, lacking this immunity, are found among survivors of P 2 infection. These cell lines have also lost the ability to generate P 2 Hy^{1}Dis when infected with P 2. Therefore, *E. coli* B probably contains a DNA segment including the determinants for immunity and early phage development, bounded by P 2-type insertion sites, and extending far enough leftward to include an allele of the *rd* locus (Fig. 12), which was rarely rescued from B along with immunity. Thus, although the immunity segments of P 2 and P 2 Hy^{1}Dis have no recognizable homology, the overall structure of the defective prophage in B has several marks of a common genetic origin with P 2.

4.3.2. Genes for Lysis and Recombination of Coliphage λ from *E. coli* K-12

The K-12 strain of *E. coli* is lysogenic for phage λ. Most λ research has been carried out on K-12 derivatives that were cured of λ by exposure to high doses of ultraviolet irradiation. Some of these cured derivatives (and therefore the parent strain as well) contain determinants that are functionally equivalent to parts of λ.

A DNA substitution (called *p4* or *qin* for "petit" and "*Q*-independent," respectively) that replaces genes *Q*, *S*, and *R* of λ with a set of genes similarly able to induce late functions and cause cellular lysis has appeared in λ stocks either for unknown reasons (Jacob and Wollman, 1954) or (at a frequency of about 10^{-5}) as a result of selection for reversion of deletions or multiple mutants (Sato and Campbell, 1970; A. Strathearn and I. Herskowitz, private communication). Rescue appears to occur by homologous recombination and virtually disappears when phage and bacterial recombination systems have been inactivated by mutation. Whereas both λ and λ *qin* have *Q* genes, their specificity is different. The *Q* of λ will turn on the late genes of λ, but not those of λ *qin*, and vice versa (Herskowitz and Signer, 1974). As in the case of P 2 Hy^{1}Dis (Section 4.3.1), the substitution includes not only the gene for a specific regulatory protein, but also its site of action—which in this case is between *Q* and *S*. λ and λ *qin* differ by a single substitution loop seen in heteroduplexes, although other irrelevant substitutions and insertions accumulate in lytically grown stocks of λ *qin* (Fiandt *et al.*, 1971).

Mutants and transducing phages lacking the recombination (*red*) genes of λ do not plate on *recA*⁻ or *polA*⁻ hosts (Zissler *et al.*, 1971). Selection for reversion of these mutations yields a λ derivative (λ reverse) with an extensive DNA substitution including the *red* locus (Gottesman *et al.*, 1974).

The host-derived DNA segment in λ reverse appears to be closely associated in the cell with that in λ *qin*. Both determinants have been spontaneously lost from various common K-12 stocks, and no case has been observed where one is present without the other (A. Strathearn and I. Herskowitz, private communication).

5. DEFECTIVE PHAGES AND POPULATION DYNAMICS

Since the studies of von Magnus on influenza virus, animal virologists have been aware that defective virions accumulate when viruses are propagated by high-multiplicity passage. Recently, many investigators have provided evidence that the selective replacement of active by defective interfering particles may be important in natural infections (Huang and Baltimore, 1970; see also Volume 9, Chapter 3, of *Comprehensive Virology*). The defective particles are frequently deletion mutants which, for unknown reasons, have some competitive advantage over normal virus in the early stages of intracellular replication (Baltimore *et al.*, 1974).

The selective pressures operating on animal viruses do not differ in any obvious quantitative manner from those experienced by bacteriophages. One may therefore ask whether any phage behaves in the same manner.

The only phage for which comparable results have been reported is the filamentous DNA phage M 13. In M 13 lysates that had been through several high multiplicity passages, Griffith and Kornberg (1974) observed that a large fraction of the virions contained DNA 0.2–0.5 the normal M 13 length. These virions were defective, but had a competitive advantage in mixed infection with wild type M 13. The miniphage DNA was circular and single-stranded, like that of M 13. The miniphage molecules are not random segments of M 13 DNA. All include a common portion of the M 13 genome, which perhaps contains the replication origin.

Another set of defective phage variants that resemble the defective animal viruses in their smaller length and competitive advantage under some circumstances are the minivariants of small RNA phages such as Qβ (Kacian *et al.*, 1972). From cultures in which Qβ RNA is serially propagated *in vitro*, derivative RNAs have been selected that are shortened by deletion and therefore can complete their replication cycle more quickly than the complete viral DNA. The smallest RNAs that were isolated from *in vitro* cloning had about 550 of the original 3600 nucleotides.

From cells infected with Qβ, even smaller molecular species have been recovered that can be replicated *in vitro* by Qβ RNA polymerase (Banerjee *et al.*, 1969). One such species (218 nucleotides in length) has been purified and the complete nucleotide sequence determined (Mills *et al.*, 1973). The sequence is similar or identical to the terminal sequences of Qβ RNA itself, as expected if the replicase recognizes these termini as a specific replication orign (Haruna and Spiegelman, 1965).

These Qβ RNA fragments are not known to be packaged into virions and therefore may not be transmissible from cell to cell. Their exact function is unknown.

6. DEFECTIVE PHAGES AND EVOLUTION

6.1. Genomes as Complexes of Separable Modules

Several authors (Hershey, 1971; Botstein and Herskowitz, 1974; Simon *et al.*, 1971; Szybalski and Szybalski, 1974) have commented on various aspects of a modular, or segmental, hypothesis of viral evolution. The concept has not been defined with great logical rigor. Roughly, it views the viral genome as composed of differentiated segments (modules) specifying particular functions, and suggests that recombinational reassortment of whole modules among distantly related varieties has been important in natural evolution.

The modular hypothesis was extended by Campbell (1972) to include not only exchanges of modules between viral varieties but also between viral chromosome and host chromosome, or between viruses and other extrachromosomal replicons. His idea was that among the basic processes that a virus like λ carries out (replication, regulation, virion formation, lysis, insertion and recombination), several are types of functions needed by the host cell and by nonviral replicons. The clustering by function of genes within λ was thus imagined to reflect the evolutionary origin of the phage genome from its component segments rather than design for optimal regulatory efficiency.

In this view, some of the entities described in this chapter, including the DNA-containing virions, need not be virus-related in the usual sense of the word. A packaging module may or may not have been part of an autonomous, packageable replicon at any time in its history.

Although the modular hypothesis might help explain the origin of viruses, that is not its primary concern. Rather it postulates that over the relatively recent evolutionary time period since a module acquired more or less its present characteristics, it has been associated with diverse combinations of other modules, perhaps only viral, perhaps host and plasmid as well. For example, the replication module of λ comprises two genes O and P that interact to promote replication from a specific origin site *ori* on λ. Related phages have different O and P products which recognize their own *ori* sites. The modular hypothesis would suppose that the recognition of λ's O and P for each other and for the λ *ori* considerably antedates the existence of λ as such.

The simplest alternative to a modular hypothesis is that groups of related viruses have developed by linear evolutionary divergence from a common viral progenitor that contained base sequences ancestral to all those present in every member of the group, and that recombination between distant relatives, though possible in the laboratory, is inconsequential in nature.

6.1.1. Gene Clustering

One prediction of the modular hypothesis is that genes belonging to the same module, and therefore related in function, should be close to each other. Such clustering is very common in prokaryotes, bacteriophages included. In λ and related phages, the spatial distribution of genes along the DNA corresponds almost completely with their functions in the viral life cycle.

Clustering is also seen between those genes exerting a direct influence on DNA behavior—transcription, replication, insertion, packaging—and the DNA target sites for their gene products. The A gene is at one terminus of the λ genetic map (Fig. 2), thus adjacent to the *cos* sequences that are cut by the A product to form the molecular ends of packaged DNA. The *int* and *xis* genes, which cause insertion of phage DNA into the host chromosome and excision from it, lie next to the site (*att*) at which insertion takes place. The products of three genes *cI*, *cro*, and *N* all influence the extent of transcription originating from two promoters, p_L and p_R, that lie between *N* and *cI* and between *cI* and *cro*, respectively. Gene *cII* stimulates transcription from a nearby promoter. Genes *O* and *P*, needed for DNA replication, lie directly to the right of the replication origin. The product of gene *Q* turns on late transcription emanating from a promoter p_{R2} between *Q* and *S*.

Two reasons have been suggested for the clustering of related genes. First, clustering may simplify regulation by allowing coordinate control of genes whose products act in concert. Alternatively, proximity between interacting loci minimizes the frequency of recombination between them, thus favoring coevolution of alleles that are well suited to each other. The first explanation does not address the proximity of genes and target sites, whereas the second does.

The modular hypothesis represents a form of the second alternative that emphasizes its intrapopulational rather than interpopulational aspects. In interpopulational competition, those species on whose genetic maps related factors are clustered should have a selective advantage, because recombination generates fewer deleterious combinations. In prokaryotes, where there is no obligatory coupling of recombination and reproduction, this is likely to be a minor concern. In the modular hypothesis, linkage between related determinants is rather a consequence of the fact that individual DNA segments have been transferred between distant varieties that have diverged from each other, surviving combinations had a selective advantage within the population, and sufficient time has not elapsed for further evolution in which components of the transferred segment have been separated from each other. Only in those cases where interacting loci were close enough to lie on the same segment has effective transfer been possible.

6.1.2. Genetic Exchange and Base Sequence Homology between Different Varieties of Phage

If each species or variety of virus comprises a set of modules concerned with different functions, various natural isolates should have different combinations of modules, and other types should be constructible by recombination between them. The results of such recombination studies, plus the detection of homologies from electron microscopy of DNA heteroduplexes between different isolates, comprise the principal data base for the modular hypothesis.

Jacob and Wollman (1961) isolated from nature a number of temperate coliphages, some of which could grow in mixed infection with λ and recombine with it. These phage lines differed from λ and from one another in numerous respects. Some obvious characteristics were the specificity of repression and the site of insertion into the bacterial chromosome. More recently, the spectrum of λ-related phages was further extended by the isolation of viable hybrids between λ and the *Salmonella* phage P 22 (Botstein and Herskowitz, 1974).

One of the earliest uses to which these strains were put was in localizing determinants of repressor specificity. Phage λ was crossed with another phage, 434. Both λ and 434 are temperate. Lysogens of either one are immune to superinfection by phage of the carried type. However, in the heterologous infections (434 infecting a λ lysogen, or λ infecting a 434 lysogen), there is no such immunity. λ repressor does not impede 434 development (nor does it bind to 434 DNA *in vitro*), or vice versa.

From a series of backcrosses between 434 and λ, in which recombinants were selected to derive their immunity characteristics from 434 and other properties from λ, Kaiser and Jacob (1957) obtained a hybrid phage differing from λ only in a small segment of the genetic map that includes the repressor gene *cI*, the two operator sites at which repressor acts, and some adjacent DNA. Within the latter is the *cro* gene (see Section 2.1.2), whose product acts on the same sites as the repressor itself. The *cro* product, like the repressor, is immunity specific; i.e., λ *cro* product is effective only on λ sites, not on 434 sites.

Electron microscopy of DNA heteroduplexes between λ and the hybrid phage show that the two differ from each other in two almost adjacent segments of the DNA molecule, in which the similarity, if any, in base sequence is insufficient to allow double helix formation (Westmoreland *et al.*, 1969). Genetic recombination, like duplex formation, is undetectable throughout most of the region, but in the small segment of homology separating the two nonhomology loops, rare recombination can occur (Wilgus *et al.*, 1973).

Further studies of other λ-related phages, and of other parts of the genome, either by genetic recombination (Liedke-Kulke and Kaiser, 1967) or by electron microscopy of heteroduplexes (Simon *et al.*, 1971; Fiandt *et al.*, 1971; Fig. 13) have extended this picture. Within any two phages, such as λ and 434 and λ and 21, there are some segments of homology, frequently gene-sized or larger, separated by segments where, as in the immunity region, there is no sign of similarity between the two partners. Looking at any such pair, we can therefore picture the genome as composed of several separate modules, within which recombination cannot occur in heterologous matings. Stretches of homology might then either lie between modules or comprise modules common to the two partners.

In terms of combining modules from functionally diverse parents, the situation is especially striking for the λ-P 22 hybrids. These viruses differ from each other not only in that they have no known common host in nature, but also in their mechanism of DNA packaging. λ packages DNA by recognition of specific *cos* sequences that become DNA termini, and P 22 by cutting headful lengths of DNA that may terminate in any base sequence. Their tail morphology is also distinct: λ has a long flexible tail with no base plate or fibers; P 22 has a base plate composed of six spikes, connected directly to the phage head. We might expect that two such dissimilar structures must have a long history of separate evolution and in fact need not have derived from a common source. Nevertheless, recombinants with the immunity and replication determinants of P 22 and the packaging genes of λ are formed and are viable. Furthermore, the specificity of the P 22 repressor is identical to that of the λ-related coliphage 21, and the immunity regions of the two phages are also identical by heteroduplex analysis (Botstein and Herskowitz, 1974).

Whereas these observations may suggest a modular hypothesis,

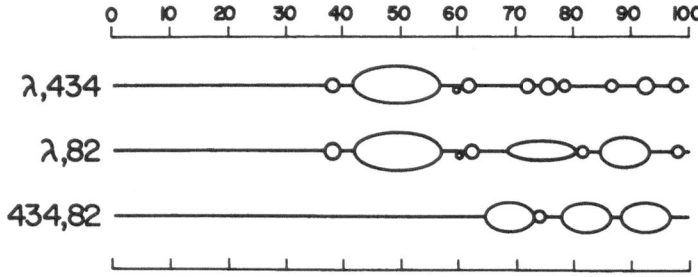

Fig. 13. Heteroduplex maps of λ and the related phages 82 and 434. Lines are segments of homology; loops indicate nonhomology. Redrawn from Simon *et al.*, 1971.

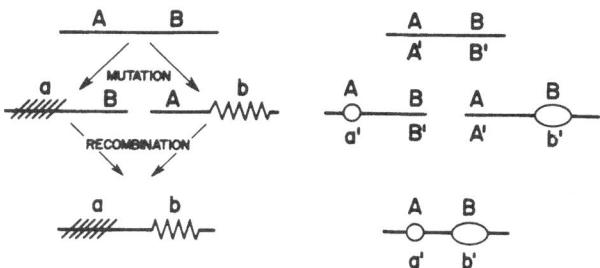

Fig. 14. Generation of changes in DNA base sequence by multiple successive mutations in localized segments of different genetic lines, followed by recombination to yield the fourth possible combination (ab). At the right are shown the appearance of heteroduplexes between the starting strain AB and each of the four genotypes.

nothing said so far eliminates the alternative explanation that different natural lines may have each evolved directly from a common ancestor, evolution having been more rapid in some lines than in others, and within a line, more rapid in some genome segments than in others.

The most convincing evidence for a recombinational origin of existing strains would be the occurrence of enough different combinations of specific determinants at two or more loci so that no family tree can be drawn relating all individuals except by postulating independent recurrence of one or more specific changes from the ancestral type. The latter possibility is tenable for any change that can happen in a few mutational steps but becomes less credible for extensive base-sequence differences recognizable in heteroduplexes, or for characters such as immunity or insertion site that require a high degree of specificity. For two loci it suffices to find natural strains that exhibit all four possible combinations of two allelic alternatives in two separate regions of the genome (Fig. 14). Any three of the four is inadequate to demonstrate the conclusion. Formalization of the requirements for more than two loci, or more than two alleles per locus, will not be developed here.

Figure 13 shows the alternating segments of homology and non-homology observed in pairwise heteroduplexes of the three phages λ, 82, and 434. The phages are obviously related, and λ clearly differs from 434 and 82 in the central part of the genome (37–64%). In the right half of the genome, all three have undergone extensive changes. It is not obvious by the two-locus criterion of Fig. 14 whether these differences are ascribable to recombination between "yet other races" of phage (Hershey, 1971) or by extensive mutational divergence (which must have taken place at some stage under most hypotheses). A fourth phage, which is homologous to 82 and 434 from 0–60% and to λ from

80 to 100%, would assure us of recombinational origin but is not yet available.

The alternative of examining natural strains for combinations of specific characteristics seems inherently less direct than heteroduplex analysis. We might be mistaken in thinking that the same insertion specificity is unlikely to have evolved twice, for example. It has the compensating advantage that the characters studied represent functionally homologous participants in the viral life cycle, whereas heterology in base sequence can originate by insertion of nonfunctional DNA (Section 3.2).

Hershey and Dove (1971) tabulated the properties of λ and some related phages. Table 5 has been modified from their table by omitting such characters as host range and tail antigen that might easily change by mutation and including Q gene specificity and *Salmonella* phage P 22. Though showing considerable variety, nothing in these data requires a recombinational origin. All phages shown could be derived by divergence from an ancestor with the insertion and N specificity of λ, the Q of phage 80, and any immunity.

The modular hypothesis, as we have stated it, implies not only that related viruses should have structurally distinct but functionally exchangeable segments, but also that each of these segments should comprise some sort of natural unit—like the *int* and *xis* genes and their target site, *att*. We might then expect that the termini of these natural units would show up in Fig. 13 as specific boundaries between homologous and heterologous regions that would be common to all phage

TABLE 5

Specific Properties of Lambda and Some Relatives

Phage	Prophage location	N gene product	Immunity	Replication	Q gene product
λ	gal-bio[a]	—	unique[a]	—	—
21	near trp[a]	unlike λ[a]	–	?	?
φ80	tlk-trp[a]	unlike λ[a]	unique[a]	?	unlike λ[e]
φ81	gal-bio[a]	?	unique[a]	?	like φ80[e]
82	gal-bio[a]	?	unique[a]	?	like φ80
424	near his[a]	?	unique[a]	?	unlike λ or φ80[e]
434	gal-bio[a]	like λ[a]	unique[a]	?	unlike λ or φ80[e]
P 22	near pro[b]	unlike λ or 21[c]	like 21[d]	unlike λ	like λ[d]

[a] Taken from Hershey and Dove, 1971.
[b] In *Salmonella typhimurium*. See Fig. 1.
[c] Botstein and Susskind, 1974.
[d] Botstein and Herskowitz, 1974.
[e] Schleif, 1972.

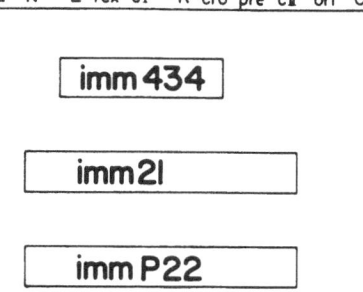

cⅢ N P_L rex cI P_R cro pre cⅡ ori O

Fig. 15. Immunity regions of λ-related phages. In each case, the bar represents the DNA segment within which viable recombinants are not found when the phage in question is crossed with λ. The top line shows the λ genes in this part of the genome (cf. Fig. 2). Data from Botstein and Herskowitz (1974).

pairs. On the other hand, progressive differentiation from a common ancestor that already had the overall gene order of λ might conserve certain segments in one line and different ones in others.

The latter statement seems more descriptive of the actual results. Furthermore, in genetic studies of functional clusters such as the immunity region, the extent of heterology among various phages differs with the phage pair studied. Figure 15 shows the results of such investigation for phage λ tested against the three partners 434, 21 and P 22. Experimentally, heterology is here defined by the absence of viable recombinants. The very rare recombination within the λ-434 immunity region (Wilgus *et al.*, 1973) is ignored. Four facts may be noted: (1) The determinants *cI* and *cro* and their target sites are inseparable in heterologous crosses. (2) With some pairs, but not others, recombination can take place between P_L and the *N* gene, whose product acts on the same target sites as those of *cI* and *cro*. (3) In some cases, gene *cII* is also not exchangeable with *cI*, though its target site is distinct from *cI*. (4) Within the sensitivity of the methods employed, regions of heterology always terminate between rather than within genes. This last property cannot hold throughout the λ genome because recombination between λ and φ80 within the exonuclease gene yields a hybrid exonuclease different from that of either parent (Radding *et al.*, 1967).

Taken together, these results indicate that, if the modular hypothesis has any significance, nature must subdivide the regulatory region into at least three separable modules.

Hershey and Dove (1971) pointed out another feature of Fig. 13 and similar data: "Identical segments occupy positions characteristic of the pair and lie approximately equidistant from the left molecular end in both members of the pair." This fits the notion that all the phages of Fig. 13 are descended from a common ancestor that was already a virus with the same general gene arrangement seen today. At the least, it places strong restrictions on alternative hypotheses (see Section 6.1.3). The inference that the λ-related phages have evolved by local differentiation followed by recombination between different members of the group was qualified by Hershey (1971) to the extent that "DNA segments brought in from outside the λ family by illegitimate recombination may have played a dominant role in the family history."

The difficulty with this concept is that the more dominant the role of such introgression is, the harder it becomes to regard the λ family as an isolatable phylogenetic unit, whose members are connectable by a family tree that remains within the family. The problem is illustrated by phage P 22, whose relation with λ was unknown in 1971. Technically, P 22 does not belong to the λ family, whose defining character is the base sequence of the molecular ends. Less formally, and more to the point, P 22's morphology and packaging mechanism differ so much from those of λ that functional homology between the genes determining virion components seems highly unlikely. Nevertheless, heteroduplex analysis between λ-related coliphages and a hybrid (λ *imm* P 22) of λ and P 22 (Fig. 16) shows that the early genes of the two phages must derive from a common source.

In the right third of the molecule this λ *imm* P 22 hybrid differs from λ in four separate heterologous regions. Therefore, at least five distinct segments of P 22 must be homologous to corresponding segments in λ. Furthermore, the repressor of P 22 is functionally identical to that of phage 21, which is remarkable considering the diversity of specific repressors among members of the λ family itself (Table 5). P 22 and 21 are homologous in part, but not all, of the immunity region (Fig. 16).

P 22 could of course be a rather recent recombinant between an unknown member of the λ family (not necessarily a coliphage) and a phage of some other family. This solution dodges the basic problem that if such interfamilial recombination is frequent, separation of viruses into distinct families becomes impossible. In any event, the properties of P 22 *vis-à-vis* λ and 21 comprise the most direct evidence for thinking that recombination has played an important part in the evolution of these phages.

Whatever the mechanism, local diversification with similarity of overall gene order is characteristic not only of phages but of their bacterial hosts as well (Zinder, 1960).

6.1.3. Genetic Exchange between Virus and Host

Campbell (1972) noted that "a phage such as lambda could have been constructed in a few steps by heterologous fusion between a non-viral plasmid and host genes determining DNA packaging, autolysis,

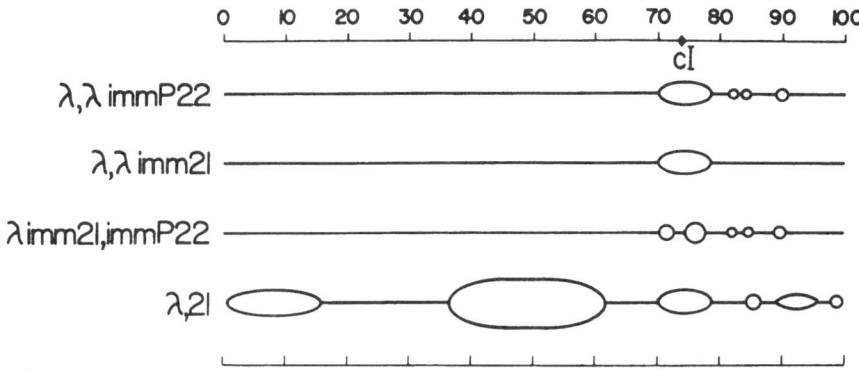

Fig. 16. Heteroduplex analysis of a hybrid phage, λ *imm* P 22, with the immunity of P 22 and the late genes of λ. λ *imm*21 is a hybrid between λ and 21 that contains the immunity region of 21 and remaining genes from λ. Based on Botstein and Herskowitz (1974). For comparison, heteroduplex analysis of λ and 21 (based on Simon *et al.*, 1971) is shown in the last line.

and recombination, respectively." This hypothesis is pertinent to the present discussion only if such fusions have recurred rather frequently, so that some of the differences among related phages arise from incorporation of different segments of foreign DNA.

As mentioned in Section 6.1.2, the fact that different members of the λ family are homologous at several corresponding positions along their lengths is understandable if they are all descended from a common viral ancestor with the same general gene order as present members of the family. If that conclusion could be stated with certainty, then any postulated origin of λ by association of modules gleaned from nonviral sources must antedate the formation of the common ancestor and hence becomes irrelevant to the data under discussion. However, all the results really require is that, if λ arose by association of nonviral elements, then other members of the family were generated by combination of some of the same elements (or of other elements partially homologous to them) arranged in the same order. This condition is restrictive, but by no means absurd. As discussed in Section 6.1.2, the idea of an ancestral λ determining a fixed matrix into which foreign DNA can be intercalated is unlikely to be uniquely superior to other hypotheses when a wider data base is available, even if recombination is entirely interviral.

This being said, we may ask whether the hypothesis that viruses do arise from, and/or give rise to, host modules helps to explain any of the facts recorded in this chapter. In general, the answer is no.

If viruses are in fact collections of host modules, we might expect that a category like "defective phages" would ultimately break down completely, because as knowledge increases we should encounter more and more objects whose classification as viral rather than host would be ambiguous. In fact, the most consistent feature of the examples presented in Section 4 is that the reverse is true. Objects that seemed phage-related by one criterion have proved, on subsequent analysis, to share so many other characteristics with known phages that they become readily classified as viral rather than nonviral. At least in these cases the ambiguity decreases with increasing knowledge, as expected if the distinction between viral and nonviral is natural rather than forced.

For example, the initial observation that phage P 2 can generate a heteroimmune recombinant in *E. coli* B (Section 4.3.1) might have meant that an immunity module homologous to those of temperate phages had been integrated into the circuitry of some host operon. However, subsequent analysis indicates a DNA segment with considerable homology to P 2 itself, as judged by rescue of alleles to various P 2 mutations, and ability to be cured by P 2 infection. Similarly, the defective *Vibrio* phages (Section 4.1.3), whose most obvious property is production of phagelike tails, prove also to form some phage heads and to exhibit immunity to active *Vibrio* phages of similar morphology.

The statistics are not large and are subject to many biases. Recombinational rescue is easier to detect where homology is large. Virion components are more readily observed and studied where their syntheses, like that of phages, is inducible by UV or mitomycin. The data do not exclude the hypothesis but as yet provide little encouragement for it.

6.2. Relation to Normal Cellular Components and Cellular Evolution

Most viruses combine at least two essential properties: autonomous replication and virion formation. Some temperate bacteriophages have at least two more: chromosomal insertion and regulated repression. Among the lambdoid phages of the *Enterobacteriaceae*, these four functions are coded by determinants that are spatially separated along the chromosome, so that, in principle, the segments of the genome determining these four separate functions might survive apart from each other. This situation is experimentally approximated with deletion derivatives of the phage.

Of these four major functions, many nonviral agents are known that replicate in a similar manner but have none of the other phage functions. Insertion into the chromosome is likewise an attribute of various elements. Some of these are unlikely, because of their limited coding capacity, to replicate autonomously (see Section 3.2). At present, the regulatory region of the DNA temperate phages is unique. Whereas we can imagine this module being recombined into new associations, no other prokaryotic genes are known to be controlled in a very similar manner.

As discussed in Section 4.1, virion formation presents special difficulties. The only place in nature where it serves a demonstrated function is in the viral life cycle. It therefore may be simplest to regard determinants of virion formation as degenerate viruses (Bradley, 1967) unless proven otherwise. Actual cases (Sections 4 and 6.1.3) indicate that the resemblance between natural defective phages and the products of phage or host mutations deranging normal lysogeny is more than superficial. However, we would hardly classify any replicon, or any inserting element, as a degenerate virus merely because it shares one property of a typical temperate phage. The frequency with which virion determinants occur and their persistence through evolutionary time (Sections 4.1.1 and 4.1.3) suggest that they serve some function, viral or otherwise, not yet fully appreciated by virologists.

It is perhaps worth mentioning that there are two distinct kinds of genetic degeneration. On the one hand, there is the simple accumulation of genetic defects, which creates progressively more useless relics of elements that were once functional. Alternatively, there is the evolution of entities with more specialized and restricted activities than their parents had. The evolution of biosynthetic capacities, which must at some time have tended toward increasing nutritional independence, seems in recent times to have proceeded mainly by loss of function (Lwoff, 1944). The reason that the rate of loss exceeds the rate of gain, rather than achieving a steady-state balance, is that there is increasing specialization within communities of interdependent species. Some of the same considerations should apply to viruses, though direct evidence is lacking.

ACKNOWLEDGMENT

This work was supported in part by the National Institute of Allergy and Infectious Diseases, National Institutes of Health Grant AI08573.

7. REFERENCES

Adhya, S., and Campbell, A., 1970, Crypticogenicity of bacteriophage λ, *J. Mol. Biol.* **50**, 481.

Adhya, S., Gottesman, M., and de Crombrugghe, B., 1974, Release of polarity in *Escherichia coli* by gene *N* of phage λ: termination and antitermination of transcription, *Proc. Nat. Acad. Sci. USA* **71**, 2534.

Appleyard, R. K., 1956, The transfer of defective λ lysogeny between strains of *Escherichia coli*, *J. Gen. Microbiol.* **14**, 573.

Arai, T., and Clowes, R., 1975, Replication of stringent and relaxed plasmids, *in* "Microbiology—1974" (D. Schlessinger, ed.), pp. 141–155, Amer. Soc. Microbiol., Washington, D.C.

Baltimore, D., Cole, D. N., Villa-Komaroff, L., and Spector, D., 1974, Poliovirus defective interfering particles, *in* "Mechanisms of Virus Disease" (W. S. Robinson and C. F. Fox, eds.), pp. 117–130, Benjamin, Menlo Park, California.

Banerjee, A. K., Rensing, U., and August, J. T., 1969, Replication of a natural 6 S RNA by the Qβ RNA polymerase, *J. Mol. Biol.* **45**, 181.

Baron, L. S., Penido, E., Rymon, I. R., and Falkow, S., 1970, Behavior of coliphage lambda in hybrids between *Escherichia coli* and *Salmonella*, *J. Bacteriol.* **102**, 221.

Barrett, K., Barclay, S., Calendar, R., Lindqvist, B., and Six, E., 1974, Reciprocal *trans*activation in a two chromosome phage system, *in* "Mechanisms of Virus Disease" (W. S. Robinson and C. F. Fox, eds.), pp. 385–402, Benjamin, Menlo Park, California.

Beckwith, J. R., 1970, Lac: the genetic system, *in* "The Lactose Operon" (J. R. Beckwith and D. Zipser, eds.), pp. 5–26, Cold Spring Harbor Laboratory, Cold Spring Harbor, New York.

Benzer, S., 1957, The elementary units of heredity, *in* "The Chemical Basis of Heredity" (W. D. McElroy and B. Glass, eds.), pp. 70–98, Johns Hopkins Press, Baltimore, Maryland.

Berg, C. M., and Curtiss, R., III, 1967, Transposition derivatives of an Hfr strain of *Escherichia coli* K-12, *Genetics* **56**, 503.

Berg, D. E., 1974*a*, Genes of phage λ essential for λ *dv* plasmids, *Virology* **62**, 224.

Berg, D. E., 1974*b*, Genetic evidence for two types of gene arrangements in new λ *dv* plasmid mutants, *J. Mol. Biol.* **86**, 59.

Berg, D. E., and Kellenberger-Gujer, G., 1974, *N* protein causes the λ *dv* plasmid to inhibit heteroimmune phage λ *imm* 434 growth and stimulates λ *dv* replication, *Virology* **62**, 234.

Bernstein, A., Rolfe, B., and Onodero, K., 1973, The *E. coli* cell surface: isolation of λ transducing phages carrying the *tol PAB* cluster, *Mol. Gen. Genet.* **121**, 325.

Bertani, L. E., and Bertani, G., 1971, Genetics of P 2 and related phages, *Adv. Genet.* **16**, 199.

Botstein, D., and Herskowitz, I., 1974, Properties of hybrids between *Salmonella* phage P 22 and coliphage λ, *Nature* (*London*) **251**, 584.

Botstein, D., and Suskind, M., 1974, Regulation of lysogeny and the evolution of temperate bacterial viruses, *in* "Mechanisms of Virus Disease" (W. S. Robinson and C. F. Fox, eds.), pp. 363–384, Benjamin, Menlo Park, California.

Boon, T., 1972, Inactivation of ribosomes *in vitro* by colicin E3, *Proc. Nat. Acad. Sci. USA* **68**, 2421.

Brachet, P., and Green, B. R., 1970, Functional analysis of early defective mutants of coliphage λ, *Virology* **40**, 792.

Bradley, D. E., 1967, Ultrastructure of bacteriophages and bacteriocins, *Bacteriol. Rev.* **31**, 230.

Busse, H. G., and Baldwin, R. L., 1972, Tandem genetic duplications of phage lambda. II. Evidence for structure and location of endpoints, *J. Mol. Biol.* **65**, 401.

Calendar, R., Lindqvist, B., Sironi, G., and Clark, A. J., 1970, Characterization of REP⁻ mutants and their interaction with P2 phage, *Virology* **40**, 72.

Campbell, A., 1968, Techniques for studying defective bacteriophages, *in* "Methods of Virology" (K. Maramorosch and H. Koprowski, eds.), Vol. 4, pp. 279–320, Academic Press, New York.

Campbell, A., 1971, Genetic structure, *in* "The Bacteriophage λ" (A. D. Hershey, ed.), pp. 13–44, Cold Spring Harbor Laboratories, Cold Spring Harbor, New York.

Campbell, A., 1972, Episomes in evolution, *in* "Evolution of Genetic Systems" (H. H. Smith, ed.), pp. 534–562 (Brookhaven Symposia in Biology, Volume 23).

Chakrabarty, A. M., and Gunsalus, I. C., 1969, Defective phage and chromosome mobilization in *Pseudomonas putida*, *Proc. Nat. Acad. Sci. USA* **64**, 1217.

Chan, R. K., Botstein, D., Watanabe, T., and Okada, Y., 1972, Specialized transduction of tetracycline resistance by phage P 22 in *Salmonella typhimurium*. II. Properties of a high-frequency-transducing lysate, *Virology* **50**, 883.

Chattoraj, D. K., and Inman, R. B., 1974, Location of DNA ends in P 2, 186, P 4, and lambda bacteriophage heads, *J. Mol. Biol.* **87**, 11.

Chattoraj, D. K., and Inman, R. D., 1975, Electron microscope heteroduplex mapping of P 2 Hy *Dis* bacteriophage DNA, *Virology* **55**, 174.

Chow, L. T., and Davidson, N., 1973, Electron microscopy of the structures of the *Bacillus subtilis* prophages SPO 2 and ϕ105, *J. Mol. Biol.* **75**, 257.

Chow, L. T., Davidson, N., and Berg, D., 1974, Electron microscope study of the structure of λ *dv* DNA's, *J. Mol. Biol.* **86**, 69.

Cohen, D., 1959, A variant of phage P 2 originating in *Escherichia coli*, strain B, *Virology* **7**, 112.

Cohen, S. S., and Barner, H. D., 1954, Studies of unbalanced growth in *Escherichia coli*, *Proc. Nat. Acad. Sci. USA* **40**, 885.

Coppo, A., Manzi, A., Pulitzer, J. F., and Takahashi, H., 1973, Abortive bacteriophage T 4 head assembly in mutants of *Escherichia coli*, *J. Mol. Biol.* **76**, 61.

Cross, R. A., and Lieb, M., 1967, Heat inducible λ phage, V. Induction of prophage with mutations in genes *O, P,* and *R*, *Genetics* **57**, 549.

Davidson, N., Deonier, R. C., Hu, S., and Ohtsubo, E., 1975, Electron microscope heteroduplex studies of sequence relations among plasmids of *Escherichia coli*. X. Deoxyribonucleic acid sequence organization of F and of F-primes and the sequences involved in Hfr formation, *in* "Microbiology-1974" (D. Schlessinger, ed.), Amer. Soc. Microbiol., Washington, D.C.

Doermann, A. H., 1973, T 4 and the rolling circle mode of replication, *Annu. Rev. Genet.* **7**, 325.

Doermann, A. H., Chase, M., and Stahl, F. W., 1955, Genetic recombination and replication in bacteriophage, *J. Cell. Comp. Physiol.* **45**, Suppl. 2, 51.

Doermann, A. H., Eiserling, F. A., and Boehner, L., 1973, Genetic control of capsid length in bacteriophage T 4. I. Isolation and preliminary description of four new mutants, *J. Virol.* **12**, 374.

Dove, W., Inokuchi, H., and Stevens, W. F., 1971, Replication control in phage lambda, *in* "The Bacteriophage Lambda" (A. D. Hershey, ed.), Cold Spring Harbor Laboratory, Cold Spring Harbor, New York.

Duckworth, D. H., 1970, The metabolism of T 4 phage infected cells. I. Macromolecular synthesis and transport of nucleic acids and protein precursors, *Virology* **40**, 673.

Echols, H., 1971, Regulation of lytic development, *in* "The Bacteriophage Lambda" (A. D. Hershey, ed.), pp. 247–270, Cold Spring Harbor Laboratory, Cold Spring Harbor, New York.

Echols, H., 1972, Developmental pathways for the temperate phage: lysis vs. lysogeny, *Annu. Rev. Genet.* **6**, 157.

Eisen, H., and Ptashne, M., 1971, Regulation of repressor synthesis, *in* "The Bacteriophage Lambda" (A. D. Hershey, ed.), pp. 239–245, Cold Spring Harbor, New York.

Eisen, H. A., Fuerst, C. R., Siminovitch, L., Thomas, R., Lambert, L., Pereira da Silva, L., and Jacob, F., 1966, Genetics and physiology of defective lysogeny in K 12 (λ): Studies of early mutants, *Virology* **30**, 224.

Emmons, S. W., and Thomas, J. O., 1975, Tandem genetic duplications in phage lambda. IV. The location of spontaneously arising tandem duplications, *J. Mol. Biol.* **91**, 147.

Fiandt, M., Hradecna, Z., Lozeron, H. A., and Szybalski, W., 1971, Electron micrographic mapping of deletions, insertions, inversions, and homologies in the DNAs of coliphage lambda and phi 80, *in* "The Bacteriophage Lambda" (A. D. Hershey, ed.), pp. 329–354, Cold Spring Harbor Laboratory, Cold Spring Harbor, New York.

Fiandt, M., Szybalski, W., and Malamy, M. H., 1972, Polar mutations in *lac, gal* and λ consist of a few IS-DNA sequences inserted with either orientation, *Mol. Gen. Genet.* **119**, 223.

Fischer-Fantuzzi, L., and Calef, E., 1964, A type of λ prophage unable to confer immunity, *Virology* **23**, 209.

Franklin, N. C., 1971, Illegitimate recombination, *in* "The Bacteriophage Lambda" (A. D. Hershey, ed.), pp. 175–194, Cold Spring Harbor Laboratory, Cold Spring Harbor, New York.

Franklin, N. C., 1974, Altered reading of genetic signals fused to the *N* operon of bacteriophage λ: Genetic evidence for modification of polymerase by the protein product of the *N* gene, *J. Mol. Biol.* **89**, 33.

Franklin, N. C., Dove, W. F., and Yanofsky, C., 1965, A linear insertion of a prophage into the chromosome of *E. coli* shown by deletion mapping, *Biochem, Biophys. Res. Commun.* **18**, 910.

Friedman, D. I., 1971, A bacterial mutant affecting lambda development, *in* "The Bacteriophage λ" (A. D. Hershey, ed.), pp. 733–738, Cold Spring Harbor Laboratory, Cold Spring Harbor, New York.

Garro, A. J., and Marmur, J., 1970, Defective bacteriophages, *J. Cell Physiol.* **76**, 253.

Georgopolous, C. P., 1971, A bacterial mutation affecting *N* function, *in* "The Bacteriophage λ" (A. D. Hershey, ed.), pp. 639–646, Cold Spring Harbor Laboratory, Cold Spring Harbor, New York.

Georgopolous, C. P., and Herskowitz, I., 1971, *Escherichia coli* mutants blocked in lambda DNA synthesis, *in* "The Bacteriophage λ" (A. D. Hershey, ed.), pp. 553–564, Cold Spring Harbor Laboratory, Cold Spring Harbor, New York.

Georgopolous, C. P., Hendrix, R. W., Casjens, S. R., and Kaiser, A. D., 1973, Host participation in bacteriophage lambda head assembly, *J. Mol. Biol.* **76**, 45.

Gerdes, J. C., and Romig, W. R., 1975, Complete and defective bacteriophages of classical *Vibrio cholerae*: Relationship to the kappa type bacteriophages, *J. Virol.* **15,** 1231.

Ghysen, A., and Pironio, M., 1972, Relationship between the *N* function of bacteriophage λ and host RNA polymerase, *J. Mol. Biol.* **65,** 259.

Goldstein, R., Lengyel, J., Pruss, G., Barrett, K., Calendar, R., and Six, E., 1974, Head size determination and the morphogenesis of satellite phage P4, *Curr. Top. Microbiol. Immunol.* **68,** 59.

Gottesman, M. M., Gottesman, M. E., Gottesman, S., and Gellert, M., 1974, Characterization of λ reverse as an *E. coli* phage carrying a unique set of host-derived recombination functions, *J. Mol. Biol.* **88,** 471.

Gottesman, S., and Beckwith, J. R., 1969, Directed transposition of the arabinose operon: a technique for the isolation of specialized transducing bacteriophages for any *Escherichia coli* gene, *J. Mol. Biol.* **44,** 117.

Grady, L. J., Cowie, D. B., and Campbell, W. P., 1971, Deoxyribonucleic acid hybridization analysis of the defective bacteriophage carried by strain 15 of *Escherichia coli*, *J. Virol.* **8,** 850.

Greer, H. A., 1975, The *kil* gene of bacteriophage lambda, Ph.D. thesis, Massachusetts Institute of Technology, Cambridge, Massachusetts.

Griffith, J., and Kornberg, A., 1974, Mini M 13 bacteriophage: circular fragments of M 13 DNA are replicated and packaged during normal infections, *Virology* **59,** 139.

Grodzicker, T., Arditti, R. R., and Eisen, H., 1972, Establishment of repression by lambdoid phage in catabolite activator protein and adenylate cyclase mutants of *Escherichia coli*, *Proc. Nat. Acad. Sci. USA* **69,** 366.

Guthrie, C., Nashimoto, H., and Nomura, M., 1969, Studies on the assembly of ribosomes in vivo, *Cold Spring Harbor Symp. Quant. Biol.* **34,** 69.

Haas, M., and Yoshikawa, H., 1969, Defective bacteriophage PBSH in *Bacillus subtilis*. II. Intracellular development of the induced prophage, *J. Virol.* **3,** 248.

Haruna, I., and Spiegelman, S., 1965, Specific template requirements of RNA replicase, *Proc. Nat. Acad. Sci. USA* **54,** 1189.

Hendrix, R. W., and Casjens, S. R., 1975, Assembly of bacteriophage lambda heads: protein processing and its genetic control in petit λ assembly, *J. Mol. Biol.* **91,** 187.

Hershey, A. D., 1971, Comparative molecular structure among related phage DNAs, *Carnegie Inst. Wash. Yearb.* **70,** 3.

Hershey, A. D., and Dove, W., 1971, Introduction to lambda, *in* "The Bacteriophage Lambda" (A. D. Hershey, ed.), Cold Spring Harbor Laboratories, Cold Spring Harbor, New York.

Herskowitz, I., and Signer, E., 1970, a site essential for expression of all late genes in bacteriophage λ, *J. Mol. Biol.* **47,** 545.

Herskowitz, I., and Signer, E. R., 1974, Substitution mutations in bacteriophage λ with new specificity for late gene expression, *Virology* **61,** 112.

Hobom, G., and Hogness, D. S., 1974, The role of recombination in the formation of circular oligomers of the λ *dv*1 plasmid, *J. Mol. Biol.* **88,** 65.

Huang, A. S., and Baltimore, D., 1970, Defective virus particles and viral disease processes, *Nature (London)* **226,** 325.

Ikeda, H., Inuzuka, M., and Tomizawa, J., 1970, P 1-like plasmid in *Escherichia coli* 15, *J. Mol. Biol.* **50,** 457.

Jacob, F., and Wollman, E. L., 1954, Étude génétique d'un bacteriophage tempéré d'*Escherichia coli*. I. Le systéme génétique du bacteriophage, *Ann. Inst. Pasteur* **87,** 663.

Jacob, F., and Wollman, E. L., 1961, "Sexuality and the Genetics of Bacteria," Academic Press, New York.

Jacob, F., Fuerst, C., and Wollman, E., 1957, Recherches sur les bacteries lysogénes defectives. II. Les types physiologíques liés aux mutations du prophage, *Ann. Inst. Pasteur* **93**, 724.

Jakes, J., Zinder, N. D., and Boon, T., 1974, Purification and properties of the colicin E3 immunity protein, *J. Mol. Biol.* **249**, 438.

Kacian, D. L., Mills, D. R., Kramer, F. R., and Spiegelman, S., 1972, A replicating RNA molecule suitable for a detailed analysis of extracellular evolution and replication, *Proc. Nat. Acad. Sci. USA* **69**, 3038.

Kaiser, A. D., and Jacob, F., 1957, Recombination between related temperate bacteriophages and the genetic control of immunity and prophage localization, *Virology* **4**, 509.

Kaiser, D., 1971, Lambda DNA replication, *in* "The Bacteriophage Lambda" (A. D. Hershey, ed.), pp. 195–210, Cold Spring Harbor Laboratory, Cold Spring Harbor, New York.

Kaiser, D., Syvanen, M., and Masuda, T., 1975, DNA packaging steps in bacteriophage lambda head assembly, *J. Mol. Biol.* **91**, 175.

Kayajanian, G., 1970, *Gal* transduction by phage λ: On the origin and nature of LFT transducing genomes, *Mol. Gen. Genet.* **108**, 338.

Kellenberger-Gujer, G., Boy de la Tour, E., and Berg, D. E., 1974, Transfer of the λ *dv* plasmid to new bacterial hosts, *Virology* **58**, 576.

Kondo, E., and Mitsuhashi, S., 1964, Drug resistance of enteric bacteria. IV. Active transducing bacteriophage P 1 CM produced by the combination of R factor with bacteriophage P 1, *J. Bacteriol.* **88**, 1266.

Kondo, E., and Mitsuhashi, S., 1966, Drug resistance of enteric bacteria. VI. Introduction of bacteriophage P 1 CM into *Salmonella typhi* and formation of P 1 *dCM* and F-*CM* elements, *J. Bacteriol.* **91**, 1787.

Konisky, J., 1975, Interaction of colicin la with *Escherichia coli*, *in* "Microbiology-1974" (D. Schlessinger, ed.), pp. 213–218, Amer. Soc. Microbiol., Washington, D.C.

Kopecko, D. J., and Cohen, S. N., 1975, Site-specific *recA*-independent recombination between bacterial plasmids: Involvement of palindromes at the recombinational loci, *Proc. Nat. Acad. Sci. USA* **72**, 1373.

Kumar, S., and Szybalski, W., 1970, Transcription of the λ *dv* plasmid and inhibition of λ phages in λ *dv* carrier cells of *Escherichia coli*, *Virology* **41**, 665.

Lee, C. S., Davis, R. W., and Davidson, N., 1970, A physical study by electron microscopy of the terminally repetitious, circularly permuted DNA from the coliphage particles of *Escherichia coli* 15, *J. Mol. Biol.* **48**, 1.

Lieb, M., 1972, Properties of polylysogens containing derepressed λ N⁻ prophage: Interference with the replication of superinfecting λ, *Virology* **49**, 582.

Liedke-Kulke, M., and Kaiser, A. D., 1967, Genetic control of prophage insertion specificity in bacteriophages λ and 21, *Virology* **32**, 465.

Lindqvist, B. H., 1974, Expression of phage transcription in P 2 lysogens infected with helper-dependent coliphage P 4, *Proc. Nat. Acad. Sci. USA* **71**, 2752.

Little, J. W., and Gottesman, M., 1971, Defective lambda particles whose DNA carries only a single cohesive end, *in* "The Bacteriophage Lambda" (A. D. Hershey, ed.), pp. 371–394, Cold Spring Harbor Laboratory, Cold Spring Harbor, New York.

Luria, S. E., and Delbrück, M., 1943, Mutations of bacteria from virus sensitivity to virus resistance, *Genetics* **28**, 491.

Luria, S. E., Adams, M. J., and Ting, R. C., 1960, Transduction of lactose-utilizing ability among strains of *E. coli* and *S. dysenteriae, Virology* **12**, 348.

Lwoff, A., 1944, "L'évolution physiologique, Étude des pertes de fonction chez les microorganismes," Hermann et Cie, Paris.

Lwoff, A., 1953, Lysogeny, *Bacteriol. Rev.* **17**, 269.

Manly, K. F., Signer, E. R., and Radding, C. M., 1969, Non-essential functions of bacteriophage λ, *Virology* **37**, 177.

Marchelli, C., Pica, L., and Soller, A., 1968, The cryptogenic factor in λ, *Virology* **34**, 650.

Matsubara, K., 1972a, Plasmid formation from bacteriophage λ as a result of interference by resident plasmid λ *dv, Virology* **47**, 618.

Matsubara, K. M., 1972b, Interference in phage growth by a resident plasmid λ *dv*. I The role of interference, *Virology* **50**, 713.

Matsubara, K., 1974, Preparation of plasmid λ*dv* from bacteriophage λ: role of promoter-operator in the plasmid replicon, *J. Virol.* **13**, 596.

Matsubara, K., and Kaiser, A. D., 1968, λ *dv*: An autonomously replicating DNA fragment, *Cold Spring Harbor Symp. Quant. Biol.* **33**, 769.

Mills, D. R., Kramer, F. R., and Spiegelman, S., 1973, Complete nucleotide sequence of a replicating RNA molecule, *Science* **180**, 916.

Mosig, G., Carnighan, J. R., Bibring, J. B., Cole, R., Bock, H.-G.O., and Bock, S., 1972, Coordinate variation in lengths of deoxyribonucleic acid molecules and head lengths in morphological variants of bacteriophage T 4, *J. Virol.* **9**, 857.

Mount, D. W. A., Harris, A. W., Fuerst, C. R., and Siminovitch, L., 1968, Mutations in bacteriophage λ affecting particle morphogenesis, *Virology* **35**, 134.

Murialdo, H., and Siminovitch, L., 1972, The morphogenesis of phage lambda. V. Form-determining function of the genes required for the assembly of the head, *Virology* **48**, 824.

Novick, R., 1967, Properties of a cryptic high-frequency transducing phage in *Staphylococcus aureus, Virology* **33**, 155.

Novick, R., Wyman, L., Bouanchoud, D., and Murphy, E., 1975, Plasmid life cycles in *Staphylococcus aureus, in* "Microbiology-1974" (D. Schlessinger, ed.), pp. 115–129, Amer. Soc. Microbiol., Washington, D.C.

Parma, D. H., 1969, The structure of genomes of individual petit particles of the bacteriophage T 4D mutant E920/96/41, *Genetics* **63**, 247.

Press, R., Glansdorff, N., Miner, P., de Vries, J., Kadner, R., and Maas, W. K., 1971, Isolation of transducing particles of φ80 bacteriophage that carry different regions of the *Escherichia coli* genome, *Proc. Nat. Acad. Sci. USA* **68**, 795.

Pruss, G., Barrett, K., Lengyel, J., Goldstein, R., and Calendar, R., 1974, Phage head size determination and head protein cleavage *in vitro, J. Supramol. Struct.* **2**, 337.

Ptashne, K., and Cohen, S. N., 1975, Occurrence of insertion sequences (IS) regions on plasmid deoxyribonucleic acid as direct and inverted nucleotide sequence duplications, *J. Bacteriol.* **122**, 776.

Ptashne, M., 1971, Repressor and its action, *in* "The Bacteriophage Lambda" (A. D. Hershey, ed.), pp. 221–237, Cold Spring Harbor Laboratory, Cold Spring Harbor, New York.

Radding, C. M., Szpirer, J., and Thomas, R., 1967, The structural gene for λ exonuclease, *Proc. Nat. Acad. Sci. USA* **57**, 277.

Reichardt, L. F., 1975a, Control of bacteriophage lambda repressor synthesis after phage infection. The role of the *N, cII, cIII* and *cro* products, *J. Mol. Biol.* **93**, 267.

Reichardt, L. F., 1975*b*, Control of bacteriophage lambda repressor synthesis: Regulation of the maintenance pathway by the *cro* and *cI* products, *J. Mol. Biol.* **93**, 289.

Reif, H. J., and Saedler, H., 1975, IS1 is involved in deletion formation in the *gal* region of *E. coli* K 12, *Mol. Gen. Genet.* **137**, 17.

Rosner, J. L., 1972, Formation, induction and curing of bacteriophage P 1 lysogens, *Virology* **48**, 679.

Saedler, H., Reif, H. J., Hu, S., and Davidson, N., 1974, IS2, a genetic element for turn-off and turn-on of gene activity in *E. coli, Mol. Gen. Genet.* **132**, 265.

Sato, K., and Campbell, A., 1970, Specialized transduction of galactose by lambda phage from a deletion lysogen, *Virology* **41**, 474.

Schleif, R., 1972, The specificity of lamboid phage late gene induction (lamboid phage late gene specificity), *Virology* **40**, 610.

Schmieger, H., and Backhaus, H., 1973, The origin of DNA in transducing particles in P 22-mutants with increased transduction-frequencies (HT-mutants), *Mol. Gen. Genet.* **120**, 181.

Scott, J. R., 1970*a*, A defective P 1 prophage with a chromosomal location, *Virology* **40**, 144.

Scott, J. R., 1970*b*, Clear plaque mutants of phage P 1, *Virology* **41**, 66.

Scott, J. R., 1973, Phage P 1 cryptic. II. Location and regulation of prophage genes, *Virology* **53**, 327.

Scott, J. R., 1975, Superinfection immunity and prophage repression in phage P 1, *Virology* **65**, 173.

Shimada, K., and Campbell, A., 1974, Int-constitutive mutants of bacteriophage lambda, *Proc. Nat. Acad. Sci. USA* **71**, 237.

Shimada, K., Weisberg, R. A., and Gottesman, M. E., 1972, Prophage lambda at unusual chromosomal locations. I. Location of the secondary attachment sites and properties of the lysogens, *J. Mol. Biol.* **63**, 483.

Shimada, K., Weisberg, R. A., and Gottesman, M. E., 1973, Prophage lambda at unusual chromosomal locations. II. Mutations induced by bacteriophage lambda in *Escherichia coli* K 12, *J. Mol. Biol.* **80**, 297.

Shimada, K., Weisberg, R. A., and Gottesman, M. E., 1975, Prophage lambda at unusual chromosomal locations. III. The components of secondary attachment sites, *J. Mol. Biol.* **93**, 415.

Sidikaro, J., and Nomura, M., 1974, E3 immunity substance, a protein from E3-colicinogenic cells that accounts for their immunity to colicin E3, *J. Biol. Chem.* **249**, 445.

Signer, E. R., 1969, Plasmid formation: A new mode of lysogeny by phage λ, *Nature* (*London*) **223**, 158.

Simon, M. N., Davis, R. W., and Davidson, N., 1971, Heteroduplexes of DNA molecules of lambdoid phages: Physical mapping of their base sequence relationships by electron microscopy, *in* "The Bacteriophage Lambda" (A. D. Hershey, ed.), Cold Spring Harbor Laboratory, Cold Spring Harbor, New York.

Six, E. W., and Klug, C. A. C., 1973, Bacteriophage P 4: A satellite virus depending on a helper such as P 2, *Virology* **51**, 327.

Skalka, A., and Enquist, L. W., 1974, Overlapping pathways for replication, recombination and repair in bacteriophage lambda, *in* "Mechanism of DNA replication" (A. R. Kolber and M. Kohiyama, eds.), pp. 181–200, Plenum Press, New York.

Starlinger, P., and Saedler, H., 1972, Insertion mutations in microorganisms, *Biochimie* **54**, 177.

Sternberg, N., 1973, Properties of a mutant of *Escherichia coli* defective in bacteriophage λ head formation. II. The propagation of phage λ, *J. Mol. Biol.* **76,** 35.

Stodolsky, M., 1973, Bacteriophage P 1 derivatives with bacterial genes: A heterozygote enrichment method for the selection of P 1 *dpro* lysogens, *Virology* **53,** 471.

Streisinger, G., Emrich, J., and Stahl, M. M., 1967, Chromosome structure in phage T 4. III. Terminal redundancy and length determination, *Proc. Nat. Acad. Sci. USA* **57,** 292.

Sunshine, M. G. and Sauer, B., 1975, A bacterial mutation blocking P2 phage late gene expression, *Proc. Nat. Acad. Sci. USA* **72,** 2770.

Szybalski, W., 1974, Bacteriophage Lambda, *in* "Handbook of Genetics" (R. C. King, ed.), pp. 309–322, Plenum Press, New York.

Szybalski, W., and Szybalski, E. H., 1974, Visualization of the evolution of viral genomes, *in* "Viruses, Evolution and Cancer" (E. Kurstak and K. Maramorosch, eds.), pp. 563–582, Academic Press, New York.

Takeda, Y., Matsubara, K., and Ogata, K., 1975, Regulation of early gene expression in bacteriophage lambda: Effect of *tof* mutations on strand-specific transcription, *Virology* **65,** 374.

Thurm, P., and Garro, A. J., 1975*a*, Bacteriophage-specific protein synthesis during induction of the defective *Bacillus subtilis* bacteriophage PBSX, *J. Virol.* **16,** 179.

Thurm, P., and Garro, A. J., 1975*b*, Isolation and characterization of prophage mutants of the defective *Bacillus subtilis* bacteriophage PBSX, *J. Virol.* **16,** 184.

Toussaint, A., 1969, Insertion of phage Mu-1 within prophage λ: A new approach for studying the control of late functions in bacteriophage λ, *Mol. Gen. Genet.* **106,** 89.

Tye, B., Chan, R. K., and Botstein, D., 1974*a*, Packaging of an oversize transducing genome by phage P 22, *J. Mol. Biol.* **85,** 485.

Tye, B., Huberman, J., and Botstein, 1974*b*, Nonrandom circular permutation of phage P 22 DNA, *J. Mol. Biol.* **85,** 501.

Vallee, M., and Cornett, J. B., 1972, A new gene of bacteriophage T 4 determining immunity against superinfecting ghosts and phages in T 4-infected *Escherichia coli,* *Virology* **48,** 777.

Vallee, M., Cornett, J. B., and Bernstein, H., 1972, The action of bacteriophage T 4 ghosts on *Escherichia coli* and the immunity to this action developed in cells preinfected with T 4, *Virology* **48,** 766.

Walker, D. H., and Anderson, T. F., 1970, Morphological variants of coliphage P 1, *J. Virol.* **5,** 765.

Walker, D. H., Mosig, G., and Bajer, M. E., 1972, Bacteriophage T 4 head models based on icosahedral symmetry, *J. Virol.* **9,** 872.

Wang, J. C., and Kaiser, A. D., 1973, Evidence that the cohesive ends of mature λ DNA are generated by the gene A product, *Nature (London) New Biol.* **241,** 16.

Weil, J., deWein, N., and Casale, A., 1975, Morphogenesis of λ with genomes containing excess DNA: Functional particles containing 12 and 15% excess DNA, *Virology* **63,** 352.

Weisberg, R., and Gottesman, M., 1969, The integration and excision defect of bacteriophage λ *dg, J. Mol. Biol.* **46,** 565.

Westmoreland, B. C., Szybalski, W., and Ris, H., 1969, Mapping of deletions and substitutions in heteroduplex DNA molecules of bacteriophage lambda by electron microscopy, *Science* **163,** 1343.

Wilgus, G. S., Mural, R. J., Friedman, D. I., Fiandt, M., and Szybalski, W., 1973, λ *imm* λ 434: A phage with a hybrid immunity region, *Virology* **56,** 46.

Zavada, V., and Calef, E., 1969, Chromosomal rearrangements in *Escherichia coli* strains carrying the cryptic lambda prophage, *Genetics* **61,** 9.

Zinder, N. D., 1960, Hybrids of *Escherichia* and *Salmonella, Science* **131,** 813.

Zissler, J., Signer, E., and Schaefer, F., 1971, The role of recombination in growth of bacteriophage lambda. I. The gamma gene, *in* "The Bacteriophage Lambda" (A. D. Hershey, ed.), pp. 455 468, Cold Spring Harbor Iaboratory Cold Spring Harbor. New York.

CHAPTER 4

The P 2–P 4 *Trans*activation System

Richard Calendar and Janet Geisselsoder

Molecular Biology Department
University of California
Berkeley, California

Melvin G. Sunshine and Erich W. Six

Microbiology Department
University of Iowa
Iowa City, Iowa

and

Björn H. Lindqvist

Institute of Medical Biology
University of Tromso
Norway

1. INTRODUCTION

The interactions between bacteriophage P 2 and its satellite phage P 4 comprise a promising model system for studying the control of transcription because the two phage have relatively small and therefore tractable genomes and yet exhibit a surprisingly complex set of effects upon one another. Each phage possesses the ability to drastically alter the expression of the other's genome.

P 2 is a noninducible temperate phage of *Escherichia coli* isolated by Bertani (1951). P 4 was isolated by Six (1963) who showed that it re-

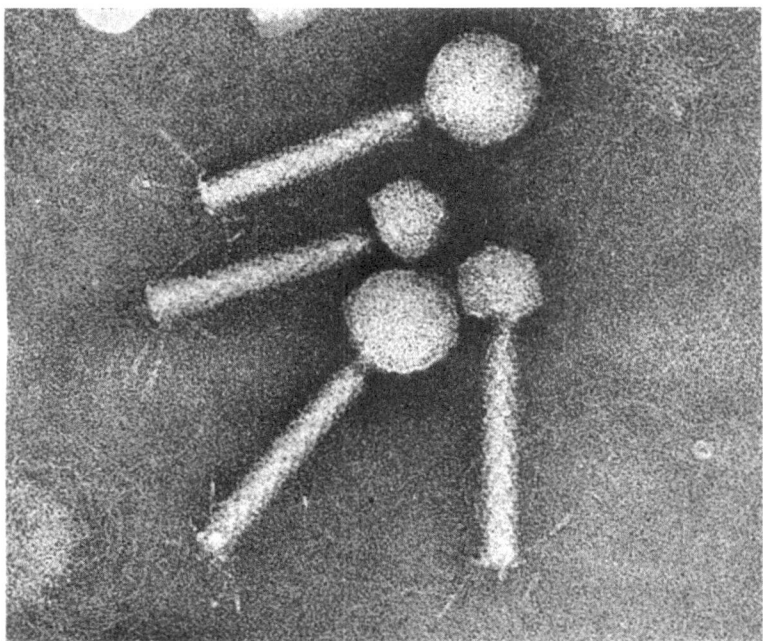

Fig. 1.   Temperate phage P 2 (large heads) and satellite phage P 4 (small heads) at 240,000-fold magnification, stained with phosphotungstic acid and photographed using minimal beam exposure (Williams and Fisher, 1970). Courtesy of Robley C. Williams.

quires a helper phage such as P 2 for its successful lytic multiplication. Figure 1 is an electron micrograph of the two phage. P 4 is able to activate genes of its helper under conditions which do not allow the helper to express those genes. This activation of genes on one phage genome by the product(s) of a different phage genome has been termed *trans*activation (Thomas, 1970). P 4 can also inactivate the immunity system of P 2, causing expression of previously repressed genes. Finally, P 2, the helper phage, *trans*activates some of the genes of its satellite, P 4.

2.   THE LIFE CYCLE OF PHAGE P 2

The temperate response of P 2 yields lysogenic strains in which the P 2 genome has been inserted into the bacterial genome (Calendar and Lindahl, 1969). Expression of the prophage is limited to three short regions of the genome (reviewed by Bertani and Bertani, 1971). DNA replication is not required for efficient lysogenization (Lindahl, 1974). The lytic response consists of an ordered expression of the genome pro-

ducing DNA replication, synthesis, and assembly of the proteins of the
phage particle enclosing a phage genome, and lysis of the cell (Geisselsoder *et al.*, 1973; Lengyel and Calendar, 1974; Pruss *et al.*, 1974*a*).

2.1. P 2 DNA Replication

Schnös and Inman (1971) have shown in density-shift experiments
that the first round of P 2 DNA replication proceeds from a defined
origin in only one direction yielding sigma or "rolling circle"
molecules. This mode of replication continues through late stages of infection (Geisselsoder, unpublished data). The products of the bacterial
genes *dna*B, *dna*E, and *dna*G are required (Bowden *et al.*, 1975; Six.
1975). In a strain carrying a *rep* mutation, P 2 DNA replication begins
but is limited to less than one full round (Calendar *et al.*, 1970; Geisselsoder, 1976).

2.2. Expression of Phage Genes

The P 2 genome is a double-stranded DNA molecule of 22×10^6
daltons (Inman and Bertani, 1969). It has 5′-terminated single-stranded
complementary ends which are 19 bases long (Wang *et al.*, 1973).
Twenty genes essential for the lytic response have been defined by both
amber and temperature-sensitive mutations (Lindahl, 1974; Fig. 2). Of

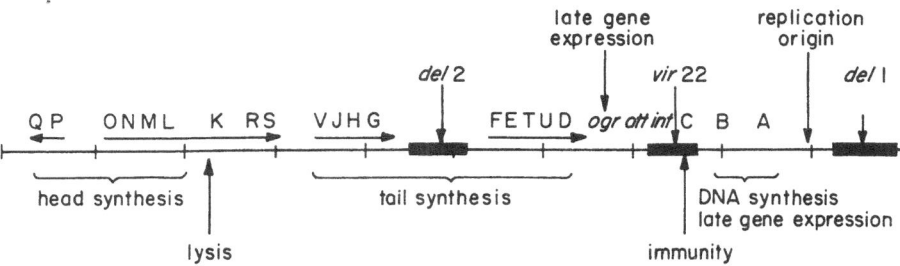

Fig. 2.   Genetic map of temperate phage P 2. Isolation and classification of mutants in
essential genes A, B, D–H, and J–V are described by Lindahl (1969*a*, 1970, 1971, 1974)
and by Sunshine *et al.* (1971). Transcription units are indicated by arrows (Lindahl,
1971; Sunshine *et al.*, 1971). Transcription is thought to begin at the base of each arrow
and proceed to the tip. Gene *ogr* is described by Sunshine and Sauer (1975). The attachment site was located by Lindahl (1969*b*) and by Calendar and Lindahl (1969). Deletion sites were located by Chattoraj and Inman (1972, 1974), and the replication
origin by Schnös and Inman (1971).

these, genes A and B are required for both DNA replication and normal expression of the other genes (Lindqvist, 1971; Lindahl, 1971). These genes are located on the physical right half of the DNA molecule, and their expression can be detected early in infection (Geisselsoder *et al.*, 1973; Lengyel and Calendar, 1974).

The P 2 genes necessary for phage particle construction and cell lysis are located predominantly on the left half of the DNA. Early in infection, no RNA hybridizable to the left half of the DNA is detectable. As the lytic cycle proceeds, however, left half transcription begins and later predominates. Lengyel and Calendar (1974) demonstrated that the kinetics of production of several late gene products are commensurate with RNA synthesis kinetics.

3. THE LIFE CYCLE OF P 4

P 4 was first identified as one of two phages released by *E. coli* strain K235 (Six, 1963). One of these phages, called PK, formed plaques on nonlysogenic strains of *Shigella* and *E. coli* C. The other, now called P 4, formed smaller plaques, and only on strains lysogenic for either P 2 or PK, not on nonlysogenic strains. Further experiments revealed that P 4 infection of strains lysogenic for P 2 or PK produced only P 4 phage. (The level of helper phage found in the lysate could be accounted for by normal spontaneous induction.) P 4-infected nonlysogens, on the other hand, did not lyse and released no progeny virus. However, P 4 was able to lysogenize them. This was demonstrated by the fact that when these lysogenic strains were further lysogenized by P 2, they released both types of virus but could not be lytically infected by P 4. It was also found that when nonlysogenic strains were coinfected with P 2 and P 4, both types appeared in the burst, though P 4 was predominant.

3.1. P 4 DNA Replication

The pathway by which P 4 replicates its DNA differs in several particulars from that utilized by P 2. First, in contrast to P 2, P 4 DNA replicates bidirectionally from a defined origin, generating theta or "Cairns" molecules (R. B. Inman, personal communication). Second, P 4 DNA replication differs from P 2 replication in its utilization of bacterial gene products: The *dna*H product is required, but not the *dna*B nor *dna* G products (Bowden *et al.*, 1975). Finally, P 4 can replicate its DNA normally in a strain carrying a *rep* mutation (Lindqvist and Six, 1971), in which P 2 DNA replication is restricted (Calendar *et al.*, 1970).

3.2. Requirement for Essential Helper Genes

Electron-microscopic observation of high-titer stocks of P 4 grown on a P 2 lysogenic strain revealed particles with a tail apparently identical to that of P 2 and an isometric head comprising one-third the volume of the P 2 head (Inman *et al.*, 1971; Fig. 1). The similarity of the P 2 and P 4 tails suggested that P 4 relies upon its helper to supply tail gene proteins. This proposal was first supported by demonstrating that the antigenicity of P 4 reflects that of its most recent helper (Six and Klug, 1973). Using strains lysogenic for various conditional lethal mutants of P 2, Six (1975) has now demonstrated that P 4 requires all of the known late genes of P 2: those utilized to construct the phage particle and lyse the cell. Barrett *et al.* (1974, 1976) have shown that all of the protein bands detectable in P 2 phage particles by SDS-polyacryl-amide gel electrophoresis are also present in P 4 phage particles. These experiments show that the P 4 phage particle is constructed from proteins coded for by the P 2 genome. The P 4 phage head, however, is much smaller than the P 2 head, so P 4 must in some way cause the helper-coded head proteins to condense into a smaller structure. The evidence for P 4 determination of small head size is that only small heads are found in P 4-infected P 2 lysogenic cells, whereas no P 4 sized heads are found in P 2-infected nonlysogenic cells (Gibbs *et al.*, 1973; Goldstein *et al.*, 1974).

P 4 does not need the products of P 2 early genes A and B, which are required for P 2 DNA replication and for P 2-mediated P 2 late gene expression (Lindahl, 1970; Lindqvist, 1971; Geisselsoder *et al.*, 1973; Six, 1975). The delay of lysis observed in P 4-infected hosts carrying a P 2 A*am* or P 2 B*am* prophage is probably a gene dosage effect (see Section 6). Three large regions of the P 2 genome can be deleted without affecting its ability to grow lytically (see Fig. 2). None of these regions contain information essential for P 4 growth (Six, 1975).

4. GENES OF THE SATELLITE PHAGE

The P 4 genome is a double-stranded DNA molecule of 7×10^6 daltons (Inman *et al.*, 1971), carrying single-stranded complementary ends which have the same sequence as the cohesive ends of P 2 DNA

(Wang *et al.*, 1973). The cohesive ends of P 2 and P 4 DNAs would have to be similar, if not identical, for the same P 2 gene products to recognize the appropriate sequence and generate both. Five P 2 gene products (M, N, O, P, and Q) involved in head formation are required to produce these scissions in vegetative P 2 DNA (Pruss and Geisselsoder, unpublished data). Outside of the cohesive end sites, there is very little, if any, homology between P 4 and P 2 DNA. Using DNA–DNA hybridization techniques the sensitivity of which would have allowed the detection of less than 1% homology, Lindqvist (1974) found none. This result, plus the differences between P 2 and P 4 DNA replication, implies that P 4 is not simply an extensively deleted helper genome.

A chromosome of the size of P 4's has the potential to code for 10 average-sized proteins. Some of the functions that these 10 genes should participate in are DNA replication, lysogenization and specification of the integration site, activation of the helper late genes, and direction of the formation of small heads. The smaller size of P 4 DNA is probably not the reason for small heads to form, since Pruss *et al.* (1974*b*, 1975) have shown that in an extract of P 2-infected cells, exogenously added P 4 DNA is not packaged into small heads. Instead trimers of P 4 DNA are packaged into large P 2-sized heads.

Electrophoretic analysis of extracts made from UV-irradiated P 4-infected cells has revealed four P 4-specified proteins. When P 4 alone infects a nonlysogenic strain, two of these proteins appear early (within 5 min), one is partially shut down subsequently, and two other proteins appear after a 40-min time lag. Hence, there is a temporal shift in the expression of the P 4 genome, implying the existence of control systems mediated by P 4 itself (Barrett et al., 1974, 1976).

As yet the genetic map of P 4 is ill-defined. Despite an extensive search, Gibbs *et al.* (1973) were able to find amber mutants in only one gene, and one temperature-sensitive mutant in a second gene. A third essential gene has been defined by one temperature-sensitive mutant (Souza *et al.*, 1977). The known P 4 essential genes are:

1. Gene α. Over 20 P 4 amber mutants have now been isolated, all of them in gene α. These mutants fail to synthesize phage DNA under nonpermissive conditions (Gibbs *et al.*, 1973). In addition, they do not synthesize a transcribing activity which copies poly(dG) · poly(dC), producing polyriboguanylic acid from GTP (Barrett *et al.*, 1972). One of the two P 4-specified early proteins detected by Barrett *et al.* (1976) is missing after infection with these mutants, and this protein has the same molecular weight as the transcribing enzyme. Thus, the two may be identical. P 4 α mutants also lack the ability of P 4+ to

suppress the polarity associated with amber mutants in P 2 late gene operons (Sunshine *et al.*, 1976; see Section 5.2).

2. Gene β. Gibbs *et al.* (1973) found one P 4 temperature-sensitive mutant which complements gene α mutants. This mutant is partially dominant to wild type and greatly reduces the overall rate of DNA, RNA, and protein synthesis in infected nonlysogenic cells, although it can replicate its own DNA (Gibbs *et al.*, 1973). The function of the wild-type allele of gene β is not known.

3. Gene C. The letter C is reserved for mutants affecting the P 4 protein which causes immunity to superinfecting P 4 (Six and Klug, 1973).

4. Gene δ. The P 4 mutation *ts*6 defines a third essential gene since it can complement and be complemented by mutants in genes α and β. DNA synthesis at the nonpermissive temperature is not impaired (Souza *et al.*, 1977). This mutant is quite leaky at the nonpermissive temperature when grown on a lysogen carrying a P 2 prophage with a functional A gene. If the prophage is mutant in gene A, however, no measurable burst is produced. Since the P 2 A gene is required for late P 2 mRNA synthesis (Geisselsoder *et al.*, 1973), these results suggested that after P 2 infection a wild-type prophage could itself be producing a small amount of P 2 late gene products, allowing a few P 4 phage particles to be made. When the P 2 prophage A gene is defective, even this avenue is blocked. Thus, P 4 δts6 mutant is defective in an alternative mechanism for activation of P 2 late genes. In support of this idea, it was found that P 4 δts6 growing on a P 2 A*am* lysogenic strain produced 10-fold reduced levels of P 2 late mRNA (Souza *et al.*, 1977).

If P 4 δts6 is defective in activating P 2 late genes, then its growth should not be impaired when those late genes can be activated in some other way. This situation is realized when P 2 and P 4 coinfect a nonlysogenic cell. Souza *et al.* (1977) found that coinfection with P 4 δts6 and P 2 did indeed yield a normal burst of P 4 δts6 at the nonpermissive temperature, but only if the conifecting P 2 carried a functional A gene and was therefore able to induce late P 2 mRNA. Formally speaking, the P 4 δ gene "replaces" the P 2 A gene in activating P 2 late genes. However, the P 4 δ gene function is not identical to the P 2 A gene function since the product of P 2 A gene acts only on the genome which coded for it (Lindahl, 1970), while the P 4 δ gene product acts in *trans*, allowing P 2 late gene transcription. Furthermore, P 4 δ gene product does not replace the P 2 A gene product in allowing P2 DNA replication (G. Pruss, personal communication).

Figure 3a is a map of P 4 genes derived from two factor crosses.

(a) P4 Genetic Map

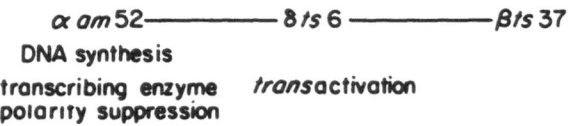

(b) Physical Map of the P4 Genome

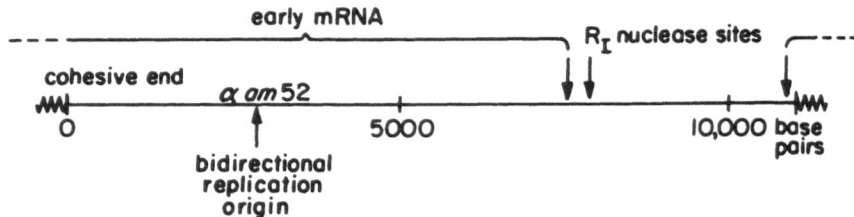

Fig. 3.   Maps of the P 4 genome. (a) Map of P 4 essential genes derived from two-fac-
tor crosses, indicating which phage functions are eliminated by the indicated mutants.
Distances between the genes have not been agreed upon, although there is a minimum
of 1% recombination between each pair of genes. (b) Physical map of the DNA, showing the cohesive ends, 3 *Eco* RI nuclease cleavage sites (arrows pointing down), the RI
nuclease fragment which hybridizes to early P 4 mRNA (under brackets), and the bidirectional replication origin located by R. B. Inman (unpublished data). References
describing the experimental findings are given in the text. The wild type allele of the P 4
αam 52 mutation can be rescued from the section of P 4 DNA which codes for early
mRNA (Yost and Souza, personal communication), and the δ gene and size determination gene also map in this segment.

5.   TWO DIFFERENT MECHANISMS FOR ACTIVATION OF HELPER LATE GENES

P 4 utilizes a mechanism for activating P 2 late genes which differs
in two major particulars from the mechanism utilized by P 2.

5.1.   Relation between DNA Replication and Late-Gene Expression

Mutations in P 2 genes A and B block both DNA replication and
late mRNA synthesis (Lindahl, 1970; Lindqvist, 1971; Geisselsoder *et
al.*, 1973). These findings suggest that DNA replication and late-gene
expression are in some way "coupled." Replicating or replicated P 2
DNA might be a competent template for late-gene expression, whereas

unreplicated DNA might be incompetent or sequestered. A similar phenomenon has been reported for phage T 4 (Hosoda and Levinthal, 1968; Riva *et al.*, 1970). Wu and Geiduschek (1975) and Coppo *et al.* (1975) have demonstrated that the absolute coupling is mediated by two separable functions of the product of T 4 gene 45. One of P 4's most interesting properties is its ability to cause the expression of P 2 late genes when P 2 DNA replication has been blocked by mutation of gene A. Also, *trans*activation of P 2 late genes by P 4 occurs normally in a *rep* host in which P 2 replication is initiated normally but a block is imposed on strand elongation (Geisselsoder, 1976).

5.2. Suppression of Polarity by the Satellite Phage

Certain P 2 late-gene amber mutations exhibit strong polar effects on large groups of contiguous genes (Lindahl, 1971; Sunshine *et al.*, 1971). Combining all data on these polar effects, it has been possible to assign every known P 2 late gene into one of the four "transcription" units delineated by arrows in Fig. 2. Polar mutations have been shown to eliminate mRNA from the downstream region of the transcription unit, e.g., in the tryptophan operon of *E. coli* (Imamoto and Yanofsky, 1967), and also in P 2 (Geisselsoder *et al.*, 1973). Sunshine *et al.* (1976) have shown that P 4 is able to relieve this polarity, by complementation studies, and also by electrophoretic analysis of the proteins coded for by genes downstream of known P 2 late-gene polar mutants. Whether this observation reflects a general change in the kinetics (or mechanism) of mRNA degradation is not known. Polar effects are now thought to be caused, at least in part, by the action of the transcription termination factor *rho* (Richardson *et al.*, 1975). It is postulated that *rho* can recognize sites within transcripts of structural genes on nascent mRNA and cause termination of transcription there if these sites are not protected by the translation apparatus following closely behind RNA polymerase. P 4 may therefore code for a protein which interferes with this action of *rho*. Another possibility is that P 4 causes new transcription initiations in the late region of P 2 DNA.

As mentioned previously, P 4 α amber mutants do not suppress polarity. The fact that the α gene appears to act in P 4 DNA synthesis (Barrett *et al.*, 1972, 1976) suggests that the α gene product does not directly cause suppression of polarity. It could be that lack of (replicated) template in P 4 αam infected cells precludes the production of sufficient amounts of the suppression product to produce this effect. Alternatively, the amber mutant in the α gene might exert a polar effect on the polarity-suppressing gene, or might itself activate the polarity-suppressing gene.

6. DEREPRESSION OF THE HELPER PROPHAGE

When P 4 α amber mutants infect a nonpermissive strain lysogenic for P 2, P 2 late mRNA is made and the cells synthesize empty head capsids (Lindqvist, 1974; Gibbs *et al.,* 1973). Also, the cells lose their immunity, allowing superinfecting immunity-sensitive P 2 to grow lytically (Lindqvist and Six, manuscript in preparation). Under these conditions, an active P 2 A gene is required for late mRNA synthesis, and thus the P 4 *trans*activation process appears to be inoperative. The same loss of immunity is inferred to take place after a P 4+ infection, since P 2 prophage DNA synthesis, dependent again on a functional A gene, can be detected under these conditions (Lindqvist and Six, manuscript in preparation). There is a paradox implicit in these findings: As measured by complementation, P 4 α mutants are able to express gene δ (Souza *et al.,* 1977), which is thought to *trans*activate P 2 late genes; yet when P 4 αam mutants derepress a P 2 A defective prophage, late genes are not *trans*activated. Also, when P 4 αam and P 2 A*am* coinfect a nonlysogenic strain, the level of P 2 late-gene expression is very low, as measured by amounts of P 2 late proteins produced (Barrett et al., 1976). The most likely explanation for these data has been previously mentioned; lack of a replicated P 4 template probably prevents sufficient synthesis of P 4's *trans*activating protein(s).

It also remains unexplained why P 2 superinfecting phages fail to grow in P 2 + P 4 coinfection of P 2 lysogens (Six and Klug, 1973), although the existence of a P 4-coded P 2 growth inhibitor seems likely.

Given the complexity of the P 2–P 4 interactions, it is important to distinguish between different modes for activating gene expression in this system. Two such modes can be clearly differentiated, one direct, the other indirect. The direct, or *trans*activation, mode is exemplified by the situation in which a P 4 virulent mutant causes expression of the late genes of a P 2 virulent A amber mutant. This P 2 mutant cannot activate its own genes; therefore, their expression is solely due to P 4. Clearly, derepression cannot be a factor in this case since neither phage can establish immunity. We assume that polarity suppression as well as replication-independent expression of P 2 late genes are associated with this *trans*activation mode, though critical evidence on this point is lacking.

The indirect mode is exemplified by the situation in which a P 4 α

amber mutant causes a P 2 prophage to be derepressed, and the subsequent P 2 late-gene activation is under the control of the P 2 A gene.
In this case, the predominant late-gene activation is due to derepression
of P 2 prophage by P 4.

7. RECIPROCAL *TRANSACTIVATION*

P 4 elicits the production of at least two classes of proteins,
"early" and "late." The two "early" proteins can be detected within 2
min after infection (Barrett *et al.*, 1974). When P 4 infects a cell with
no helper present, two other proteins make their appearance after 40
min (Barrett *et al.*, 1976). When a P 2 helper is present, the kinetics of
synthesis of the two "early" proteins is not altered, but the two "late"
proteins now appear at 15 min after infection and are synthesized in 5–
10 fold greater quantity. This effect of the helper on satellite phage late-
gene expression has been called "reciprocal *trans*activation" (Barrett *et
al.*, 1974).

The reciprocal *trans*activation effect is not abolished by mutations
in any of the known P 2 essential genes nor by the deletions *del*1, *del*2
and *vir*22, which, among them, delete 18% of the P 2 genome (Fig. 2).
It could be that a P 2 early gene, as yet not defined by mutation or
perhaps not essential to P 2 in the normal laboratory hosts, is responsible for this effect. If so, this gene probably lies between the right end of
the *vir*22 deletion and gene A: P 4 does not form visible plaques on
strains in which this region of the prophage genome is deleted. The
latent period is extended threefold in these strains, although the size of
the burst is normal (S. Barclay, E. Six, and L. E. Bertani, unpublished
results; partial data presented in Barrett *et al.*, 1974). There is coding
capacity for approximately 150,000 daltons of protein between the
Bam116 mutation and the left end of the deletion *del*1 (Chattoraj and
Inman, 1972, 1974), which would well accommodate part of gene B (B
gene product is 19,000 daltons; Lengyel and Calendar, 1974), gene A
(the molecular weight of its product is unknown), and another gene.
Whether the prophage deletions abolish reciprocal *trans*activation is
not yet known since P 4 proteins have only been detected after heavy
ultraviolet irradiation of the host, which would destroy a prophage
helper. It has not been possible to grow such P 2 prophage deletion
mutants lytically and to test them as coinfecting helpers.

Tests for late P 4 mRNA have now become possible, since the P 4
genome has been divided into several major sections by cleavage with
EcoRI endonuclease (Thomas *et al.*, 1974; Goldstein *et al.*, 1975; Fig.

3b). Two such segments have been incorporated into lambda phages which can be propagated lytically. The strands of the DNA of these phages can be separated, and hence individual strands of the P 4 DNA segments can be obtained in large quantity. P 4 early mRNA anneals to only one strand of one segment of P 4 DNA (J. Harris, personal communication; Fig. 3b), while late P 4 mRNA anneals to both strands of both major fragments. The appearance of late mRNA could be measured in order to detect reciprocal *trans*activation and to decide whether the above-mentioned prophage deletions abolish *trans*activation of P 4 late genes.

Another P 2 gene involved in late gene activation is defined by P 2 *ogr* mutations, which map between tail gene D and *att* (Fig. 2). P 2 *ogr* mutants are able to grow on an *E. coli* mutant called *gro*109, which does not allow late-gene expression from a P 2+ genome although DNA replication proceeds at a normal rate (Sunshine and Sauer, 1975). The *gro*109 mutation maps in the *str-spc* gene cluster at 64 min on the *E. coli* map and produces a leucine to histidine substitution in the alpha subunit of DNA-dependent RNA polymerase (Jaskunas *et al.*, 1975; Fujiki *et al.*, 1976). That the *E. coli* RNA polymerase is utilized to synthesize P 2 and P 4 mRNA can be inferred from the data of Lindqvist (1974). Rifampicin was shown to prevent the synthesis of mRNA of both phage in a rifampicin sensitive host, but not in a resistant host. P 2 *ogr* mutants can promote the expression of coinfecting P 2+ late genes in *gro*109 strain (Sunshine and Sauer, 1975); therefore, the *ogr* gene must code for a diffusible product. The product probably interacts with the transcription apparatus to allow P 2 late-gene expression. The *gro*109 allele also inhibits P 4 production unless a P 2 *ogr* phage serves as helper (Sunshine, unpublished results) and the burst of P 4 is delayed (Sauer, personal communication; Sunshine, unpublished results). Neither P 2 nor P 4 late proteins can be detected after coinfection of a *gro*109 strain (Sauer, personal communication). It appears, then, that the P 2 *ogr* is involved in reciprocal *trans*activation.

It is not yet known whether *ogr* is an essential P 2 gene in wild-type *E. coli* host strains, although a mutation in it is clearly required for P 2 late genes to be expressed in a strain carrying the *gro*109 allele. No amber or temperature-sensitive mutants in *ogr* have yet been identified. However, circumstantial evidence suggests that it is essential: S. Hocking and J. B. Egan (personal communication) have isolated a conditional lethal amber mutation of the related phage 186 which maps in a position equivalent to P 2 *ogr*.

8. SUMMARY

The late genes of temperate phage P 2 are normally under the control of P 2 early genes A and B. In an *E. coli gro* mutant host a newly defined P 2 gene, called *ogr*, is also needed for P 2 late-gene expression. Other P 2 regulatory genes may exist, since the early region of the genetic map is not saturated with mutations. In the absence of P 2 genes A and B, satellite phage P 4 can *trans*activate P 2 late-gene expression. P 4 can also, by derepressing a prophage, induce P 2 to activate its own genes. This mode of activation requires a functional P 2 A gene. P 4 gene δ appears to replace P 2 gene A in activating P 2 late genes. The simple replacement of P 2 A and B genes by P 4 does not completely explain the *trans*activation of P 2 late genes, since the polarity associated with certain P 2 late amber mutations is suppressed by P 4. Moreover, P 4 can cause P 2 late genes to be expressed without concomitant DNA replication, suggesting that P 4 alters the mechanism for P 2 late-gene expression.

P 4 itself has late genes, which may participate in the determination of small head size. Expression of these late genes is stimulated by the presence of a helper phage. Thus, *trans*activation is reciprocal in the P 2–P 4 system: P 2 turns on P 4 late genes, and P 4 turns on P 2 late genes. The P 2 *ogr* gene is needed for reciprocal *trans*activation of P 4 late genes, and another P 2 early gene may also be required. These actions and interactions are summarized in Fig. 4.

The interactions between P 2 and P 4 may aid in elucidating similar interactions found in eukaryotic systems, such as the turn-on of

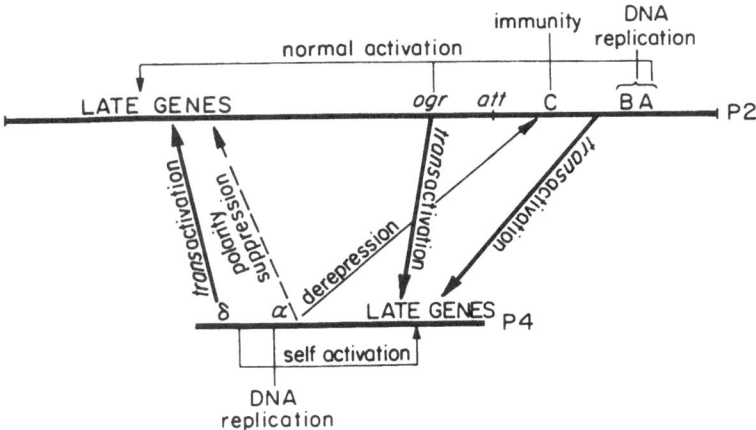

Fig. 4.   Schematic representation of P 2–P 4 actions and interactions.

adeno-associated satellite virus genes by adenovirus (Blacklow *et al.,* 1971) and the activation of a variety of autosomal genes in the presence of a Y chromosome (Hamerton, 1968; Mittwoch, 1973).

ACKNOWLEDGMENTS

The recent research upon which this article is based was supported by NIH research grants AI 08722 (to R.C.), AI 04043 (to E.W.S.), and postdoctoral fellowship AI 45852 (to J.G.) from the National Institute of Allergy and Infectious Diseases; by NIH research grant CA 14097 and training grant CA 05028 from the National Cancer Institute; by National Science Foundation grants Nos. BMS-74-19607 (to R.C.) and GU 2591 (to the University of Iowa); by Norwegian Research Council for Science and Humanities (C.07.14-5; to B.H.L.); by grant VC-188 (to R.C.) from the American Cancer Society; and by Cancer Research Funds of the University of California.

9. REFERENCES

Barrett, K., Gibbs, W., and Calendar, R., 1972, A transcribing activity induced by satellite phage P 4, *Proc. Nat. Acad. Sci. USA* **69,** 2986.

Barrett, K., Barclay, S., Calendar, R., Lindqvist, B., and Six, E. W., 1974, Reciprocal *trans*activation in a two-chromosome phage system, *in* "Mechanism of Virus Disease" (W. S. Robinson and C. F. Fox, eds.), p. 385, W. A. Benjamin, Inc., Menlo Park, California.

Barrett, K. J., Marsh, M. L., and Calendar, R., 1976, Interactions between a satellite phage and its helper, *J. Mol. Biol.* **106,** 683.

Bertani, G., 1951, Studies on lysogenesis. I. The mode of phage liberation by lysogenic *E. coli, J. Bacteriol.* **62,** 293.

Bertani, L. E., 1968, Abortive induction of bacteriophage P 2, *Virology* **36,** 87.

Bertani, L. E., and Bertani, G., 1971, Genetics of P 2 and related phages, *Adv. Genet.* **16,** 199.

Blacklow, N. R., Dolin, R., and Hoggan, M. D., 1971, Studies of the enhancement of an adenovirus-associated virus by herpes simplex virus, *J. Gen. Virol.* **10,** 29.

Bowden, D. W., Twersky, R. S., and Calendar, R., 1975, *Escherichia coli* deoxyribonucleic acid synthesis mutants: Their effect upon bacteriophage P 2 and satellite bacteriophage P 4 deoxyribonucleic acid synthesis, *J. Bacteriol.* **124,** 167.

Calendar, R., and Lindahl, G., 1969, Attachment of prophage P 2: Gene order at different host chromosomal sites, *Virology* **39,** 867.

Calendar, R., Lindqvist, B. H., Sironi, G., and Clark, A. J., 1970, Characterization of REP⁻ mutants and their interactions with P 2 phage, *Virology* **40,** 72.

Chattoraj, D. K., and Inman, R. B., 1972, Position of two deletion mutations on the physical map of bacteriophage P 2, *J. Mol. Biol.* **66,** 423.

Chattoraj, D. K., and Inman, R. B., 1974, Tandem duplication in bacteriophage P 2: Electron microscopic mapping, *Proc. Nat. Acad. Sci. USA* **71**, 311.

Coppo, A., Manzi, A., Pulitzer, J. F., and Takahashi, H., 1975, Host mutant (*tab*D) induced inhibition of bacteriophage T 4 late transcription. I. Isolation and phenotypic characterization of the mutants, *J. Mol. Biol.* **96**, 579.

Fujiki, H., Palm, P., Zillig, W., Calendar, R., and Sunshine, M., 1976, Identification of a mutation within the structural gene for the α subunit of DNA-dependent RNA polymerase of *E. coli, Molec. Gen. Genet.* **145**, 19.

Geisselsoder, J., 1976, Strand-specific discontinuity in replicating P 2 DNA, *J. Mol. Biol.* **100**, 13–22.

Geisselsoder, J., Mandel, M., Calendar, R., and Chattoraj, D. K., 1973, Patterns of transcription after infection with bacteriophage P 2, *J. Mol. Biol.* **77**, 405.

Gibbs, W., Goldstein, R., Wiener, R., Lindqvist, B., and Calendar, R., 1973, Satellite bacteriophage P 4: Characterization of mutants in two essential genes, *Virology* **53**, 24.

Goldstein, L., Thomas, M., and Davis, R. W., 1975, EcoRI endonuclease cleavage map of bacteriophage P 4 DNA, *Virology* **66**, 420.

Goldstein, R. N., Lengyel, J., Pruss, G., Barrett, K., Calendar, R., and Six, E. W., 1974, Phage head size determination and the morphogenesis of satellite phage P 4, *Curr. Top. Microbiol. Immunol.* **68**, 59.

Hamerton, J. L., 1968, Significance of sex chromosome derived heterochromatin in mammals, *Nature (London)* **219**, 910.

Hosoda, J., and Levinthal, C., 1968, Protein synthesis by *Escherichia coli* infected with bacteriophage T 4D, *Virology* **34**, 709.

Imamoto, F., and Yanofsky, C., 1967, Transcription of the tryptophan operon in polarity mutants of *E. coli.* I. Characterization of the tryptophan messenger RNA of polar mutants, *J. Mol. Biol.* **28**, 1.

Inman, R., and Bertani, G., 1969, Heat denaturation of P 2 DNA: Compositional heterogeneity, *J. Mol. Biol.* **44**, 533.

Inman, R., Schnös, M., Simon, L., Six, E. W., and Walker, D., 1971, Some morphological properties of P 4 bacteriophage and P 4 DNA, *Virology* **44**, 67.

Jaskunas, S. R., Burgess, R., Lindahl, L., and Nomura, M., 1975, Two clusters of genes for RNA polymerase and ribosome components in *E. coli, in* "RNA Polymerase" (M. Chamberlin and R. Losick, eds.), Cold Spring Harbor, New York.

Lengyel, J., and Calendar, R., 1974, Control of bacteriophage P 2 protein and DNA synthesis, *Virology* **57**, 305.

Lindahl, G., 1969*a*, Genetic map of bacteriophage P 2, *Virology* **39**, 839.

Lindahl, G., 1969*b*, Multiple recombination mechanisms in bacteriophage P 2, *Virology* **39**, 861.

Lindahl, G., 1970, Bacteriophage P 2: Replication of the chromosome requires a protein which acts only on the genome that coded for it, *Virology* **42**, 522.

Lindahl, G., 1971, On the control of transcription in bacteriophage P 2, *Virology* **46**, 620.

Lindahl, G., 1974, Characterization of conditional lethal mutants of bacteriophage P 2, *Mol. Gen. Genet.* **128**, 249.

Lindqvist, B. H., 1971, Vegetative DNA of temperate coliphage P 2, *Mol. Gen. Genet.* **110**, 178.

Lindqvist, B. H., 1974, Expression of phage transcription in P 2 lysogens infected with helper-dependent coliphage P 4, *Proc. Nat. Acad. Sci. USA* **71**, 2752.

Lindqvist, B. H., and Six, E. W., 1971, Replication of bacteriophage P 4 DNA in a nonlysogenic host, *Virology* **43**, 1.

Mittwoch, U., 1973, "Genetics of Sex Differentiation," Academic Press, London.

Pruss, G., Goldstein, R. N., and Calendar, R., 1974*a, In vitro* packaging of satellite phage P 4 DNA, *Proc. Nat. Acad. Sci. USA* **71**, 2367.

Pruss, G., Barrett, K., Engvel, J., Goldstein, R., and Calendar, R., 1974*b*, Phage head size determination and head protein cleavage *in vitro, J. Supramol. Struct.* **2**, 337.

Pruss, G., Wang, J. C., and Calendar, R., 1975, *In vitro* packaging of covalently closed circular monomers of bacteriophage DNA, *J. Mol. Biol.* **98**, 465.

Richardson, J. P., Grimley, C., and Lowrey, C., 1975, Transcription termination factor *rho* activity is altered *Escherichia coli* with *suA* gene mutations, *Proc. Nat. Acad. Sci. USA* **72**, 1725.

Riva, S., Cascino, A., and Geiduscheck, E. P., 1970, Coupling of late transcription of viral replication in bacteriophage T 4 development, *J. Mol. Biol.* **54**, 85.

Schnös, M., and Inman, R. B., 1971, Starting point and direction of replication in P 2 DNA, *J. Mol. Biol.* **55**, 31.

Six, E. W., 1963, A defective phage depending on phage P 2, *Bacteriol. Proc.,* p. 138.

Six, E. W., 1975, The helper dependence of satellite bacteriophage P 4: Which gene functions of bacteriophage P 2 are needed by P 4, *Virology* **67**, 249.

Six, E. W., and Lindqvist, B. H., 1971, Multiplication of bacteriophage P 4 in the absence of replication of DNA of its helper, *Virology* **43**, 8.

Six, E. W., and Klug, C., 1973, Helper-dependent bacteriophage P 4: A satellite virus dependent on a helper such as P 2, *Virology* **51**, 327.

Souza, L., Calendar, R., and Six, E. W., 1977, A *trans*activation mutant of satellite phage P 4, *Virology,* in press.

Sunshine, M., and Sauer, B., 1975, A bacterial mutation blocking P 2 phage late-gene expression, *Proc. Nat. Acad. Sci. USA* **72**, 2770.

Sunshine, M., Thorn, M., Gibbs, W., Calendar, R., and Kelly, B., 1971, P 2 phage amber mutants: Characterization by use of a polarity suppressor, *Virology* **46**, 691.

Sunshine, M., Six, E. W., Barrett, K., and Calendar, R., 1976, Relief of P 2 phage amber mutant polarity by the satellite phage P 4, *J. Mol. Biol.,* **106**, 673.

Thomas, M., Cameron, J., and Davis, R., 1974, Viable molecular hybrids of bacteriophage λ and eukaryotic DNA, *Proc. Nat. Acad. Sci. USA* **71**, 4579.

Thomas, R., 1970, Control of development in temperate phage III. Which prophage genes are and which are not *trans*activatable in the presence of immunity, *J. Mol. Biol.* **49**, 393.

Wang, J. C., Martin, K. V., and Calendar, R., 1973, On the sequence similarity of the cohesive ends of coliphage P 4, P 2, and 186 DNA, *Biochemistry* **12**, 2119.

Williams, R. C., and Fisher, H., 1970, Electron microscopy of tobacco mosaic virus under conditions of minimal beam exposure, *J. Mol. Biol.* **52**, 121.

Wu, R., and Geiduschek, E. P., 1975, The role of replication proteins in the regulation of bacteriophage T 4 transcription. I. Gene 45 and hydroxymethyl C containing DNA, *J. Mol. Biol.* **96**, 513.

Index